Einstein and the Generations of Science

rev John Heilbron, Science 185, 1974, 776-778

Einstein and the Generations of Science

SECOND EDITION

LEWIS S. FEUER

Transaction Books
New Brunswick (U.S.A.) and London (U.K.)

TO

Robin and Christopher Miller

Library of Congress Catalog Number: 82-1874
ISBN: 0-87855-899-3 (paper)
Printed in the United States of America

Library of Congress Cataloging in Publication Data
Feuer, Lewis Samuel, 1912-
Einstein and the generations of science.
Includes bibliographical references and index.
1. Einstein, Albert, 1879-1955. 2. Physicists—Biography. 3. Relativity (Physics) 4. Science—History. I. Title.
QC16.E5F48 1982 530'.092'4 [B] 82-1874
ISBN 0-87855-899-3 (pbk.) AACR2

Contents

Contents

Prologue

I.

Perhaps the kindest words concerning my *Einstein and the Generations of Science* were written by M. Serge Moscovici, the distinguished social psychologist, in his preface to its French edition. He found my efforts truly contributing to a sociological theory of science, and free from the "ossified" languages of functionalism, Marxism, and paradigmatism. He detected in me the lineaments of the old "liberal" who held up the process of scientific advance as a model for how generational conflict might be constructively channelized. I was pleased that he accepted my explanation as to why the theory of relativity issued from a circle of impecunious scientific outsiders rather than from the "social pyramid" of French science of which the Ecole Polytechnique was the center of gravity; he gently demurred, however, at what he regards as my touchingly innocent faith in science despite "the monsters it has engendered." Nonetheless, he felt that "in having treated science as a total social phenomenon in the sense of Mauss, there is a merit before which many reservations dissolve. . . . Dealing with such a subject, it is rare and it is exceptional."

The reviewer in *Science,* however, though finding my book "valuable," disliked its occasional psychological conjectures and byways. Yet, only when such hypotheses are proposed and tested in a frankly tentative and exploratory spirit can we advance our knowledge of scientific research as a human phenomenon. The study, for instance, of Mach's character and dreams seems to me to illumine the emotive, extralogical sources of his empiricism. One sociological reviewer was under the impression that my title intended to play on the word "generations" of science, though actually what was in my mind was the old biblical usage in *Genesis* which I had studied as a child: "And these were the generations of. . . ." Another reviewer in a journal of the history of science has objected to my analysis of Einstein's circle as "countercultural," that is, as much affected with a spirit of generational rebellion. Curiously, I had first met the reviewer a few years earlier when he was still actively involved with a semiterrorist sect, and filled with the zeal of an extreme generational rebellion. *Ce sont les derniers convertis qui sont les plus farouches* (Those who are converted last are the fiercest). Among Israeli critics a few resented the place of Werner Heisenberg in my book, and felt that scholarly detachment was misplaced in writing about a scientist who had once been a Nazi sympathizer. I replied at length to one such criticism in two articles in *Philosophy of the Social Sciences.* The paramount duty seemed to me not to allow the scientific judgment of the genesis of scientific theories to be thrown askew by our ethical feelings.

Since I wrote the basic draft of *Einstein and the Generations of Science,* additional facts have come to light concerning the friends of his youth who helped Einstein shape his relativist mode of thought. The personality of Friedrich Adler was further illumined for me by the late Paul Lazarsfeld, for many years professor of sociology at Columbia Unversity. Lazarsfeld, after having read the portions on Einstein published in *Annals of Science,* narrated some unusual details as we dined together in Toronto late in 1972. "My mother," said Lazarsfeld, "was responsible for destroying three men, my father, Friedrich Adler, and myself. I always say this." It appeared that for many years, Adler had lived with Sofie Lazarsfeld and her family. Indeed, on the morning of the day that Adler assassinated the prime minister of Austria-Hungary, he had left the house, and then mailed them a postcard saying he was in good spirits. Neither Sofie Lazarsfeld nor her son, then 14 years old, had any foreknowledge of the pending events. Then, when "Fritz" Adler was imprisoned, young Paul would visit the jail every day, bringing books with him that Adler requested; the political prisoner was engaged in writing a book on the theory of relativity. Lazarsfeld's own career was very much shaped by Friedrich Adler. Apart from his work for some time on behalf of the Socialist International and Social Democratic party, Lazarsfeld took a doctorate in mathematics because Adler impressed upon him that without mathematics, there can be no contribution to science. Lazarsfeld thought that my view of the role of social factors in the formation of hypotheses was valid for original scientists, not for unoriginal ones—when he was working on his doctoral problem, a formula for the orbit of Mercury, he found in a mathematical textbook the description of a function that was exactly suitable, but such work he scarcely regarded as original. As in the public lecture he then gave, he said he always observed that his best students wanted to disprove their professor (an academic rendition of "conflict of generations"). Lazarsfeld urged me to speak with his mother, Sofie, who then was 92 years old, still living in an apartment in New York's West Side, and practicing Adlerian (Alfred) individual psychology.

For several hours on December 9, 1972, I listened to this still beautiful and charming woman, and found myself transported into a world that survived only in the plays of Schnitzler and the novels of Stefan Zweig—graceful, witty, rational, liberated, poised unknowingly on the edge of debacle. Her relationship with Friedrich Adler had begun around 1915 when she met him and his father, the Austrian socialist leader Victor Adler, at a summer resort; son and father looked alike, as two brothers, though she realized that Fritz lacked his father's political genius. She had indeed not known on the fateful morning in 1916 that Fritz planned to shoot the prime minister; she could never have thought of his doing such a thing. The day he murdered the prime minister, Fritz sent a letter to her son Paul, and the day after, a postcard to her little daughter. There was an inner turmoil in Fritz's life that people who know only his selfless political idealism never realized. Many years later, Friedrich Adler sent a reproachful, self-explaining letter to his wife Katya; it was more than a hundred pages long, and he gave a copy to Sofie Lazarsfeld. She knew of course the whole story of the unselfish action in which Fritz Adler had made way for Albert Einstein to secure his first academic post. Her son Paul had introduced her to the pyschological theories of Alfred Adler. In this vein, she wrote in later years an unusual article, "Did Oedipus Have An Oedipus Complex?"[1] She accepted the conflict of generations as a universal theme in history but tried to explain the phenomenon through the son's inevitable feelings of inferiority, and his effort to compensate for them; the will to power and security, she believed, was underlying even to the formation of the Oedipus Complex. A few months later, both Paul and Sofie Lazarsfeld died, the son shortly before his mother.

In my book, I aimed nonetheless to show in part that the name "scientific revolution" is a misnomer; the traits that lead us to speak of certain types of social change as examples of social "revolution" are absent in the presumed scientific variety. One scholar, while finding my critique of "scientific revolution" useful,[2] argued that psychoanalysis had not encountered, as I thought, an "ordeal of publication." She believed that "Freud's dour remarks about the reception of his writings" misled me, and maintained that, as a matter of fact, "Freud and other psychoanalysts had no trouble getting their papers and books published before they founded their own journals." If that had been the case, however, one might have expected a contemporaneous rebuttal along these lines by critics of Freud, as well as on the part later of such former coworkers as Breuer, Adler and Jung; they would have taken Freud to task directly for being querulous, and insisted that in their view, editors of the medical journals and organizers of medical sessions had welcomed Freud's contributions.

Apart from the absence of such evidence, we have the contrary testimony of G. Stanley Hall, the president of Clark University who organized the memorable conference of psychoanalysis in 1909. Hall, a pioneer in laboratory and statistical psychology, author of *Aspects of German Culture*, knew the German milieu well. He wrote in 1909: "His [Freud's] views are now being talked of in Germany as the psychology of the future. . . . And yet, partly because so much that he taught was new and revolutionary, he has until lately had but scant recognition; and because he attempts to do justice to sex in his scheme, he was for years, socially ostracized. Happily, he is now coming into his own . . ."[3] Freud's ideas evidently did encounter an ordeal of resistance that happily the work neither of Einstein, Bohr, de Broglie, nor Heisenberg was obliged to undergo.

II

Einstein and the Prague Circle

The first edition of my book in 1973 was principally concerned, as far as Einstein's work was involved, in showing how the character of his Zurich-Berne circle and friends contributed to his discovery of the special theory of relativity. I would now add that Einstein's later sojourn in Prague probably contributed in a similar fashion to forming the emotional-intellectual standpoint that came to underly the general theory of relativity.

Einstein used to say that the special theory of relativity was in the air when he discovered it, that Paul Langevin, for instance, might have done so as well. The general theory, on the other hand, was, he thought, a unique achievement that would not have found an alternative discoverer.[4] The discovery of the general theory of relativity was not, in Einstein's view, part of an inevitable logic in the evolution of science; it did not arise from any discussion that was taking place in scientific circles, and Max Planck, among others, regarded Einstein as making a mistake in devoting his energies to this path of research.[5] Evidently unusual circumstances may well have helped lead Einstein into what we might call his cosmological stage. Indeed, even as his circle of Marxian-Machian friends in Zurich and Berne was a cultural mainstay in his formation of the special theory of relativity, so likewise Einstein's association with the Prague Circle of Jewish mystics and intellectuals may have assisted his transition to a cosmological stage, a shift from Hume to Spinoza, and the formation of the general theory of relativity.

During the Easter of 1911, Einstein left the University of Zurich to take up the duties of professor ordinarius of physics in Prague at the German University. This move, from Zurich to Prague, brought a radical change in Einstein's friendships, and his religious and social milieu. In Prague, while taking "solitary walks," as Einstein told his son-in-law, "the general theory of relativity matured and took form."[6] At Prague in June 1911, Einstein published his calculations on the deflection of light rays in the sun's gravitational field, and called urgently on astronomers to resolve the question. In February 1912, he was looking forward hopefully to an expedition by the young German astronomer Erwin Freundlich to observe the solar eclipse in the Crimea; the First World War, however, postponed that confirmation of his evolving general theory.[7] Although his sojourn in Prague lasted only a year and a half, it seems to have catalyzed or contributed to a profound alteration in Einstein's personal and philosophical outlook.

Unlike Zurich, Prague was not a city of social revolution, of Marxist political parties and sects. Even physically, it was immersed in tradition—old buildings, dark streets; "The gravestones of the old Jewish cemetery told a thousand years' story of his race. Einstein's artistic nature was stirred by all this."[8] Prague, indeed, had, as Hans Tramer wrote, "remained a 'Jewish' town as almost no other in the Western world."[9] In front of its town hall stood a monument that was itself a strange symbol, that of the legendary creator of the golem, Rabbi Löw. Three peoples lived in Prague: 415,000 Czechs, 25,000 Jews, and 10,000 Germans. The Jewish students, excluded from the Czech and German associations, had formed their own under the name of Bar Kochba (the last warrior of Jewish independence against Rome), which by 1899 adopted the Zionist ideology. At Zurich and Berne, Einstein's circle had consisted of a cosmopolitan group of rebels and socialists: the Marxist-Machian Friedrich Adler; the impecunious Rumanian May Day marcher, Maurice Solovine; the eccentric engineer of the Garibaldian family, Besso; two Slavic women mathematicians; and several struggling Swiss. Now Einstein became a member of a Prague Circle composed exclusively of Jews, who moreover were undertaking the enterprise of a Jewish renaissance.

A strong overtone of neo-mystical philosophy pervaded the Prague Circle. From 1909 to 1911, the leading young Jewish philosopher, Martin Buber, delivered three lectures which left a permanent impress upon his listeners. A half-century later the novelist Max Brod still remembered "Buber's addresses before the Prague Circle over which he exerted so enduring an influence."[10] They were regarded as having marked the beginning of modern Jewish mysticism.[11] The German University at Prague became, as the historian Hans Kohn wrote, the center from which a new German and Jewish intellectual movement began.[12] Einstein attended the Tuesday meetings of this circle regularly at the home of a highly cultured woman, Mrs. Bertha Fanta. Led by a young philosopher, Hugo Bergmann, later to become the first rector of the Hebrew University at Jerusalem,[13] the group also included among its members (and intermittent visitors) the young novelist Max Brod; his close friend, the later celebrated Franz Kafka, depictor of human bewilderment in a bureaucratic society; the writers Arnold Zweig and Jakob Wasserman; the student of Spinoza, Margarete Susmann; the philosophers and scholars, Erich Kahler, Felix Weltsch, and Hans Kohn; and the Berlin mystical anarchist, Gustav Landauer.[14] Not a single Marxist could be found among them; furthermore, all were or were becoming Zionists and were to varying degrees touched with mysticism. Even Kafka under their influence found himself drawn toward their enthusiasm. Until after midnight they would read Kant's *Prolegomena to a Future Metaphysic* or the *Critique of Pure Reason.*

Einstein brought his violin, and when they were surfeited with metaphysics, they would play chamber music, with Hugo Bergmann conducting. The latter's "outstanding gifts made him the intellectual leader of his fellow students"; a convinced Zionist, he had not only undertaken to learn the Hebrew language but also Yiddish, the language of the orthodox poor or proletarian Polish-Russian Jews, so outlandish to those of bourgeois background. Buber was introducing a whole new generation to the strange domain of "underground Judaism," the spontaneous mysticism of the prescientific Eastern European Judaism.[15] Max Brod living in the state of Israel during his last years recalled in his autobiography that Einstein was not an orthodox "Einsteinian," that his mind was venturesome, open to all sorts of ideas, including those altogether contrary to his own:

> We met at the Fanta Circle. . . . we argued with Einstein and with Philipp Frank, ardent interpreter of Einstein. While Einstein himself was no such orthodox "Einsteinian" and always delighted me with his perpetual wonder and inspiration, I then observed how he would change his standpoint in discussion as an experiment, knowing how to try out once a standpoint opposed to his own, and to regard the whole question anew from an altered point of view. He seemed to take pleasure in savoring fully and with boldness all the possible ways of scientifically treating a topic. . . . This quality of his of scientific courage and ever-new beginning I copied in my Tycho Brahe novel in my figure of Kepler . . .[16]

Now it would be a mistake to say that there was a theoretical relationship between Einstein's thought and the kind of philosophic mysticism that Martin Buber and his disciples in the Prague Circle advocated. For Buber's mysticism tended to denigrate the significance of science as an avenue to basic truth.[17] Buber, drawing his famous contrast between the I-Thou relation and that of the I-It, regarded science as also based on turning the Thou into an It, thereby distorting and depersonalizing it. Moreover, Buber and his associates in the Prague Circle all believed in the notion of free will, in the exemption, at least partially, of the human spirit from deterministic laws of nature.

Yet in his evenings with the Prague Circle Einstein did find rekindled within himself mystical leanings that he had put aside at the age of 12. As a child, Einstein had "longed for a religious life and for religious instruction," and felt in nature "the whole of

God's majesty." He had disliked the ironic remarks that his father, a freethinker, made about religion and religious ritual. The child Albert "wrote and set to music songs in praise of God," in which, in Spinozist fashion, he identified God with Nature.[18] In youth, however, Einstein had become "a professed atheist and completely indifferent to all religious problems, " describing himself on the governmental roll as "unaffiliated."[19] As late as 1908 or 1909 Einstein had in discussions with Besso defended his separating himself from the Jewish community.[20] A return to a form of religious thinking, combined with an awakened sense of linkage with the Jewish people, evidently began for Einstein during his Prague interlude. Though he still rejected Judaic and Zionist notions, he had begun to experiment with them intellectually. Generally a "choice of comrades" (in Ignazio Silone's words) indicates the direction in thought to which one is potentially, perhaps unconsciously, drawn. This was Einstein's state of mind in Prague. He began to fashion for himself what might be called a philosophy of scientific mysticism, to ask himself in evaluating a theory: "Would God have made the universe in that way?" He would, for instance, be less able to rest content with the notion that while uniform motion is relative, accelerated motion should still remain absolute; now that judgment would not simply rest on a principle of economy or an aesthetic preference; he would say, "God would not have fashioned the world so inharmoniously."[21] The notion that scientific discovery is a kind of divination in the sense that the scientific theorist is somehow seeking to retrace or share in God's thinking is one that Einstein would not have used even metaphorically in his earlier Zurich-Berne years.

In the Prague Circle Einstein also heard more of Spinoza's blend of science with mysticism. When in 1913 their society Bar Kochba published a collective volume of essays, *Von Judentum*, with such contributors as Bergmann, Buber, Brod, Erich Kahler, and Gustav Landauer, they also included an essay on Spinoza, "Spinoza und das jüdische Weltgefühl" (Spinoza and the Jewish Cosmic Emotion).[22] Einstein in later years, describing himself as sharing Spinoza's conception of God, also used the expression "cosmic religious feeling" as the favorite description for his philosophy. It was the highest stage of religious experience, Einstein wrote in 1930: "I shall call it cosmic religious feeling."[23] "[I]f those searching for knowledge had not been inspired by Spinoza's *Amor Dei Intellectualis*, they would hardly have been capable of that untiring devotion which alone enables man to attain his greatest achievements," he reiterated in 1941.[24] Through the intellectual love of God, the human preoccupation with self could be surmounted by making the mind coextensive as far as possible with a knowledge of the universal laws of nature. As he said in 1934, the scientist's "religious feeling takes the form of a rapturous amazement at the harmony of natural law, which reveals an intelligence of such superiority that, compared with it, all the systematic thinking and acting of human beings is an utterly insignificant reflection. This feeling is the guiding principle of his life and work, insofar as he succeeds in keeping himself from the shackles of selfish desire."[25] He continued, however, to reject the sort of mysticism which like Buber's tried to make our comprehension of the world "a footstool of its incomprehensibility,"[26] that would elevate the place of the human ego in the scheme of existence. Einstein refused to praise man, pamper him, or concede him an epistemological privilege. Determinism for Einstein was to be denigrated neither as an instrumental device, nor as a reductionist, exploitative, or domination-seeking attitude toward nature. No observer, no human actor, had a privilege of exemption from the determinist laws of nature; the fear of death would be conquered only insofar as our minds became identified with the eternal nature of things. Whoever has participated in helping advance the "rational unification" of the experienced manifold, wrote Einstein in later years, "is moved by profound reverence for the rationality made manifest in existence. By way of understanding he achieves a far-reaching emancipation from the shackles of personal hopes and desires, and thereby attains that humble attitude of mind towards the grandeur of reason incarnate in existence, and which in its profoundest depths, is inaccessible to man."[27]

Although Einstein explicitly denied any doctrine of a personal God, when he tried to explain the grounds of his mystic conviction in the rationality of the world, his argument took the form that a divine intelligence was made manifest in nature's laws. He explained himself most fully toward the end of his life in a letter to his old friend, Maurice Solovine:

> You find it curious that I regard the intelligibility of the world (in the measure that we are authorized to speak of such an intelligibility) as a miracle or an eternal mystery. Well, *a priori* one should expect a chaotic world, which could in no way be grasped by thought.

> One could (or *one should*) expect that the world would be rendered lawful only to the extent that we intervene with our ordering intelligence. It would be a kind of order like the alphabetical order of the words of a language. The kind of order, on the contrary, created, for example, by Newton's theory of gravitation, is of altogether different character. Even if the axioms of the theory are set by men, the success of such an endeavor presupposes in the objective world a high degree of order that we were *a priori* in no way authorized to expect. That is the "miracle" that is strengthened more and more with the development of our knowledge.
>
> Here is where the weak point is found of the positivists and dedicated atheists who regard themselves fortunate because they feel they have not only with entire success deprived the world of gods but also "despoiled" it of miracles. The curious thing is that we have to content ourselves with recognizing the "miracle" without having a legitimate way of going beyond it. . . .[28]

The slowly growing Jewish mystical component in Einstein's personality evidently also contributed to the unhappiness he came to feel in his first marriage. Einstein's wife, Mileva Maritsch, his fellow student at Zurich and loyal companion during his years of hardship, had checked his mathematical deductions in his first papers. A Serbian by nationality, Serbian Orthodox in religious background, a freethinker and radical in her ideas, Mileva a decade later on September 21, 1913, had their two small children baptized. She did so while visiting her native country, and on the suggestion of her father and brother. Einstein professed not to take it seriously.[29] Michele Besso, Einstein's old close friend, recalling in 1928 his crucial role in Einstein's intellectual and personal life, wrote him: "perhaps it was my defence of Judaism and the Jewish family that led to your family life taking a different turn, and that I had to take Mileva back from Berlin to Zurich."[30] Einstein and his wife were separated in 1914;[31] their marriage had begun to founder in Prague.[32] Their life in Prague differed from that in Zurich chiefly in the character of their social circle; it is doubtful that Mileva could have felt at home in the Prague Circle of Jewish mystics. Subsequently, Einstein married his cousin and childhood friend, Elsa. It was doubly a return to the maternal Jewish fold, for Elsa was doubly his cousin; her father was a cousin of Einstein's father, and her mother was the sister of Einstein's mother.[33] Novel tensions, however, commenced in Einstein's life;

whereas Mileva had been taciturn and averse to social amenities, Elsa was vivacious and fond of social occasions. Upon the death in 1955 of Besso, Einstein commented sadly that Michele had had the good fortune to live happily, in "constant accord" with one wife, while as for himself, "I have failed lamentably twice."[34] So casual was Einstein in his way of informing Max Born that Ilse, his second wife, was dead, "the death of my mate who was more attached to human beings than I," that Born, noting it "rather strange" added that Einstein was "totally detached from his environment and the human beings included in it."[35]

III

Einstein's Political Oscillations: The Unlovely Laws of the Social Universe

Einstein's heuristic notion, that the laws of nature were such that one could say that a divine intelligence had fashioned them, we might call the "deiform" principle; not only are the laws of nature uniform throughout all space-time but their form has a simplicity and beauty that bespeak God's thought. The deiform principle became Einstein's beacon increasingly during the latter part of his scientific career. But the deiform principle comes to grief when applied to social science, to social and political problems. If a person studying the history and workings of societies were to ask, how would God have solved these problems, he would be utterly at a loss, and give way either to hopelessness or to utopian fantasy. In 1953 Einstein enunciated a law of human stupidity: "The majority of the stupid is invincible and guaranteed for all time. The terror of their tyranny, however, is alleviated by their lack of consistency";[36] did this law of human stupidity bear the stamp of a divine intelligence? And no sociological theorist—Malthus, Marx, Weber, Durkheim, Pareto—has ever attested admiringly to a harmony, the counterpart for sociological law to that which Einstein expressed for physical law.

Indeed, his belief in the deiform character of nature's laws probably explains in part why Einstein kept oscillating between socialist-communist longings and feelings of disillusionment. With the advent and aftermath of the First World War, Einstein affiliated himself with a long series of causes and committees; almost always the upshot was disaffection with sociological law. The same temperament that would find delight in the general theory of relativity and gravitation could find no equivalent in any equally imposing general theory of society, and the attempts to construct the second were pitiful in contrast to the first. Pondering the human mutual destruction in the First World War in 1917, and citing the case of Nernst whose two sons had been killed in the fighting, Einstein asked whether Jehovah was dead: "Der alte Jehovah lebt noch?"[37] The God that replaced Jehovah was a Supreme Physicist but evidently an Incompetent Sociologist.

Einstein during the First World War had stood apart as the German intellectual class was almost completely engulfed by a military enthusiasm. For the first time in his life he joined a political society, the Bund Neues Vaterland, a "left-leaning" pacifist group, constituted in 1915 of a handful of distinguished thinkers. After the war, it evolved into the German League for Defending the Rights of Man (Deutsche Liga für Menschenrechte). The high point of his pacifism came on December 14, 1930, when Einstein declared that "even if only two percent of those assigned to perform military service should announce their refusal to fight," wars would end.[38] Less than three years later Einstein revoked his view, declaring instead that Western Europe, confronted with Hitler, had no alternative but to rearm; he hoped they would not wait, but would attack first, and he, for one, would not refuse military service.[39] Pacifists thereupon denounced Einstein as an "apostate."

Stirred by the Bolshevik Revolution in October 1917, Einstein became a sponsor to various organizations of fellow travellers. When the adroit Communist organizer, Willi Münzenberg, organized in 1927 a conference of the League for the Struggle against Imperialism, he was able to persuade Einstein, as an honorary member of its presidium, to address a message condemning the white nations who "ruled the world": "I . . . am convinced that the successful accomplishment of the task which you have undertaken is in the interest of all to whom the dignity of man is dear."[40]

Soon, however, Einstein was appalled by the evils that the new Soviet ethos generated. Authenticated documents emerged from the Soviet Union during the early twenties written by the political inmates of prisons and prison camps, describing their tortures, maltreatment, and the Soviet disregard for law. Gathered together in a volume entitled *Letters From Russian Prisons* by the anarchist Alexander Berkman, the civil libertarian Roger Baldwin and the staunch anti-Communist newspaperman Isaac Don Levine, this volume a half-century ago was in every sense the forerunner of Alexander Solzhenitsyn's *The Gulag Archipelago*. To it in 1925 Einstein contributed an introduction:

> You will contemplate with horror this tragedy of frightfulness in Russia. . . . And it is just the best, the most altruistic individuals who are being tortured and slaughtered. . . .
>
> All serious people should be under obligation to the editor of these documents. Their publication should contribute to affecting a change in this terrible state of affairs. For the powers that be in Russia will be compelled to alter their methods after the appearance of these letters in print, if they desire to continue their attempt to acquire moral standing among the civilized peoples. They will lose the last shred of sympathy they now enjoy if they are not able to demonstrate through a great and courageous act of liberation that they do not require this bloody terror to put their political ideas in force.[41]

When, in the autumn of 1930, the Soviet regime staged one of its first fraudulent trials, that of the Menshevik economic and technological "specialists," Einstein once more joined in signing a protest.[42] The next year, in 1931, he signed an appeal to the Soviet government protesting "the harsh and restrictive rules" that were being used to persecute the "teachers and students of the Hebrew language."[43] This was an era of appeals and protests; it is doubtful that any of them were to avail. Einstein had greeted the advent of the German Republic in November 1918 by signing the call for the German Democratic Party, a liberal, middle-class political organization whose most famous member was Walther Rathenau, a man whom Einstein deeply admired for his combination of political wisdom, philosophic depth, organizing ability and scientific talent: Einstein was soon mourning his assassination.[44] Fourteen years later, early in 1933, with the Weimar Republic near shipwreck, Einstein signed a call for an immediate union of the Social Democratic and Communist parties to halt Adolf Hitler's drive to power. But the Marxist and liberal parties were broken in spirit, lacking in resilience, and overwhelmed by a sense of inexorable catastrophe. Einstein had also joined in 1932 a communist-led group, the Organizing Committee of the League against War and Fascism, that denounced capitalism as the cause of war, and called upon all to defend the Soviet Union. Both

Friedrich Adler and Leon Trotsky assailed this Stalinist organization. Trotsky trenchantly observed: "the 'left' intelligentsia of the West has gone down on its knees before the Soviet Union."[45] Einstein's own anger with Karl Kautsky, the inveterate Social Democratic critic of Soviet communism, still remained so sharp that in 1938 he refused to endorse his nomination for a Nobel Peace Prize; Kautsky, he maintained, had as editor of the German war documents not exposed the German guilt, especially in its invasion of Belgium.[46]

Toward the end of the Second World War, with the Soviet battle against Hitler fresh in people's minds in 1945, Einstein declared that the Soviet dictatorship had been a historical necessity because the Russian people had lacked political education. "If I had been born a Russian," he said, "I believe I could have adjusted myself to this situation."[47] This last proposition indicates the extent to which Einstein's deiform principle could mislead him on social and political questions. For Einstein also insisted that without freedom "there would have been no Shakespeare, no Goethe, no Newton, no Faraday, no Pasteur, and no Lister," that "only a free individual can make a discovery," and he had asked sardonically: "Can you imagine an organization of scientists making the discoveries of Charles Darwin?"[48] Above all, Einstein knew the bizarre story of the suppression of the theory of relativity in the Soviet Union. He had corresponded with Max Born in 1929 concerning the persecution of adherents to the theory of relativity in the Soviet Union, and had seen that the situation had gotten much worse in the thirties.[49] But with the advent of Hitler's power, Einstein refused to criticize the Soviet Union; he declined, for instance, to criticize Stalin's execution of political prisoners after a Politburo member was assassinated, writing to Isaac Don Levine in 1934 that while in Germany "many thousands of Jewish workers" were being "driven to death . . . without causing a stir in the non-Jewish world. . . . The Russians have proved that their only aim is really the improvement of the lot of the Russian people. . . . Why direct the attention of public opinion in other countries solely to the blunders of this regime?"[50]

The so-called "Moscow trials" of the late thirties exhibited strikingly how Einstein's social perception could be warped by the desire to remake the world as God should have made it. At first, he felt that the trials were plainly frame-ups, "a case of a dictator's despotic acts, based on lies and deception." The fellow travellers in Einstein's confidence persuaded him otherwise, "that the Russian trials are not faked, but that there is a plot among those who look upon Stalin as a stupid reactionary who has betrayed the ideas of the revolution. . . . [T]hose who know Russia best are all more or less of the same opinion."[51] He oscillated between projective fantasy and empirical perception; if his idealistic longings made him suggestible to emotional appeals, his sense of reality would later reassert itself. Einstein would never have accepted such claims to knowledge in his own scientific thinking. Learning traumatically from his social experience, he denounced the communist purge trials in Prague in 1952 as bereft of any shred of legality.[52] In 1954, in the course of a correspondence on the Rosenberg case, Einstein stated his final view of the Moscow trials: he did not believe that the leaders executed by Stalin in the thirties were guilty of the alleged crimes.[53]

Personalities whose utterances had a prophetic accent exercised an inordinate attraction upon Einstein. Unfortunately their prophecies were less of the self-fulfilling than the self-confuting kind. Occasionally too Einstein himself experienced the impact of Soviet agencies of repression. Angered when the Soviet delegates at the World Congress of Intellectuals in 1948 at Wroclaw prevented his paper from being read, Einstein declared: "Such practices cannot contribute to the creation of an atmosphere of mutual trust."[54] Toward the end of his life Einstein's standards of integrity separated him from both the left-wing intellectuals and the dominant American majority. He thought, for instance, that the verdict that Julius and Ethel Rosenberg were guilty of espionage was indeed just, but he added that others were more guilty.[55] Even as he took a strong public stand against the questioning of intellectuals by congressional committees, he perceived as well in 1953 the corruption among the intellectuals themselves in having recourse to "the well-known subterfuge of invoking the Fifth Amendment against possible self-incrimination."[56] When the overwhelming majority of leftist intellectuals pleaded the Fifth Amendment, the inference was plainly in Einstein's eyes that they were not "blameless." Nonetheless, Einstein, in the years after the Second World War "gave his name unhesitatingly to communist fronts, twenty-six of them" (in Max Eastman's words), and was assailed as "among the most willing of the liberal tools of the communist conspiracy."[57]

That Einstein would have survived, as he believed, in a communist environment is altogether doubtful. Those like himself who valued freedom of science perished. He himself had said in 1933 that he could never live "in Russia under the rule of the secret police."[58] But Einstein clung to the curious faith that the genuine scientist could survive unnoticed in the interstices of bureaucracy. He had what might be called a "lighthouse complex"; he felt that such places as lighthouses should be filled "with young people who wish to think out scientific problems, especially of a mathematical or philosophical nature." Einstein was apparently never inside a lighthouse, nor did he know its working conditions.[59] Nevertheless, he felt that under such arrangements, young scientists, free from the pressure to produce results quickly, would grow serenely in creativity. He seemed to return in social fantasy to the golden years from 1902 to 1909 when as a functionary in the Berne patent office, he and Besso, his fellow "patent boy," worked a couple of hours a day and then discussed questions of physics and philosophy. No academic committees, no intrigues, no theses, no promotions, no envious colleagues. Franz Kafka had a similar post in Prague as a clerk in the Workers' Accident Insurance Institute. Kafka's reaction, however, to existence in the bureaucratic interstices was altogether different from Einstein's.[60] Kafka saw the individual overwhelmed in the bureaucratic complexity. Max Born observed the flaw in Einstein's conception: "To be able to practice science as a hobby, one has to be an Einstein."[61] Perhaps it all went back to Einstein's reading of the *Ethics of the Fathers*, a book he adored as a child, with its maxim: "The study of Torah is beautiful when combined with an ordinary occupation." Before the "lighthouse complex" congealed in his mind, Einstein used to advise young scientists to get a "shoemaker's job."[62] When the physicist Leo Szilard arrived in the United States, he was favored by Einstein with a discourse on the beauties of lighthouse-keeping. That might be just the thing for Einstein, Szilard concluded, but it would never do for himself.

Einstein's "lighthouse complex" brought a happy conclusion, for it finally won him an honorary membership in the Plumbers' Union. Asked in November 1954 what he thought of the situation of American scientists, then being interrogated by the hostile congressional committees, Einstein made his oft-cited reply: "If I were a young man again and had to decide how to make a living, I would not try to become a scientist or scholar or teacher. I would rather choose to be a plumber or a peddler, in the hope of finding that modest degree of independence still available under present circumstances."[63]

Although *The New York Times* argued over-powerfully that Dr. Einstein's counsel was a "grievous error," the president of the Plumbers and Steamfitters Union, A.F. of L., Mr. Martin P. Durkin, greeted graciously Einstein's proletarian enthusiasm; Einstein would be welcome to a union card, he said, even if it was a little late to start. In any case, he promised him an honorary union membership.[64] The following week Einstein received a working card and membership in the Chicago Local 130, Plumbers Union.[65] "Plumber Einstein Happy," read the news report. Never has a greater solidarity been achieved between science and the working class. The plumber president, it is pleasant to note, had only recently served as the secretary of labor in the Republican administration of President Eisenhower; he had shown himself honorably militant.

IV

Einstein's Determinism

During the last thirty years of his life Einstein became and avowed himself a complete determinist. This avowal I have suggested, was not a logical consequence of his scientific researches. Einstein, the theorist of Brownian movement, and with his name given to a branch of statistics, might have been expected to welcome the probabilistic-indeterminacy mode of analysis that arose in the mid-1920s. Einstein's mysticism was not necessarily bound to a determinist postulate. Arthur Eddington and James Jeans, one, a Quaker, were equally mystical in their standpoint, yet they finally found an exhilaration in the personal freedom which the principle of indeterminacy seemed to allow.

The sources of Einstein's determinism were, I have tried to show, indeed primarily psychological, sociological, nonlogical. His determinist standpoint had much in common with that of his older contemporaries, the psychoanalyst Sigmund Freud and the biochemist Jacques Loeb. For all these three men, the doctrine of determinism was in large part a defensive-retaliative standpoint against human irrationalities that sometimes seemed overwhelming. All three had experienced intensely the impact of anti-Semitism. Einstein indeed was the only scientist in modern times to have experienced the threats of repeated plots to assassinate him. They were serious. Shortly after the Nazis took power in 1933, Scotland Yard received information that Einstein would be assassinated on October 3rd when he was scheduled to speak at London's Albert Hall together with such men as Jeans, Rutherford, and Austen Chamberlain. The Hall was heavily guarded, though Einstein at the time was in cheerful ignorance.[66] Twelve years earlier such threats of assassination led Einstein to terminate his lectures at Berlin permanently. He wrote to Solovine in July 1922: "Our days have been anxious since the frightful assassination of Rathenau. I am also always on guard, I have discontinued my course and am officially absent, though actually I'm always here. The anti-Semitism is very great."[67] To Max Planck he wrote: "I have been informed independently by serious persons that it would be dangerous for me in the near future to stay in Berlin or, for that matter, to appear anywhere in public in Germany, for I am supposed to belong to that group of persons whom the people are planning to assassinate. . . . [T]he situation obtaining at present allows it to appear highly credible."[68] A person living under the perpetual menace of assassins is probably apt to regard the laws of the much-vaunted human nature as special cases of biophysical and biochemical laws.

Now from the psychological standpoint, there are at least three varieties of determinism. One is like that of Karl Marx, where the thinker exults in an identification with determinist forces that are supposed to guarantee his own victory over the forces of evil. The second is the determinism of the resigned and defeated who regard themselves as destined inevitably to be overwhelmed. The third, which was indeed Einstein's, uses the doctrine of determinism in retaliative denigration against evil, irrational men; it shows that the "free" actions on which they preen themselves are every whit as determined as any chemical reaction. As William James probably said, nothing insults a haughty individual more than subsuming each act of his under a causal law. Einstein, Freud, and Loeb belonged to a group that could not strike back aggressively at their oppressors; the alternative they chose was to espouse a retaliative determinism, aimed at reducing the deeds of their persecutors into the resultants of polygons of social forces, subsuming them to stupidities and envies flattering to no human ego. Bitter against the agencies of evil, natural and human, Einstein finally deemed it fortunate that there was no immortality of the individual self: "It is despite all a good thing that our individual life with all its problems and tensions has an end. . . . Those who invented the belief in individual life after death must have been miserable people!"[69] He too might take comfort from the belief that there would be no afterlife to perpetuate the likes of Hitler.[70]

V

Jewish Loyalties and the Determinist Postulate

Precisely during those years that Einstein's determinist conviction deepened, his concern for Zionist activities grew, a response to the rising hatreds that he saw festering in Europe. The question had been far from his mind while he lived in Switzerland, when, in Einstein's words, "I did not realize my Judaism. There was nothing that called forth any Jewish sentiments in me. When I moved to Berlin all that changed." He was dismayed especially by "the difficulties with which many young Jews were confronted. I saw how, amid anti-Semitic surroundings, systematic study . . . was made impossible for them. This refers especially to the Eastern-born Jews in Germany, who were continually exposed to provocation. . . ." When the German government considered expelling them, "I stood up for them, wrote Einstein, "and pointed out . . . the inhumanity and folly of such a measure." Together with other colleagues, Jews and Gentiles, he added, "I started University courses for these Eastern-born Jews. . . ." He played his violin at gatherings to raise funds for the *Hilfsverein* that was organized to assist impoverished Jewish immigrants. He admired liberal England, where the sense of objectivity was not disturbed by political views, and the theory of relativity had awakened a tremendous enthusiasm, whereas "in Germany the judgment of my theories depended on the party politics of the press. . . ." What perturbed Einstein most deeply was that the German intellectuals especially had made themselves the proponents of anti-Semitism: "And it is just the educated circles who have set themselves up as carriers of the anti-Semitic disease. They have built up for themselves a 'culture' of anti-Semitism. . . ." He was chagrined that these attitudes evidently spread to many German Jews who then proclaimed that they had "nothing to do" with the poor East European Jews.[71] Such reflections led Einstein to agree to visit the United States in 1921 in company with Chaim Weizmann, later the first president of Israel, to advocate the causes of Zionism, the Hebrew University, and Weizmann's leadership. Reluctant at first to undertake such a mission, Einstein finally agreed because he felt the Zionist enterprise required such an individual sacrifice.[72] Later he wrote: "It was in America that I first discovered the Jewish people—I have seen any number of Jews, but the Jewish people I had never met either in Berlin or elsewhere in Germany. This Jewish people, which I found in America, came from Russia, Poland, and Eastern Europe generally. These men and women still retain a healthy national feeling. . . ."[73] When he involved himself, however, in practical Zionist activities on behalf of the Hebrew University, Einstein's methodology, so heuristic in physical theory, showed itself baneful. He became obstinate, perverse, childish, vacillatory, susceptible to poor advice. "A rotten letter of Einstein" in 1933 provoked Chaim Weizmann to say: "The man and his adviser are determined to provoke . . . a great Chillul Hashem (blasphemy)." "A great nuisance," Weizmann called him, "continuing his campaign of bitterness and slander always prompted by his 'new friend' . . . I'm afraid seriously that Einstein must be mentally deranged." The following year in 1934 Weizmann noted that Einstein was "changing his mind for the fourth time during the past year and a half," and he was concerned at the harm Einstein's poor judgment was doing to the University.[74]

Withal, Einstein was indeed about the only one among the great German Jewish physicists who evinced strong Jewish loyalties at this time. This identification persisted until the end of his life. "[M]y relationship to the Jewish people has become my strongest human bond," wrote Einstein in 1952. None among his scientific friends—Max Born, Philipp Frank, Paul Ehrenfest, or even his older personal ones such as Adler, Besso, or Solovine—shared Einstein's degree of self-association with the Jewish people. Max Born, for instance, despite long discussions with Weizmann who proposed that he emigrate in 1933 to Palestine, told Einstein, "I myself have never felt particularly Jewish," and his wife and children even less so.[75] Philipp Frank, later Einstein's biographer, was ironically hostile toward Zionism. German Jewish scientists, who were Nobel laureates, were devoid of such Jewish feelings as Einstein's.[76]

A reconstructed atheist, Einstein agreed to serve as chairman of the Management Committee of the Akademie für die Wissenschaft des Judentums, a society dedicated to the cultivation of Jewish learning.[77] Einstein indeed came to think that the hatred for the intellect was the underlying motive of anti-Semites: "Hence the hatred of the Jews by those who have reason to shun popular enlightenment. More than anything else in the world, they fear the influence of men of intellectual independence. I see in this the essential cause for the savage hatred of Jews raging in present-day Germany."[78] His pessimism as to the lot of man deepened as he saw the virulence of the regressive, anti-civilizational feelings. When the Western democracies made their culminating appeasement of Hitler in 1938, Einstein wrote that were it not for his work, he would no longer want to live.[79]

Beset by the circumpressures of an irrational human world, the postulate of determinism seemed to Einstein essential to preserve one's own rationality. His attraction to Spinoza's character and philosophy focused therefore on Spinoza's doctrine of determinism, his thoroughgoing rejection of human free will. In Einstein's conception of the laws of nature as God's laws, the determinist principle became at least as significant as that of the splendor of the laws themselves; it expressed Einstein's pessimism. Like Spinoza, he added that "a deterministic attitude leads to a more lenient and sympathetic attitude towards one's fellow creatures, to a mitigation of feelings of hatred. This is perhaps the greatest blessing that Spinoza has given to humanity."[80] Discussing with his converted friend Besso in 1948 the precept of Jesus to love one's enemies, Einstein declared: "For me, however, the intellectual foundation lies in a confidence in unlimited causality. 'I am not able to hate him [the enemy] because he *has* to do what he does.' I am thus closer to Spinoza than the prophets. That's why 'sin' doesn't exist for me."[81]

Probably Einstein's tragically unhappy relationship with his younger son, Eduard, strengthened the father's disposition toward a pessimistic determinism. The son, brilliant though without the father's genius, had a tendency to depression that was not tempered by his father's humor and creativity. He came to hate Einstein for having deserted mother and children; he reproached him fiercely. Eduard sustained a breakdown in 1930, from which he never fully recovered. Einstein brooded over his guilt.[82] Had he sacrificed the well-being of his family to his fuller pursuit of scientific work? Was his a new version of the Abraham-Isaac story? Had he, obedient to the vocation of science, constructed an altar upon which to sacrifice his own son? Jehovah had intervened to save Isaac, but no God had intervened to prevent the sacrifice of Eduard Einstein. In anguish in 1917 over the condition of his seven-year-old son, Einstein wrote to Besso his innermost feelings: "Who knows, perhaps it would be better for him to take his departure from this world before really having known life? I really feel guilty towards him, and for the first time in my life hurl reproaches against myself."[83] The guiltless Einstein of the Prague year was replaced by an Einstein who brooded on the conflict of fathers and sons. He read Dostoyevsky's novel *The Brothers Karamazov* in 1920 "with great enthusiasm," writing Paul Ehrenfest that it was "the most remarkable book I have ever read."[84] Probably the tragedy of Eduard led Einstein to change his mind on the validity of Freud's theories. For a long time Einstein had been skeptical of them, not finding them validated in his own dreams. But on April 21, 1936, on the occasion of Freud's eightieth birthday, Einstein declared that he had "had the opportunity of hearing about a few instances . . . which in my judgment excludes any other interpretation than that provided by the theory of repression." He added that "it is always delightful when a great and beautiful conception is proven to be consonant with reality."[85] Evidently, Freud, the theorist of the universality of man's guilt, and the eternal drama of the conflict between father and son, acquired a new significance for Einstein. The determinist philosophy, too, in Spinoza's form, was an answer to the anxieties of guilt, for Spinoza had written that guilt (remorse) was one of the two archenemies of mankind, and that the determinist principle alone could alleviate the burden of guilt by reminding one that the causal sequence of events was a necessary, inevitable one. Thus Einstein's determinism had a therapeutic quality, but it was a strangely nonlogical formation in his thinking.

There can be little doubt, however, that Einstein's pessimism deepened in his last years, even as did perhaps his sense of humor. He told one correspondent that the passions that governed men were predominantly those of hatred and shortsighted selfishness, and in the mood of Ecclesiastes: "Things always remain essentially the same. Nations continue to fall into the same trap, because atavistic drives are more powerful than either reason or acquired convictions." And with cosmic finality: "If all our efforts are in vain and man goes down in self-destruction, the universe will shed no tears."[86] At the same time Einstein continued to urge the merits of world government. A visitor to Einstein's house was once much puzzled why this great thinker who thought so ill of the human race should still be so concerned with such projects as world government. He asked: "Why, Professor Einstein, are you so deeply opposed to the disappearance of the human race?"[87] Einstein was much disturbed by the question. Despite the portrait of Schopenhauer that hung on the wall of one of Einstein's bedrooms, he evidently felt a great resistance toward drawing the last consequence of his determinist premises. He still maintained that if people willed a goal, they could change the impossible into the possible,[88] though this working hypothesis seemed out of keeping with his sociological determinism.

Einstein would refer somewhat ironically to his working conviction in a cosmological, unified field theory, derived from a few mathematical premises that had almost a divine, necessary character, as "Jewish physics." When Max Born wrote an essay against this a priori deductivism as evinced, for instance, in the books of the British scientists Eddington and E.A. Milne, Einstein defended this "quixotic," "Hegelian" element; when it is altogether missing, he wrote, "the inveterate philistine rules. I am therefore confident that 'Jewish physics' is not to be killed."[89] This humorous sally of Einstein's was especially perverse because in September 1944 (when he wrote this letter) the war against the Nazis was at its height. Nazi physicists had introduced the term "Jewish physics" precisely to attack the "abstract," mathematical method of Einstein. From the sociological standpoint, the designation was indeed off the mark: the experimentalist A.A. Michelson, also Jewish, found the theory of relativity uncongenial all his life precisely because it offended his "concrete" common sense; the French Jewish philosopher Bergson argued that the theory lacked the stamp of intuitive reality; moreover, such mystic Christians as Eddington and Milne far exceeded Einstein in their mathematical apriorism. Nevertheless, Einstein borrowed the Hitlerite term, converting it into a slogan for his own mystical conviction that in the highest acts of human knowing there is an ingredient of divination.

VI

Einstein's Friendship Toward Eccentrics

Einstein had a warm, understanding sympathy with the eccentric fringe to the "scientific community," those whom the overwhelming majority of scientists regard as crackpots, or semi-sane, and would even excommunicate. Nobody since William James has acted with such consideration toward these strange errant characters. The memory of those days when he was a penniless, jobless graduate of the Federal Polytechnic never left Einstein—those years when he had felt utterly deserted, and when he was sustained by the companionship of such eccentrics as Michele Besso. Thus he was ready to listen to the aberrant hypotheses of Immanuel Velikovsky and Wilhelm Reich (as he had been to Upton Sinclair), to correspond and discuss with them, and to explain courteously why he thought they were wrong. He wrote Velikovsky in 1955 that he admired the dramatic talent of his book, but appealed to him to show more of a sense of humor in his clash with astronomer Harlow Shapley: "I would feel happy if you could savor the whole episode for its humorous side."[90] Einstein indeed had known Velikovsky since the mid-twenties in Germany when he had helped him in editing the translation of scientific monographs.[91] Again, during the Second World War's most tense days, Einstein on January 13, 1941, received at Princeton for a long talk the bizarre Wilhelm Reich. He listend to Reich expound the powers of a box he had contrived that would heighten and concentrate the alleged "orgone energy" of its resident. Einstein subsequently wrote Reich a letter explaining where his theorizing had gone wrong, and would not pursue the matter further. He incurred in this episode a nuisance loss of time as well as outlandish recriminations. The unfortunate Reich, persuading himself that Einstein had been drawn into a conspiracy instigated by communists against his work, published a brochure of complaint, *The Einstein Affair*.[92] Upton Sinclair once remarked: "All kinds of cranks accept me as a brother."[93] The reverse was just as true for Einstein. Einstein was likewise gently understanding when his friend Besso underwent a religious conversion, leaving his early Machian convictions far behind.

That affection that Einstein had shown for the academic marginal men of the Olympia Circle remained with him until the end of his life. When Einstein died, it transpired that he had named Dr. Otto Nathan, an economist living in Greenwich Village, as trustee for his estate. Nathan was the kind of man with whom Einstein felt most at home, and to whom he looked for information and opinion on all social and political subjects. Nathan's academic career had been a troubled one. German-born, a left socialist, he had found himself rendered unwelcome in a series of American colleges and universities—Princeton, Vassar, New York University, Howard. Responding to a subpoena of the House Un-American Activities Committee, Nathan subsequently went to the courts to secure his right to a passport. His name recurred unfailingly in fellow-travelling and pacifist manifestos and manifestations. Einstein's designation of Nathan as trustee came as a surprise to such an old friend as his biographer, Philipp Frank, who had expected that Einstein would name a scientific companion as himself. And Frank indeed recalled that Einstein had been much displeased with Nathan's behavior at the Soviet-sponsored Wroclaw conference of intellectuals in 1948; Dr. Nathan, acting as the bearer of Einstein's paper, refrained from its reading when the Soviet spokesmen threatened to retaliate with a worldwide campaign against Einstein. Einstein then commented that it would have made no difference to him if they did.

Nonetheless, Otto Nathan was the kind of unattached, protesting intellectual that Einstein prized—a member of the order of presumably free spirits who had failed to come to terms with institutions. This was ultimately the basis of Einstein's choice of comrades. All of Einstein's closest lifelong personal friends had something in them of the lost, isolated eccentric. The inordinate trust that Einstein placed in an Orientalist, Professor Abraham Shalom Yahuda, appalled Chaim Weizmann, in whose view it led to serious damage to the Hebrew University. Weizmann and the famed chemist, Fritz Haber, almost believed that Einstein had taken leave of his senses. Actually, this trait of attraction to the eccentric was an invariant one in Einstein's character.

The name of Michele Angelo Besso, who was Einstein's closest confidant all his life, was thus inscribed in Einstein's first classical paper in 1905 on the theory of relativity. Besso changed very little in later years from the restless searcher whom Einstein knew as a young man; his curiosity was boundless and multidirectional. Enjoying the Bohemian experience of a perpetual student, always enrolling in university courses, he could endure ignorance in no branch of learning. He studied law, celestial mechanics, English literature, the theory of groups, and the physiology of the nervous system. He conceived many intellectual projects but could never complete any. As an electrical engineer, he worked to devise safety measures for industrial installations. But he also lectured at the Federal Polytechnic from 1916 to 1938 on patent law. Through the marriages of their sisters into the same family, Einstein and Besso were, moreover, relatives.[94] Besso had that selfless character, that freedom from the drive for personal aggrandizement, from the competitive compulsion that was the trait Einstein most valued in persons. When a scientist lacked that trait, even such a man as J. Robert Oppenheimer, Einstein could never really feel warm toward him. To Besso, by contrast, he could write: "I have never seen you do or say anything other than what was intended to do good."[95] To him Einstein turned during the trying days of his separation from Mileva and their two sons, relying on Besso to visit them, reassure them, and tell Einstein the latest news.

Einstein received news of Besso in his last years: "Uncle Michele is a gentle, little man who sits in Bern, Switzerland, and looks out into the world, leaning on a white beard that descends from almost under his blue eyes to the end of his necktie. Every night for twenty years, in the company of a friend, he has looked into *The Divine Comedy*, taking time off to look into his soul with a fierce, puritanical spirit tempered by a great deal of natural goodness; he has looked into the field of economics, trying to find mathematical formulae to solve the crises of the world . . ."[96] Michele, said Einstein gallantly, had been too much of a "universal spirit" to get results in science.

Most at home with the defeated, the lost, the bewildered in this world, with those prizing the things of the spirit, knowledge, art and unable to acclimatize themselves to the terms of competitive existence, Einstein too found competitive activities traumatic. He even, therefore, disliked playing chess. "I have to confess that I have always disliked the fierce competitive spirit embodied in that highly intellectual game," Einstein wrote.[97] He took many walks in later years with the world's chess champion, Emanuel Lasker, but reflected that Spinoza had been "luckier" than Lasker, because Spinoza, as a lens-grinder, followed a trade that left his mind "free and independent," whereas Lasker's brilliant philosophic and scientific intelligence was enthralled by the games yet to be won.

Probably Einstein rather enjoyed the isolation of his last twenty scientific years because his pursuit of a unified deterministic field theory left him in a domain with virtually few competitors. He continued to seek out independent, nonacademic intellectuals for free discussion. In the spring of 1937 for instance, he chanced to meet the gifted Max Eastman, the great editor of the prewar *Masses*, and friend and American proponent of Leon Trotsky. Eastman, the interpreter of humor, was reputedly the handsomest man in America, and also its foremost anti-Stalinist intellectual. Einstein was taken with Eastman's unusual argument against determinism, that if the mind were determined in its judgment by antecedent causes, then the validity of all scientific judgments was destroyed because the mind then cannot be determined by the reasons on which it allegedly bases its judgments.[98] Twice that spring Einstein invited Eastman back to Princeton to continue that discussion. Einstein seemingly conveyed in Kantian language that he was indeed preparing to abandon his old universal determinism, and to accept the view that there were certain kinds of phenomena that are uncaused.[99] Eastman was elated by his presumable role in the history of science in persuading Einstein to reverse himself on the determinism issue, and recorded with that charming egotism which he indulged: "Just why I should have been chosen to receive the first news of this revolution, I can not imagine."[100] No doubt this was another intellectual experiment in which Einstein was engaged. What Eastman overlooked was Einstein's readiness to engage in such intellectual experiments with ideas opposed to his own, and then having reconsidered them from all sides, probably to reject them. But Eastman treasured the memory of "the naive generosity with which Einstein gave the gift of attention to anything that anybody said,"[101] his laughter at Eastman's sallies against Marx, dialectical materialism, and John Dewey: "'this man is wicked! He is really wicked!'" And then parting at the door: "'Hasn't it been a delightful discussion!'"[102]

VII

Einstein's Humor and the Nature of Things

Einstein's sense of humor tended to be ironic; he enjoyed reducing high-sounding pomposity to indeed low, perhaps vulgar dimensions. He seems to have enjoyed telling jokes in which "old whores" figured prominently. He even told Max Born that the latter's animosity toward aprioristic physics "reminds me of the beautiful proverb: Young whores—old bigots, particularly when I think of Max Born."[103]

His views on women's abilities aroused indignation; he disturbed Mrs. Born, a writer and enthusiast for Asian mysticism, by referring to "the well-documented truth that the centre of gravity for creative activity is located in different parts of the body in men and women."[104] Mrs. Born in her reply to Einstein referred to "all your shameful sayings fixed in my memory."[105] He was flirtatious with women, as Vera Weizmann has written, a trait that evidently led to deep resentment on the part of Mileva, Einstein's first wife.[106]

Reducing things to their lowest common denominator was the central aspect of Einstein's sense of humor. (Even his special theory of relativity, he said, had been inspired in the spirit of a joke.) Perhaps there was an element of personal malice involved—the retaliation against authority, the overthrow of the stuffed shirt.[107] Katya Adler, the wife of Friedrich Adler, recalled Einstein's first reaction to toilet tissue when they all lived in the same house in Zurich in 1909. Einstein once saw her holding rolls of toilet tissue. Much puzzled, he asked what those were. Mrs. Adler explained their purpose. Einstein replied that his family would continue to use newspapers.[108]

In many ways, Einstein was not a "modern." His attitude toward women, we might note, was much like that of Freud and Schopenhauer. Like Freud, he felt America was afflicted with petticoat government, a view that involved Einstein in an unpleasant interchange, widely publicized, shortly after his visit to the United States. On July 8, 1921, *The New York Times* carried a dispatch from Berlin with arresting headlines: Einstein Declares Women Rule Here / Scientist Says He Found American Men the Toy Dogs of the Other Sex / People Colossally Bored / Showed Excessive Enthusiasm Over Him for Lack of Other Things, He Thinks. Among its direct quotations was one that sounded like an echo of Sinclair Lewis's then popular *Main Street* and *Babbitt*: "There are cities with 1,000,000 inhabitants, despite which what poverty, intellectual poverty! . . . Above all things there are the women who, as a literal fact, dominate the entire life in America.

The men take an interest in absolutely nothing at all. They work and work, the like of which I have never seen anywhere yet. For the rest they are the toy dogs of the women, who spend the money in a most unmeasurable, illimitable way and wrap themselves in a fog of extravagance . . . now quite by chance they have thrown themselves on the Einstein fashion." "The magic power of mystery," "of what they cannot conceive," he thought, had an allure for them. And he concluded with the remark: "[T]o compare the general scientific life in America with Europe is nonsense."

These ungenerous remarks were perhaps just another of the anti-American deprecations customary among European visitors from Charles Dickens to the worst of them, Bertrand Russell. *The New York Times* firmly rebuked Einstein by writing that "it is a well-known fact that high development in one direction has underdevelopment elsewhere to balance it,"[109] that a great physicist could be a superficial sociological observer, and that Einstein's remark was "ludicrously and offensively false." Perhaps Einstein was irritated, the editorial suggested, because he and Weizmann had largely failed in their American mission, having indeed aroused "antagonism" rather than approval. Nonetheless, Einstein's bad manners and sociological ignorance said the *Times*, would not make them any "less ready" to honor his scientific achievement. Especially effective in rebutting Einstein was the editor of the *Popular Science Monthly*, Kenneth W. Payne. Americans had a higher average education than Europeans, Germans in particular he noted, and their interest in science was much greater. Payne pointed to the "over million and a half readers of the popular scientific magazines" and the three and a half million of technical magazines: "What European nation can even approach such figures showing widespread popular interest in science? . . . And where are the European papers that give the same consistent play to scientific developments that we do?" Einstein "has completely misinterpreted the popular sensation accompanying his reception here," he concluded.[110]

Probably Einstein ignored the sociological truth that unless there is a widespread popular interest in scientific work, and a reasoned support for its achievement, the scientific institutions will exist precariously. The German people may have led the world in terms of the number of its Nobel Prize winners, but there was no popular scientific reagent to neutralize the anti-intellectualism that spread in Hitler's era. The high estate of the scientific community concealed perhaps the fact that it existed in a socially suspended state.

The women of Chicago were understandably indignant with Einstein's Schopenhauerian denigration of their sex. "Perfectly ridiculous," said one woman spokesman, though the press reported that the Chicago men seemed to agree with Einstein's assessment. On the other hand, a professor from Northwestern's chemistry department judged it "incredible" that a man of Einstein's attainments could have made such statements. Millikan at Chicago opined charitably that Einstein's judgment had been impaired by the "hectic time" he had had.[111] Einstein, discomfitted, tried unconvincingly to disavow or mitigate the original interview. Unimpressed, the *Times* still felt that he had "made but a poor return" for the cordial reception he had been accorded.[112] A later news story did, however, find the actual text of the interview to have been "not as harsh" as the original cabled report, since it included a tribute to the younger generation, especially *The New Republic* group.[113] Probably Einstein had indeed met several of the editors and writers of that organ of liberal intellectuals—Herbert Croly, Alvin Johnson, Morris R. Cohen, John Dewey. *The New Republic*, in articles published before and after Einstein's visit, declared that the theory of relativity provided the basis for a new American liberal social philosophy. Leaving aside the dubious ideological reasoning in this claim, it did make for a sense of comradeship between Einstein and the American radicals, so that he could repeat in trusting faith the radical indictments of American civilization. Whether the Einstein-Freud standpoint toward women is a trait linked with the highest scientific genius has not been sufficiently explored by students of the scientific character. Einstein himself seems to have been fonder of his sister Maja than of any other person in his life; in the years of her last illness, he would sit by her bedside and read to her for hours from such books as the Greek classics, Sophocles, Aeschylus, Thucydides, and with much pleasure, Bertrand Russell's new *A History of Western Philosophy* ("He has remained a kind of scoundrel [Lausbub] till his oldest years").[114] Perhaps, in a sense, so had Einstein.

VIII

Revolutionist or Frontiersman?

Einstein has been much characterized as the maker of a scientific revolution. The analogy of social revolution, extended to the development of scientific ideas, did strongly appeal to him in early manhood. In his later years, however, the image of the revolutionist in science became for him a recessive one. Rather, he was the explorer, held in wonder and awe, by the mysterious laws of the universe.

Two metaphors indeed have competed for dominance among scientists: on the one hand, the one now highly popular of the "scientific revolutionist"; but there frequently recurs that of the pioneer, the frontiersman, the explorer. Bridgman and Szilard thought of themselves as frontiersmen, "advancing into new territory," in Bridgman's words. Szilard offered a course not on scientific revolutions but on frontiers of science. When Ernest Rutherford died in 1937, Chadwick wrote: "He was indeed a pioneer—a word he often used—at his best in exploring an unknown country, pointing out the really important features and leaving the rest for others to survey at leisure." And Vannevar Bush in 1940 wrote his celebrated memorandum to the United States government, *Science: The Endless Frontier.* He did not write *Science: The Permanent Revolution* with an unconscious Trotskyism.[115]

Undoubtedly the notion of the scientist as revolutionist has illumined certain traits and moments in the lives of scientists. But Francis Bacon, we might note, who grasped most prophetically the spirit of the emerging science, in explicating his Tables of Inventions, invoked the example of Columbus, seeking out "new lands and continents."[116] And Bacon too specifically thought that the character of the scientific and inventive activity differed from that of the social revolutionist.[117]

From the psychological standpoint, the pioneer frontiersman has much in comon with the social revolutionist. Both are usually young men who have felt it necessary to break with an established way of life. In Europe very often frustrated revolutionists became emigrants, while such a votary as the anarchist Bakunin said nature had actually intended for him to be a bold pioneer on the American frontier. The metaphor of the frontier, however, adds a cluster of traits omitted by that of revolution. For one, the pioneer lives in loneliness, works in isolation, and in that sense the work of Einstein was very much, as he insisted, the product of a man who had sought solitude all his life.[118] Newton and Darwin shared very much this propensity toward isolation. In Einstein's case, at a time when people were asserting their identifications with parties, classes, movements, there came instead his statement: "I am a horse for single harness, not cut out for tandem or team work. I have never belonged wholeheartedly to country or state, to my circle of friends, or even to my own family."[119]

Revolutions on the other hand are the work of party committees, caucuses, factions, but as Einstein had asked sarcastically, what organization could have produced the discoveries of a Charles Darwin?[120] Moreover, the pioneer, once he has opened the frontier, founds cultural institutions that transmit all that he most prizes from his original land; the great frontier societies, from the time of the Hellenic to the British colonies, have been marked by a cultural continuity. In a similar spirit, Ludwig Boltzmann, the guide to kinetic theory at the century's turn, explained that "the development of experimental physics proceeds almost continuously and without any leaps that are too sudden and this is why it is never visited by great revolutions or commotions."[121] Einstein too came especially to emphasize the continuity of the theory of relativity with the great scientific tradition.

Lastly, the pioneer has a sense of the unknown that is never vouchsafed to the revolutionist. The maker of revolutions is generally self-confident, dogmatic, with an assurance that all history is moving inevitably with him; he believes the law of history is clear and that there are indeed no scientific social mysteries. The political conflict itself, moreover, involving the would-be revolutionists in a quest for personal power, makes their sense of relative status always strong among the revolutionists. From Robespierre to Marx and Lenin, none has a sense of mystery, of the humility of man's groping intellect. This sense of the unknown, however, of transcendent realities, of the microcosmic status of man, was common to Newton, Darwin, and Einstein. As Einstein wrote: "In my long life I have learned one thing: it is damn hard to get close to Him, as soon as one wishes to go beyond the surface."[122] Einstein, indeed, on his fiftieth birthday, asserted that "the psychological urge" lying at ths source of the scientist's work (as well as of the artist), is "the individual pioneering effort."

Every devotee of some transient crackpot fashion now reiterates that he is bringing science a new "paradigm"; the devoted labor and mastery of the previous science which characterized such men as Darwin, Planck, and Einstein is forgotten, and the concept of "scientific revolution" is thus transmuted into an ideology of antiscience. Perhaps this generation of growing students and teachers may have had their surfeit of metaphors of revolution. To them Einstein offers the example of the life of the pioneer in science in the fields of the unknown, an activity the least sullied by selfishness that the world has known.

NOTES

*Parts of this Prologue were read at the Symposia to commemorate Einstein's centenary anniversary convened in 1979 by the Royal Society of Canada in Toronto, and the American Chemical Society in Washington, D.C.

1. Sofie Lazarsfeld, "Did Oedipus Have an Oedipus Complex?," *The American Journal of Orthopsychiatry*," Vol. XIV, 1944, pp. 226-29.
2. Hannah S. Decker, *Freud in Germany: Revolution and Reaction in Science, 1893-1907*, New York, 1977, pp. 13-14.
3. Cited in the authoritative book, Nathan G. Hale, Jr., *Freud and the Americans: The Beginning of Psychoanalysis in the United States, 1876-1917*, New York, 1917, p. 20.
4. Leopold Infeld, *Albert Einstein: His Work and Its Influence on Our World*, revised edition, New York, 1950, p. 46.
5. Albert Einstein and Michele Besso, *Correspondance 1903-1955*, tr. Pierre Speziali, Paris, 1972, p. 51.
6. Anton Reiser, *Albert Einstein: A Biographical Portrait*, New York, 1930, pp. 85-86. Philipp Frank, *Einstein: His Life and Times*, trans. George Rosen, New York, 1947, pp. 98.
7. Einstein and Besso, *Correspondance 1903-1955*, p. 49. Eric G. Forbes, "Erwin Finlay Freundlich," *Dictionary of Scientific Biography*, New York, 1972, vol. V, p. 182.
8. Reiser, *Albert Einstein*, p. 85.
9. Hans Tramer, "Prague: City of Three Peoples," *Publications of the Leo Baeck Institute: Year Book IX*, London, 1964, p. 313.
10. Max Brod, "Judaism and Christianity in the Work of Martin Buber," in Paul Arthur Schilpp and Maurice Friedman, eds., *The Philosophy of Martin Buber*, La Salle, 1967, pp. 322-23.
11. Cf. Grete Schraeder, *The Hebrew Humanism of Martin Buber*, tr. Noah J. Jacobs, Wayne State University Press, Detroit, 1973, pp. 127-28. Hans Kohn, *Martin Buber: Sein Werk Und Seine Zeist*, Wiesbaden, 1979, p. 90.
12. Hans Kohn, *Living in a World Revolution: My Encounters with History*, New York, 1964, p. 64.
13. Curt D. Wormann, "German Jews in Israel: Their Cultural Situation since 1933," *Publications of the Leo Baeck Institute, Year Book XV*, London, 1970, p. 86. Ruth Kestenberg Gladstein, "The Jews Between Czechs and Germans in the Historic Lands, 1848-1914," in *The Jews of Czechoslovakia: Historical Strikes and Surveys*, New York, 1968, vol. I, p. 54. Hugo Bergmann later married Mrs. Fanta's daughter, Else.
14. Carl Seelig, *Albert Einstein: A Documentary Biography*, tr. Mervyn Savill, London, 1956, p. 121. Hans Tramer, "Prague," pp. 329, 317-18. Max Brod, *Der Prager Kreis*, Stuttgart, 1966, pp. 36-37, 57, 111-12. Landauer's life was in jeopardy in 1919 after his participation as minister of culture in the Munich Soviet Republic; thereupon Einstein joined with Buber and Margarete Susmann and other friends in a vain effort to save him. Ruth Link-

Salinger (Hyman), *Gustav Landauer: Philosopher of Utopia*, Indianapolis, 1977, p. 101.

15. Chaim Schatzker, "Martin Buber's Influence on the Jewish Youth Movement in Germany," *Publications of the Leo Baeck Institute, Year Book XIII*, London, 1978, p. 151.
16. Max Brod, *Streitbares Leben 1884-1968*, Munchen, 1968, pp. 201-02.
17. Charles Hartshorne, "Buber's Metaphysics," in Paul A. Schilpp and Maurice Friedman, eds., *The Philosophy of Martin Buber*, La Salle, 1967, p. 58.
18. Anton Reiser, *Albert Einstein*, pp. 28-29.
19. Carl Seelig, *Albert Einstein*, p. 113. Also, cf. Banesh Hoffman, "Albert Einstein," *Publications of the Leo Baeck Institute, Year Book XXI*, London, 1976, pp. 281, 285.
20. Einstein and Besso, *Correspondance 1903-1955*, pp. 346-47.
21. Banesh Hoffman, "Albert Einstein," p. 284.
22. Margarete Susmann, *Ich Habe Viele Leben Gelebt*, Stuttgart, 1966, p. 78.
23. Albert Einstein, *Ideas and Opinions*, New York, 1976, p. 47.
24. Ibid., p. 61.
25. Ibid., p. 50.
26. Buber's words are cited in Hugo Bergman, "Martin Buber and Mysticism," in Schilpp and Friedman, *The Philosophy of Martin Buber*, p. 298.
27. Albert Einstein, *Out of My Later Years*, New York, 1967, p. 33.
28. Albert Einstein, *Lettres à Maurice Solovine*, Paris, 1956, pp. 114-15.
29. Carl Seelig, *Albert Einstein*, p. 113. Desanka Durić-Trbuhović, *U senci Alberta Ajnštajna (In the Shadow of Albert Einstein)*, Krushevatz, Yugoslavia, 1969.
30. Einstein and Besso, *Correspondance 1903-1955*, pp. 238-39.
31. G.J. Whitrow, ed., *Einstein: The Man and his Achievement*, London, 1967, p. 30; New York, 1973.
32. Carl Seelig, *Albert Einstein*, p. 119.
33. Ibid., p. 158. Frank, *Einstein*, p. 123. Einstein was formally divorced from Mileva on February 14, 1919. According to Frank, he married his cousin Elsa "while the war was still going on." Some error in dating seems to have taken place. Cf. Seelig, *Albert Einstein*, p. 148; Frank, op. cit., p. 124.
34. Einstein and Besso, *Correspondance 1903-1955*, p. 538.
35. *The Born-Einstein Letters: Correspondence between Albert Einstein and Max and Hedwig Born from 1916 to 1955*, tr. Irene Born, New York, 1971, p. 128-30.
36. Einstein, *Ideas and Opinions*, p. 38. The document was displayed at the Einstein centenary exhibition in May 1979 at the Leo Baeck Institute in New York.
37. Einstein and Besso, *Correspondance 1903-1955*, pp. 114, 115.
38. Margarete Turnowsky-Pinner, "A Student's Friendship with Ernst Toller," *Publications of the Leo Baeck Institute, Year Book XV*, London, 1970, p. 216. Eugene Lunn, *Prophet of Community: The Romantic Socialism of Gustav Landauer*, Berkeley, 1973, pp. 248-49.
39. Otto Nathan and Heinz Norden, *Einstein on Peace*, New York 1960; reprinted 1968, pp. 117, 229, 230.
40. Babette Gross, *Willi Münzenberg: A Political Biography*, tr. Marian Jackson, Michigan State University Press, 1974, pp. 185, 189. Werner T. Angress, *Stillborn Revolution: The Communist Bid for Power in Germany, 1921-1973*, Princeton, 1963, pp. 346-47. Helmut Gruber, "Willi Münzenberg: Propagandist For and Against the Comintern," *International Review of Social History*, vol. X (1965), Part 2, p. 193.

41. Albert Einstein, Introductory Letter, in Alexander Berkman, ed., *Letters from Russian Prisons*, New York, 1925, reprinted Westport, Conn., 1977, p. 7. Also cf. Roger N. Baldwin, *Liberty Under the Soviets*, New York, 1928, p. 2. Isaac Don Levine, *Eyewitness to History*, New York, 1973, p. 169.
42. David Caute, *The Fellow-Travellers: A Postscript to the Enlightenment*, London, 1973, p. 52.
43. Margarita Pazi, "Arnold Zweig and Max Brod," *Publication of the Leo Baeck Institute, Year Book XXIII*, London, 1978, p. 259.
44. Bruce B. Frye, "The German Democratic Party and the 'Jewish Problem' in the Weimar Republic," *Publications of the Leo Baeck Institute, Year Book XXI*, London, 1976, p. 152.
45. David Caute, *The Fellow-Travellers: A Postscript to The Enlightenment*, London, 1973, pp. 133, 210.
46. George Garvy, "Albert Einstein and the Nobel Peace Prize for Karl Kautsky," *International Review of Social History*, vol. XVIII, 1973, Part 1, pp. 107-10.
47. Albert Einstein, *Out of My Later Years*, p. 174.
48. Ibid., pp. 140, 175-76.
49. *The Born-Einstein Letters*, p. 101ff.
50. Levine, *Eyewitness*, pp. 171-72.
50. *The Born-Einstein Letters*, p. 130. Max Eastman tells that in June 1937 Einstein gave his opinion of the "confessions" at the Moscow trials: "'Of course they are not true. It is impossible that twenty men being caught in a conspiracy, would react in the same way—and that in so unnatural a way as to defile themselves publicly.'" Eastman, *Great Companions: Critical Memoirs of Some Famous Friends*, New York, 1959, p. 33.
52. Nathan and Norden, *Einstein on Peace*, p. 571.
53. Letter of Albert Einstein to Irwin Edelman, September 8, 1954.
54. Nathan and Norden, *Einstein on Peace*, p. 493.
55. Seelig, *Albert Einstein*, p. 206.
56. Einstein, *Ideas and Opinions*, pp. 43-44. Nathan and Norden, *Einstein on Peace*, p. 547.
57. Eastman, *Great Companions*, pp. 34-35.
58. Nathan and Norden, *Einstein on Peace*, p. 234.
59. Einstein, *Out of My Later Years*, p. 141.
60. Max Brod, *Franz Kafka*, p. 84.
61. *The Born-Einstein Letters*, p. 108.
62. Reiser, *Albert Einstein*, pp. 65, 89.
63. Albert Einstein, "Correspondence," *The Reporter*, vol. 11, no. 9, Nov. 18, 1954, p. 8. Einstein's opinion had evidently been asked concerning two articles by Theodore H. White, "U.S. Science: The Troubled Quest," I and II, in *The Reporter*, vol. 11, no. 4, Sept. 14, 1954, pp. 12-18, and No. 5, Sept. 23, 1954, pp. 26-27, 30-34. Nathan and Norden, *Einstein on Peace*, p. 613.
64. *The New York Times*, "Dr. Einstein's Quest," November 11, 1954, p. 30. Ibid., p. 33.
65. Ibid., Nov. 20, 1954, p. 31.
66. Nathan and Norden, *Einstein on Peace*, pp. 236-37.
67. Albert Einstein, *Lettres à Maurice Solovine*, p. 43.
68. Carl Seelig, *Albert Einstein*, pp. 178-79. When Einstein went, however, to the new University of Hamburg in 1920 to deliver a lecture, he was, despite the anti-Jewish demonstrations that had taken place in 1919, greeted by the student corps with applause. Ernst R. Loewenberg, "Jakob Loewenberg: Excerpts from his Diaries and Letters," *Publications of the Leo Baeck Institute,*

Year Book XV, London, 1970, pp. 197-198.
69. Einstein and Besso, *Correspondance*, p. 474.
70. Bertrand Russell, *Religion and Science*, New York, 1935, p. 143.
71. Albert Einstein, *About Zionism: Speeches and Letters*, London, 1930, pp. 22-27. "James Simon: Industrialist, Art Collector, Philanthropist," *Publications of the Leo Baeck Institute, Year Book X*, London, 1965, p. 21.
72. Kurt Blumenfeld, "Einstein on Zionism," *Jewish Frontier*, vol. XLVI, March 1979, p. 11.
73. Einstein, *About Zionism*, p. 25.
74. Barnet Litvinoff, ed., *The Letters and Papers of Chaim Weizmann*, Vol. XVI, New Brunswick, pp. 17, 18, 36, 44, 319.
75. *Born-Einstein Letters*, p. 116ff., 176-77. Also, p. 124.
76. Chaim Weizmann, *Trial and Error: The Autobiography of Chaim Weizmann*, New York, 1949, pp. 350-55.
77. Ernst Hamburger, "Hugo Preusz: Scholar and Statesman," *Publications of the Leo Baeck Institute, Year Book XX*, London, 1975, p. 204.
78. Einstein, *Ideas and Opinions*, p. 195.
79. Einstein and Besso, *Correspondance*, pp. 330, 332.
80. Letter to Rabbi Abraham Geller, April 17, 1933, at Leo Baeck Museum, New York.
81. Einstein and Besso, *Correspondance*, p. 393.
82. Peter Michelmore, *Einstein: Profile of the Man*, pp. 146-48.
83. Einstein and Besso, *Correspondance 1903-1955*, p. 103.
84. Carl Seelig, *Albert Einstein*, p. 159. Also, cf. Ilya Ehrenburg, *Post-War Years 1945-1954*, tr. Tatiana Shebunina, Cleveland, 1967, p. 77.
85. Nathan and Norden, *Einstein on Peace*, p. 186. Ernest Jones, *The Life and Work of Sigmund Freud*, Vol. 3, New York, 1957, p. 203.
86. Cf. Nathan and Norden, *Einstein on Peace*, pp. 556, 567, 568.
87. Alfred Werner, "Albert Einstein and His God," *Congress Weekly*, vol. 22, no. 17, May 2, 1955, p. 6.
88. Nathan and Norden, *Einstein on Peace*, p. 567.
89. *The Born-Einstein Letters*, p. 149.
90. Stephen L. Talbott, et al., eds., *Velikovsky Reconsidered*, New York, n.d., pp. 64-65. Horace M. Kallen, "Shapley, Velikovsky, and the Scientific Spirit," in ibid, p. 65.
91. Immanuel Velikovsky, "Genesis of the first Jerusalem 'Scripta,'" *The Jewish Quarterly*, vol. 26, Spring, 1978, pp. 16-17.
92. Ilse Ollendorff Reich, *Wilhelm Reich: A Personal Biography*, New York, 1969, reprint, 1970, pp. 84-87.
93. Leon Harris, *Upton Sinclair: American Rebel*, New York, 1975, p. 262.
94. Einstein and Besso, *Correspondance 1903-1955*, pp. 105-06.
95. Ibid., pp. xxviii-xxix.
96. Niccolo Tucci, "A Reporter at Large: The Great Foreigner," *The New Yorker*, vol. XXIII, November 22, 1947, p. 46.
97. J. Hannock, *Emanuel Lasker: The Life of a Chess Master*, Foreword by Albert Einstein, tr. Heinrich Fraenkel, New York, 1959, p. 7.
98. Max Eastman, *Great Companions*, p. 26.
99. Ibid., p. 30.
100. Ibid., p. 31.
101. Ibid., p. 26.
102. Ibid., p. 28.
103. *The Born-Einstein Letters*, p. 149. Also cf. Peter Michelmore, *Einstein: Profile of the Man*, New York, 1962, p. 43.

104. *The Born-Einstein Letters*, p. 92.
105. Seelig, *Albert Einstein*, p. 180.
106. In 1947, I inquired of Philipp Frank and his wife as to what had led to the estrangement of the Einsteins; all he would say cryptically was that when Einstein was young, he was very handsome, and looked like a movie actor.
107. Einstein wrote Besso in 1918 that the sessions of the Berlin Academy were "curious: the majority of these old gentlemen are mixed up and stupefied.... When I think of Treitschke... a god beside the super-god Bismarck... I have the reputation of a thoroughgoing socialist [sozi]." Einstein and Besso, *Correspondance*, pp. 145-46.
108. Katya Adler arrived in America at the time of the outbreak of the Second World War. She told this story at a gathering for her at Madison, Wisconsin, where she met the distinguished labor historian Professor Selig Perlman. An Austrian socialist, Mr. Joseph Mire, later an official in the Office and Professional Employees International Union in Washington, D.C., told me this story at our table at the banquet in honor of George Meany, president of the American Federation of Labor, on March 31, 1977 at the Shoreham Americana Hotel.
109. *The New York Times*, July 9, 1921, p. 8.
110. Kenneth W. Payne, "Einstein on Americans," *The New York Times*, July 10, 1921, Section 2, p. 2.
111. *The New York Times*, July 9, 1921, p. 7.
112. *The New York Times*, July 12, 1921, p. 12.
113. *The New York Times*, July 31, 1921, section two, p. 4.
114. Tucci, "A Reporter at Large," p. 53. Einstein and Besso, *Correspondance 1903-1955*, p. 453.
115. P.W. Bridgman, *Reflections of a Physicist*, sec. ed., New York, 1953, pp. 110, 432. Abram L. Sachar, *A Host at Last*, Boston, 1976, p. 103. Bentley Glass, in his presidential address on Dec. 28, 1970, to the American Association for the Advancement of Science, compared scientists to the "explorers of a great continent who have penetrated to its margin." "Science: Endless Horizons or Golden Age," *Science*, Vol. 1711 (8 January 1971), p. 24. James Chadwick in *Nature*, October 30, 1937, cited in Sir Mark Oliphant, *Rutherford: Recollections of the Cambridge Days*, Amsterdam, 1972, p. 69.
116. Francis Bacon, *Novum Organum*, Book I, Par. 92; also Par. 113, 114.
117. "Civil reformation seldom is carried on without violence and confusion, while inventions are a blessing and a benefit without injuring or afflicting any." Solomon, says Bacon, said "it is the glory of God to conceal a thing, but the glory of a king to search it out." Ibid., Par. 129.
118. Albert Einstein, in *Living Philosophies*, New York, 1931, reprinted 1941, p. 4.
119. Loc. cit.
120. Einstein, *Out of My Later Years*, pp. 140, 175-76.
121. Ludwig Boltzmann, *Theoretical Physics and Philosophical Problems: Selected Writings*, tr. Paul Foulkes, ed. Brian McGuinness, Dordrecht, 1974, p. 160.
122. Einstein and Besso, *Correspondance 1903-1955*, pp. 439, 441.

Preface

Generational movements in modern times have given rise to the highest forms of creativity as well as the most destructive acts. The greatest scientists of our century have had the character of generational revolutionists; their bursts of scientific creative energies were symbiotic with generational movements in their contemporaneous social and political worlds. Einstein came to intellectual maturity in an environment charged with the revolutionary philosophy of his cosmopolitan, bohemian friends and fellow students; he can be regarded as the scientific embodiment of the revolutionary generation of 1905. The works of Niels Bohr, Werner Heisenberg, and Louis de Broglie likewise bear the impress of their student circles. Bohr shared a great enthusiasm for Kierkegaard with his Copenhagen *Ekliptika* circle; Heisenberg was inspired by the idealist, anti-Marxist outlook of his comrades in the German Youth Movement; de Broglie was kin to the young Bergsonians of the Third French Republic.

Only in the sciences has human society devised a means for resolving the conflict of generations. The scientific community has undergone basic reconstructions of ideas without suffering the equivalent of social revolutions. Fundamental theoretical changes have been made in a rational, constitutional spirit; a common loyalty to scientific truth has overridden divisive generational, national, political, and religious forces.

The fashion today, however, is to look at all reality under the aspect of "revolution." The word itself exerts tremendous power over our time; it is the emotional symbol with greatest allure for many intellectuals of the younger generation. I have therefore inquired into the theory of "scientific revolution" to test the notion that the advancement of science has taken place in accordance with a pattern or law of revolution. My conclusion, one might say, is a complementarist one: insofar as scientists as individuals are concerned, emotions of generational revolt have provided much of the motive energy for their greatest achievements; insofar, however, as the social workings of the scientific community are regarded, progress has generally taken place without an analogue of the revolutionary process. Subjectively, the young scientist may feel the emotions of the generational revolutionist; objectively, the scientific community is a reformist, intergenerational, constitutional republic. Thus the activities of the scientific community with its own conflicting social movements are misperceived when the observer imposes upon them a revolutionary a priori.

In previous books I have been concerned with the conflict of generations as a universal historical theme, with the character of intellectuals, and with the psychological sources of the scientific temperament. All three themes coalesce in this book. Here I offer evidence that the emotional, often irrational, strivings of men can be transmuted through scientific reasoning into the highest truths.

My approach to the sociology of science naturally bears the stamp of the experiences of my own generation. Probably no generation has searched its own presuppositions, postulates, motives, and reasonings with as much unflinching power. To Arthur O. Lovejoy whose stature in character and scholarship grows with time, I owe much in my conception of the history of thought. I am grateful to Philipp Frank for our many discussions in bygone years on some of the central questions of this book. Needless to say, I alone, however, bear the responsibility for the approach and detail of this book.

My warm thanks go to the Canada Council for its help in making possible the writing of this book. Parts of Chapter 1 were published in *Annals of Science*, volume 27, September and December, 1971. The editors have kindly granted permission for their use in this book.

LEWIS S. FEUER

1973

The "Olympia Academy" of Berne, 1902:
Conrad Habicht, Maurice Solovine, and Albert Einstein.

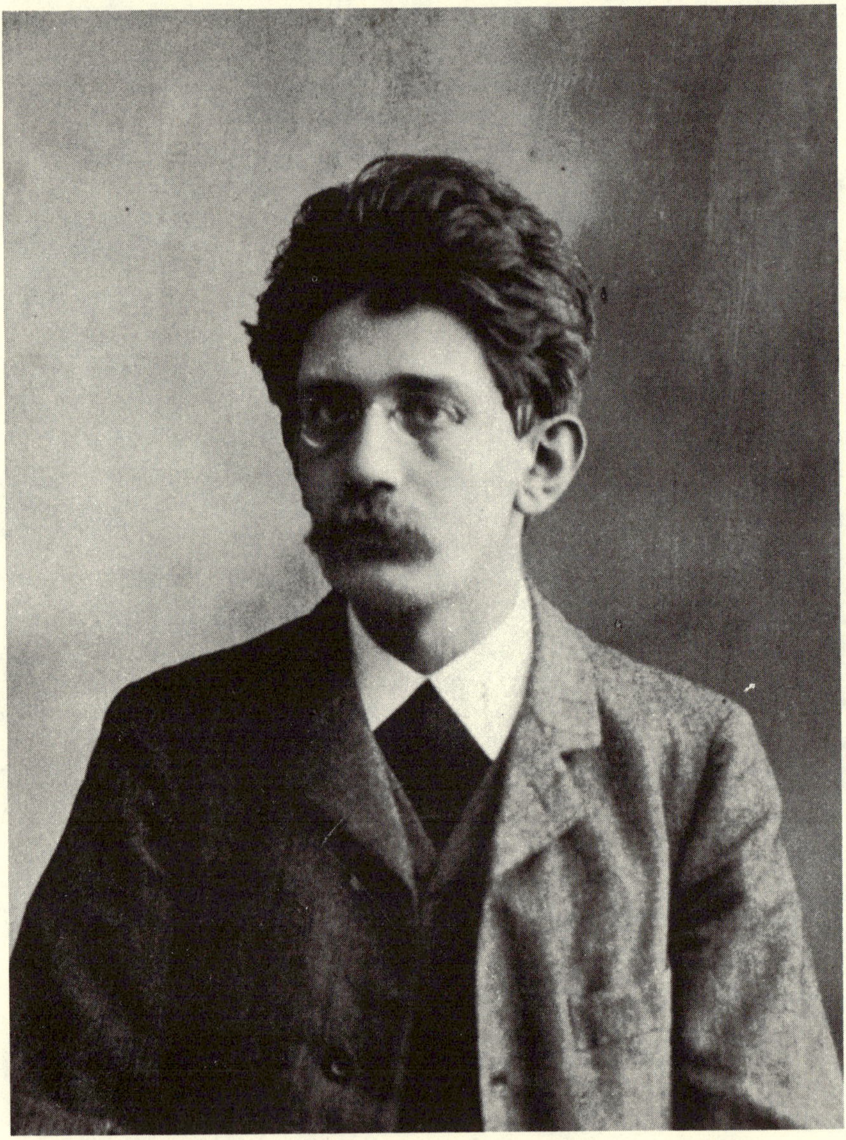

Friedrich Adler, Machian physicist and Marxist revolutionary.
REPRODUCED BY COURTESY OF FELIX T. ADLER.

Albert Einstein and Mileva Maric, married in 1903.

Einstein overjoyed as he meets the famed liberal mayor
of New York City, Fiorello M. La Guardia,
together with Rabbi Stephen S. Wise.

Einstein as drawn by the Russian painter, Leonid Pasternak.

Zurich: peaceful cradle of European revolution in the arts, science, and politics.
REPRODUCED BY PERMISSION OF DR. M. HÜRLIMANN, ZURICH.

Werner Heisenberg: young physicist and activist of the German Youth Movement.

COURTESY OF PROFESSOR WERNER HEISENBERG. PHOTO BY F. HUND.

Louis Victor, Prince de Broglie: scion of an aristocratic family, founder of wave mechanics.

Niels Bohr and his wife (photographed by George Gamow).
REPRODUCED BY PERMISSION OF BARBARA GAMOW (MRS. GEORGE GAMOW).

Einstein in old age.

1

THE Social Roots OF Einstein's Theory OF Relativity

THE great anarchist and distinguished scientist, Prince Peter Kropotkin, once wrote:

All great researches, all discoveries revolutionizing science, have been made outside academies and universities, either by men rich enough to remain independent, like Darwin and Lyell, or by men who undermined their health by working in poverty, and often in great straits, losing endless time for want of a laboratory, and unable to procure the instruments or books necessary to continue their researches, but persevering against hope, and often dying before they had reached the end in view. Their name is legion.[1]

This is what we might call "Kropotkin's Hypothesis." It is an insight that we shall find pertinent to our understanding of the social roots of Einstein's theory of relativity, though doubtless it must be qualified for a later scientific era and for such scientists as Newton in an earlier era. We shall see that Einstein's high period of original thought was sustained by a curious circle of young intellectuals who were filled with emotions of social and scientific generational rebellion. Theirs was a countercommunity of scientists and cosmopolitan bohemians outside the official scientific establishment,

I would like to express my gratitude for the kind help of Dr. Otto Nathan, trustee to the estate of Albert Einstein, and Miss Helene Dukas, Einstein's loyal secretary for many years. The greatness of Einstein's spirit is reflected in his choice of friends. I would also like to acknowledge the generous suggestions of the late Dr. Julius Braunthal, the biographer of Friedrich Adler, and the helpful criticism of Professors V. F. Lenzen and Edward Teller. Needless to say, I alone bear the responsibility for the approach and detail of this book.

that was moved in a revolutionary time to see the world in a new way.

As Philipp Frank has written, many factors, including the political and social, are involved in the making of a scientific theory; when a theory is psychologically conceived, the relevant experimental facts and logical considerations are not the exclusive causal agents.[2] Let us now turn to the components in Einstein's social environment that helped to shape his formulation of the theory of relativity.

Zurich: The Peaceful Cradle of European Revolution

To understand the social setting that gave sustenance to Einstein's revolutionary mode of thought, we must first recapture the spirit of Zurich, Switzerland, as it was at the turn of the century. For Zurich molded Einstein as Athens had shaped Socrates. It liberated his mind.

Albert Einstein was a student at the Federal Polytechnic School in Zurich from autumn 1896 to autumn 1900. In his boyhood, he had attended the Luitpold *Gymnasium* in Munich where he found life a continuous frustration; he had chafed at the rigid classical curriculum and the military-like discipline. His teacher suggested that he leave school saying that his presence in the class destroyed the respect of the other students. In Zurich, however, Einstein found sheer joy in existence. Zurich, which "always meant home to him," was where his best friends lived and where he won Swiss citizenship in 1901. For Zurich he was prepared to render a soldier's duty and was evidently disappointed that he was rejected as physically unfit.[3] During his four years at the Federal Polytechnic School, Einstein, for all his tendencies to isolation, reveled in the free atmosphere, the excellent science teachers, and the unusual student body.[4]

What manner of city was Zurich which was such a liberating force in Einstein's emotional and intellectual life? A small city of 153,000 inhabitants (according to the guidebook of 1903),[5] Zurich was above all the crossroads of European revolutionary movements.

As the marketplace of Athens witnessed the dialogues of Socrates, Parmenides, Thrasymachus, and Alcibiades, the cafés of Zurich were the stage for the debates of Russian revolutionists of all kinds: Marxists, nihilists, and Zionists. From the Middle Ages, Zurich had been regarded as the seat of beauty and wisdom; since the sixteenth century it was said: "whom God loves, He gives a house in Zurich." And in the early twentieth century tourists called it the "Athens on the Limmat." But few tourists could have realized how momentous for the history of the world were the noisy debates of the angry, gesticulating students they saw in Zurich's cafés.

As early as the 1870s, Zurich had welcomed waves of Russian women students unable to secure an education in Russia.[6] The famed Russian revolutionist Stepniak, describing the place that Switzerland and Zurich held in the consciousness of his comrades, wrote a poetic paean:

> Free Switzerland, which shuts out no one from its frontiers or its schools, was the favorite country of these new pilgrims, and the famous city of Zurich was their Jerusalem. From all ports of Russia, from the plains of the placid Volga; from the Caucasus; from distant Siberia—young girls of scarcely sixteen, with scanty luggage and slender means, went forth alone into an unknown country, eager for the knowledge which alone could insure them the independence they coveted. But on arriving in the country of their dreams, they found not only schools of medicine there, but also a great social movement of which many had no conception.[7]

Too often, however, the Russian students deserted the treatises of science for the revolutionary pamphlets of Marx, Bakunin, and Proudhon. "In a short time the city of Zurich from being a place of study was transformed into an immense permanent Club. Its fame spread throughout all Russia, and attracted to it hundreds and hundreds of persons, men and women." Therefore in 1873 the Russian government demanded that under the pain of outlawry Russian students "immediately abandon the terrible city of Zurich."

Rosa Luxemburg came to Zurich in 1889 to enroll in its university; her professor regarded her as his ablest student. She became the most original thinker of the left wing of the German Social Democratic party and was murdered as a Spartacist in the civil strife of 1919. Einstein greatly admired her and said: "She was far too good for this world." The German people, he told Philipp Frank, had not deserved a woman such as Rosa.[8] She lived in Zurich until 1898; then, her development as a socialist complete, moved to Berlin. But she continued to miss Zurich greatly.[9]

In 1898 an admirer of Rosa Luxemburg, the beautiful young Russian, Alexandra Kollontay, came to Zurich fleeing an unhappy marriage to begin a three-year study of economics at the university. Einstein had probably heard of her among his own socialist and Slavic friends and listened to discussions of her daring theories of free love, the reconstruction of the family, and the new woman. Sexual absolutes were under criticism. In later years, Alexandra became the leader of the workers' opposition in the Soviet Union and the Soviet ambassadress to Norway and Mexico. She was preeminently the exemplar of the Russian woman student revolutionary in the Zurich of Einstein's time.[10]

From the American continent came Florence Kelley, an eager young woman fresh from Cornell University and filled with a zeal for social welfare; in later years she became the first chief state factory inspector in the United States under Governor John P. Altgeld of Illinois and a famed pioneer in social legislation. She enrolled for courses at the Federal Polytechnic in 1883 and partook for several years of its socialist student life. A forest extended down from the top of the Zürichberg almost to the Polytechnic. "Here we students walked by the hour arguing in English, French, or German," wrote Florence Kelley. "The women students were almost all Russians, candidates for degrees in medicine and the sciences." There was "a joke among the polyglot students that the Russians were so busy with the future that they never knew whether the snowcaps were clear and lovely or shrouded in fog, any beauty that survived despite our modern capitalist civilization being unworthy their notice."

This international of students was confident that it could solve all "baffling, human problems . . . [H]ere in Zurich among students from many lands, was the philosophy of Socialism, its assurance flooding the minds of youth and the wage-earners with hope that, within the inevitable development of modern industry, was the coming solution." Stirred by ideology and hope, Florence proceeded to translate into English the book by Friedrich Engels, *The Condition of the Working Classes in England in 1844*, and to correspond with Marx's revered, surviving friend. She went to her first socialist meeting and "trembling with excitement," listened to the speaker, Eduard Bernstein, the exiled German socialist editor and friend of

Engels. "And here was I in the World of the Future!" she wrote. She was like an early Christian vouchsafed a meeting with one of the original Apostles. In the surrounding environment were the placid Swiss students, "having no political or social grievances" and "frankly bored by the foreigners." Nonetheless, wrote Florence Kelley, "Swiss people were the freest in the whole range of civilization."[11]

Several generations of socialist exiles and emigrants were attracted to Zurich. German socialists fled to Zurich to escape prosecution for sedition after the passage in 1878 of the Anti-Socialist Law and took courage in its free atmosphere. The Zurich section of the International Workingmen's Association survived for a decade after its parent body vanished because it provided a meeting-ground for socialists from all nations. In Zurich the intellectual spokesman of Russian populism, Peter Lavrov, gathered around him young Russian socialist students to expound his theory that the salvation of their country would come from the "critically thinking individuals," the intellectuals. Pavel Axelrod and Leo Deutsch, pioneers of Russian Marxism, also established themselves in Zurich. "Near the Polytechnic," wrote Kropotkin,

> was a corner of Russia, . . . the students lived as most Russian students do, especially the women; that is, upon very little. Tea and bread, some milk, and a thin slice of meat cooked over a spirit lamp, amidst animated discussions of the latest news from the socialistic world of the last book read,—that was their regular fare. Those who had more money than was needed for such a mode of living gave it for the common cause. . . .[12]

According to Eduard Bernstein, the later founder of revisionist Marxism, the Slavic students—Russians, Serbians, Bulgarians—lived on tea and ideology.[13]

Warring factions of Polish students brought George Plekhanov, the father of Russian Marxism, to Zurich in 1893 to compose their differences. He received such an emotional greeting that it brought tears to the eyes of Anatoly Lunacharsky, later Lenin's commissar of education.[14] Also in 1893 the Socialist International convened in Zurich, resolving bravely to "oppose with all their strength the chauvinist lusts of the ruling class," and affirming that world peace would be realized with the overthrow of capitalism.[15]

Not only Zurich but Berne, Basle, and Geneva had become the ideological capitals of Europe. Chaim Weizmann, the first president of Israel, reminisced:

Switzerland was, at the turn of the century, the crossroads of Europe's revolutionary forces. Lenin and Plekhanov made it their center. Trotzky, who was some years younger than I, was often there. The Jewish students were swayed—it might be better to say overawed—by the intellectual and moral authority of the older revolutionaries, with whose names was already associated the glamor of Siberian records.[16]

Factions and factionalist leaders disputed for possession of the soul of a generation of young Russian Jews. A new century was beginning: Would it bring to the Jewish people a destiny in their own land or a free life in socialist commonwealths of Europe?

The Zionist movement was founded by Theodor Herzl in 1897, and its first three World Zionist Congresses were held in Basle in the summers of 1897, 1898, and 1899. Student Zionists constituted the left wing of the movement. Jewish students in all the Swiss university towns were involved in ideological ferment. In Zurich and Berne in 1898, during the preparations for the Second Zionist Congress, student Zionist societies proposed, though without success, the convening of a students' conference with the aim of establishing an association of student Zionist societies.[17]

Letters written by the young Weizmann record the tensions among the Jewish students in Switzerland at this time. He was depressed by what he saw, and from Geneva in June 1900 he wrote to a student friend in Berlin:

Here among the Swiss students, quasi-socialist, poorly assimilated as regards Judaism, lacking in *moralischer Halt* (moral fiber), the atmosphere spread by Plekhanov, the General of the Russian revolution, and *tutti quanti* has a pernicious effect upon them. I have seen socialists here unable to utter two words without swearing, spitting or slandering someone three times, and I confess, my dear friends, that I am not filled with respect for these representatives of social conscience and social justice. All the socialism of most of the local people, even of the most outstanding of them, is merely an *Ausfluss* (result) of their own insignificance and individual weakness: they need the crowd to hide behind it, not to educate it. They flock to the crowd because they are frightened of themselves. Only bitterness and similar negative motives drive our semi-Jewish youth here into the socialist camp, and on the whole they bring to socialism the elements of corruption and ferment. In this sense they are the true sons of their own people, agents of decomposition! . . .

That is why they cannot be Zionists . . . They can assess Zionism only by analogy with other nationalist currents of this century: they can assess it only *relativ* (relative), by comparison with the Polish or Armenian movements, . . . and being materialists (*sic*) they can never understand the prophetic aspect of Zionism.[18]

From this point of view, Weizmann probably would have classified the circle of students with whom, as we shall see, Einstein was associated as just another group of materialists whose sense of their cultural heritage had been decomposed by their relativism.

However, Weizmann's more usual undaunted spirit was restored in the following year when shortly before the first Zionist Youth Conference in Switzerland in 1901, Weizmann debated the famed "father of Russian Marxism," George Plekhanov. For three days and nights Marxists and Zionists argued and counterargued. To Plekhanov's consternation the debate ended with 180 students joining the Zionist society. This was the so-called Berne Rebellion that reverberated throughout the Jewish student body.[19] Weizmann was overjoyed; in November 1901, he wrote:

They [the Marxists] reckoned that Plekhanov would have the last word but nothing of the sort happened. The comrade who had the floor after Plekhanov gave up his turn to me, and Mr. Plekhanov was debunked and routed, and retreated in the most ignominious manner. . . . Terrific excitement. This had never happened in Switzerland before. Just think of it: Plekhanov, the favorite, the idol who is worshipped so. . . .

I was in the seventh heaven at having knocked out Balaam. I had been looking for an opportunity to come to grips with Mr. P[lekhanov] for a long time, but as you know I would not set foot on his *terrain*, and here he thrust himself upon us, for the first and last time. I doubt whether I shall ever again be in such spirits as I was that evening. We had an almost sleepless night afterwards. The Zionists are jubilant. . . .[20]

Public debates between Zionists and anti-Zionists took place as well in Zurich in October and November 1901,[21] and Jewish student factions clashed vehemently in 1902 in Berne.[22] During this time discrimination against Jewish students in both Russia and Germany became so great that Weizmann and other Swiss student Zionists devoted much thought to the founding of a Jewish university. When the Russian minister of education was assassinated by a student in February 1901, the Jews were blamed though the assassin was a non-Jew; and strict quotas were decreed for the number of Jewish students permissible in the various university faculties. Weizmann reported on August 29, 1901 that "in all German academic circles, high and low, a strong anti-Russian (anti-Jewish) propaganda is now being conducted with great success." Therefore, he felt, it was "no longer a luxury but a necessity" to start to think of creating not only a university but several technical schools, possibly on the Swiss model.[23] In 1902 Weizmann, together with the philosopher Martin Buber and their Zurich representative, published a pamphlet, *Eine Jüdische Hochschule* (A Jewish University). Weizmann returned to this theme repeatedly, but its realization was left to the next generation. No doubt something of all this reached Einstein's brooding consciousness. In 1921, when he and Weizmann toured the United States together, they helped organize groups to cooperate with the Hebrew University Advisory Committee. And when the university finally had its dedication ceremony in 1925 Einstein became a member of its first board of governors.[24]

Zurich and Berne vied for the affection of the foreign revolutionary colonies. As Einstein went about the streets of these cities in 1903 and 1904, he might have stopped to listen to a pale, young Italian, gesticulating, exhorting small crowds of Italian workingmen. This was Benito Mussolini, then a revolutionary socialist, who lived in Zurich and Berne during those years organizing on behalf of the Italian Socialist party in Switzerland. There too he met his mentor, Angelica Balabanoff, at a Zurich congress.[25] According to his biographer and intimate friend, Mussolini "mixed much with the Russian students, women as well as men, all kinds of them—a strange, dissolute, eccentric, fantastic group, Nihilists, and Bohemians, the last word in fervid, feverish modernity." Mussolini even claimed to have frequented the Federal Polytechnic, and to have joined the Russian students, "both women and men," in their "orgies of strong talk and weak tea."[26] Mussolini, however, is said to have called Switzerland a "republic of sausages."[27]

Also haunting political meetings in Zurich in 1903, quarreling vociferously with an anarchist speaker till the audience threw him out, was a Polish Jew, Karl Radek. Radek was employed in a library and was preparing himself for a later career as the world's leading communist journalist, provost of Sun Yat-sen University, organizer of an unsuccessful insurrection in Germany, and defendant, confessed and convicted, in Stalin's staged trials in Moscow among the Bolshevik comrades.[28]

Zurich was also favored by the Mensheviks, the more democratic faction of the Russian Social Democratic party. And Lenin, too, liked Zurich more than Berne, not only because "there were a large number of revolutionary-minded young foreigners in Zurich," but also because the "Social Democratic party there was of a more Leftist tendency, and the petty-bourgeois spirit seemed to be less in evidence there." Lenin, an inveterate bookman, was especially enthusiastic about the libraries of the Swiss towns. Working in libraries in Geneva in 1903, in Berne in 1914, and in Zurich in 1916, Lenin marveled at their lack of bureaucracy, "no red tape, fine catalogues, open shelves and the exceptional interest taken in the reader." Lenin, however, unlike Einstein, never really liked the Swiss democracy. "Life there," wrote Lenin's wife Krupskaya, "is soaked in a sort of petty-bourgeois dullness. Berne is very 'democratic'—the wife of the Republic's highest official shakes her rugs out on the balcony every day . . . ," but nonetheless Krupskaya felt that a "petty-bourgeois stamp lay upon everything in Switzerland." When Lenin saw Tolstoi's play, *The Living Corpse*, he was moved by what he took to be its critique of philistinism; the Swiss citizens, however, gave their sympathies to the tormented wife in the drama. Lenin scarcely shared the enthusiastic feeling of some Russian revolutionists that Switzerland would be the center of the world's future revolution. Lenin was prepared to concede that it was democratic, that it had solved the question of nationalities, and that its library system was a model for the future Russian revolutionary society. Nevertheless, he felt this bookworms' paradise was still "a country of health resorts," without a revolutionary proletariat. In 1917, however, Bolshevik and Menshevik exiles alike departed from Zurich and Berne in a sealed train for Russia.[29]

The Swiss towns were also a mecca for sociological observers hoping to learn from their unique achievements in social welfare and political structure. The American radical, Henry Demarest Lloyd, who was renowned for documenting the sins of the Standard Oil Company, came to Switzerland in 1901 and 1902 for two long visits. "Students of politics," he wrote were "generally agreed that Switzerland contains a larger variety of instructive experiments in political and economic democracy than any other living nation. Nowhere else has the direct personal participation of the body of citizens in acts of government been applied in so many different ways. . . ." Switzerland, he said, succeeded in avoiding "the helpless stratification of existence"; it showed how our race "evolves when not overcome by monopoly."[30] Despite complaints by revolutionary socialists that Switzerland was a "bourgeois democracy," cooperative and voluntary movements flourished in its cantons; the referendum and initiative were part of its political procedure.[31] The chief socialist organization, the *Grütli-Verein*, had won an impressive body of social legislation through constitutional means. The social legislation of Zurich included laws for the protection of workingwomen, courts of arbitration, children's holiday colonies, housing ventures, and stations for migratory workers. In Zurich and Berne, the cooperative societies dominated local commerce. As Einstein walked in the winding streets of Zurich and along its gently flowing river, signs told of cooperative restaurants, savings banks, sick benefit societies, and stores; Zurich was famous for them. In 1900 Switzerland had 116 cooperative societies, 419 stores, and 83,549 members.[32] This was what Henry D. Lloyd called the "experimental socialism" of Swiss democracy, inaugurated by the people themselves, not by a bureaucracy. "The direct participation of a simple citizen in acts of government, and the application of the federal principle, are the key-notes to Swiss democracy." This was an instance of a country, "we might almost say of a civilization," wrote Lloyd, "which has escaped the most bewildering of the modern currents. Its people have not been led off into the pursuit of fabulous wealth. . . ."[33] New Zealand, in Lloyd's opinion, had gone further in socialistic experiments, the accomplishments in Switzerland were more the people's own: "The democracy in the world which cannot be betrayed is the Swiss."[34] "Happy Switzerland!" concluded the American publicist, "It has no coast, no navy, no colonies, no empire, no masses, no new wealth and very little old wealth, no trusts . . . Confined within the narrow limits of her little land she is forced to concentrate her energy upon her internal resources. Here we have a key-note to the genuineness of her democracy, the counter-foil of the infinities of the great nations."[35]

From burgeoning, expanding, powerful America, Lloyd was moved by the small-scale Swiss democracy to profound reflections on the fate of states. Were the "most admirable traits of Swiss democracy . . . directly traceable to the smallness of the nation?" Lloyd enunciated a theorem for the future: "It is a deep-seated illusion to suppose that either the future or the present lies with great empires."[36] The most highly civilized countries were also small ones: Denmark, Holland, Sweden, Switzerland, New Zealand. Lloyd was not a solitary, eccentric enthusiast in his observations of the Swiss democracy. The noted and experienced English social historian, William Harbutt Dawson, also reported in 1897 on "the boldness and originality which the Cantonal and Municipal Governments of the Swiss Republic have shown of late years in their manifold excursions in the field of social reform"; he hoped Britain would learn from the Swiss experience.[37]

The Swiss democracy seemed an oasis of freedom in Central Europe. The right of political asylum, guaranteed by the Constitution of 1848, was maintained even against the might of Bismark's German Empire. Zurich in particular had a rich tradition of socialist ideas and leadership. In 1901, the socialist discussion was especially intense as diverse groups coalesced to form the Swiss Social Democratic party.[38]

In the early 1900s Zurich's climate of freedom and zest nurtured original thought in the most diverse directions. Zurich was indeed the only place in Europe apart from Vienna where a circle of physicians arose whose interest in psychoanalysis was a friendly, exploratory one. As Freud wrote, "whilst in all the other places" there was "nothing but a very emphatic repudiation" of the subject, "in Zurich, on the contrary, agreement on general lines was the dominant note. Moreover, nowhere else did such a compact little group of adherents exist," together with a public clinic at the service of psychoanalytic investigation, as well as a clinical teacher who regarded the principles of psychoanalysis as an integral part of the teaching of psychiatry. "The Zurich group thus became the nucleus of the small band who were fighting for the recognition of psychoanalysis." Only in Zurich could one learn the new art and apply it in practice. "Most of my followers and co-workers at the present time came to me by way of Zurich, even those who were geographically much nearer to Vienna than to Switzerland. . . . For many years Vienna has been affected by strong prejudices. Representatives of all the most important nations congregate in Switzerland, where intellectual activity is so lively; a focus of infection there was bound to be of great importance for the spread of the 'psychical epidemic.' "[39] Thus Freud "repeatedly acknowledged with gratitude the great efforts of the Zurich School of Psychiatry. . . ."

In 1900 the young Carl Gustav Jung came to Zurich to work as an assistant in the Burghölzli Mental Hospital. Jung recalled: "When I came to Zurich I felt the difference at once. Zurich relates to the world not by the intellect but by commerce. Yet here the air was free . . . Here you were not weighed down by the brown fog of the centuries, even though we missed the rich background of culture."[40] Jung and his coworkers constructed, as Freud said, "the first bridge leading from experimental psychology to psychoanalysis." In Zurich too at this time lived and worked a rare spirit, Oskar Pfister, pastor of the Prediger Church, the first educator to recognize the importance of psychoanalysis, and the first religious thinker to try to use psychoanalytical insights to achieve a deeper religious understanding.[41] Zurich was also the place of origin for movements in art, politics, and science which diffused outward through the rest of Europe. "It was in a Zurich café, for instance, in 1916 that Tristan Tzara invented the word 'Dada,' " as he and his circle sought "to scandalize opinion and shake it from its lethargy."[42] From Zurich came the missionaries and manifestoes of the surrealist movement in art, the Marxist revolution in society, and the revolutionary theory of relativity in physics. The monument of Zwingli stood in a Zurich square reminding the burghers that a sermon on New Year's Day in 1519 had ignited the Protestant Reformation. A scientific reformation was similarly prepared in the discussions of a leftist, bohemian student circle at the turn of the nineteenth century.

Friedrich Adler and Albert Einstein

But how did the multifarious political and intellectual currents of the radical circles of Zurich enter into the thinking and scientific development of the young Albert Einstein? Here we are led to the unique role of his friendship with a fellow student, Friedrich Adler, who was the chief intermediary between Einstein and the new emotional-political-intellectual movement.

The Federal Polytechnic School was at this time a locus of confluence of the most advanced ideas in science and politics. With its high international reputation, it drew, as Philipp Frank writes, many students "from eastern and southeastern Europe who could not or would not study in their native countries for political reasons, and hence Zurich became a place where future revolutions were nurtured." Such a student was Friedrich Adler, "a thin, pale, blonde young man who like other students from the east united within himself an intense devotion to his studies and a fanatical faith in the revolutionary development of society."[43]

Friedrich Adler was one of the most remarkable characters in the history of the European socialist movement. He was the son of the brilliant leader of the Austrian Social Democratic party, Dr. Victor Adler (with whom Freud oddly felt himself in a kind of rivalry). When he and Einstein were the closest of friends Friedrich Adler was an ardent physicist-philosopher with the soul of a political student activist. Adler lived in the very same rooms at the student lodgings at 37 Pestalozzistrasse that Rosa Luxemburg had occupied a short time earlier.[44]

Victor Adler was deeply troubled by his son's calling to political action. He worried whether Friedrich's sensitive nervous system could bear political burdens since the shadow of mental illness was on the family. Friedrich's mother had for years suffered from severe depression, his sister had become insane, and relatives of Victor's father had also been afflicted with psychical disorders. "Victor as a psychiatrist was worried with anxiety that Friedrich's sensitive nervous system might one day break down. This danger, he said, had always hovered before his eyes."[45]

As a youngster Friedrich was excited not only by his reading in natural science but by the classics of Marxist ideology; he read Friedrich Engels' *Anti-Dühring* and Karl Kautsky's exposition of the economic doctrines of Marx. The physical sciences did not satisfy what his biographer has called his "Faustian drive"—his passionate philosophical desire to know the innermost bonds of reality.[46] Marxism became the religion of Friedrich Adler. He felt himself a "party comrade" at the age of fifteen, read faithfully every column of its daily newspaper *Arbeiter-Zeitung* and the analyses of the theoretical organ *Neue Zeit*, and attended meetings and demonstrations. But Friedrich wanted to practice his creed in its fullest sense of identification with the working class, and over this, he came into sharp conflict with his father. Victor Adler, the practical, skilful socialist leader, the man of wisdom, insisted that his son acquire a bourgeois profession and hoped that he would continue his chemistry studies. But in October 1897, during his first year at the University of Zurich, Friedrich wrote his father that he wanted to experience working-class life and intended

> as was my original plan, to go directly to a mine, and indeed for three months. I think the Falkenauer district would be most suitable . . . If I see, which I don't believe, that after a short while I can't endure it, then I

can always return and possibly do something which you define as "reasonable" . . . After I've been for three months in the mine, I shall go for nine months in a factory, possibly a machine factory . . .

The father was dismayed. His son's idea seemed to him a flight of romantic fantasy. What sense did it make, he pleaded with Friedrich, for him to take upon himself the misery of proletarian existence if it helped no one? If he really wished to serve the working-class movement, he must be economically independent of it and acquire the competence of a practical vocation. The son answered with irritation that the issue was then that as a worker, he couldn't lead so comfortable a life as he could as a bourgeois. Friedrich knew that his father had made bitter sacrifices in his own life. Would he, however, if he had the same resources and could choose again, become a quiet, comfortable, private citizen? "How can you demand of me that I fail to take the same step in my setting which you took in yours?" He was furious at his father's demand that he spend three years at the university to insure his becoming "a member of bourgeois society." "And I tell you that I had rather be the editor of the smallest party sheet, and were I too stupid for that, a factory worker than do anything whatever to achieve a bourgeois status . . ." Friedrich was then in his nineteenth year.

The son, however, yielded to his father's authority. Soon afterward he acknowledged that his father had perhaps been right. "You will be pleased", he wrote, "that I have come partially to your standpoint. I went to Zurich because I recognized that without the knowledge of physical *science* one can't begin to do anything; today however I also see that real work has an immense meaning, and you are entirely right when you say that mere bookmen are worth nothing. . . ." The son, however, still clung to his wish to dedicate himself to the working-class movement. Therefore, at the end of the second semester, he wanted to study history, economics, and philosophy instead of chemistry. But Victor Adler, chief of the Austrian Social Democratic party, would have none of that. They argued it out for three days during vacation time. "The youngster is brave, but foolish and young, and intransigent to extremes," wrote Victor to his sister-in-law. "I said no categorically. Then he proposed a compromise. He will become a Doctor of Physics because he expects from physics and mathematics a 'more spiritual education' than from chemistry."[47] Thus in the third semester of 1898

Friedrich Adler began the study of physics, finding in it an immense spiritual and ideological satisfaction.

In later years, the political activist in Adler became dominant and he finally abandoned his professional career as a scientist. He retained all the emotions of a student revolutionary, including the predilection for political assassination—a tactic abhorred by the orthodox socialist movement. In 1916, Friedrich Adler startled the world by assassinating the Austrian prime minister, Count Carl Stürgh. Adler felt it was the only protest possible against the prime minister's dictatorial refusal to convene the Austrian Parliament and the suppression of political freedom.

When Adler and Einstein first met as students, evidently in 1897, the future seemed more promising. Friedrich Adler made a deep personal impression on Einstein. Anton Reiser, Einstein's son-in-law, tells the story:

> With Albert, in Zurich, at the same time, was a young blonde sensitive student, the purest and most idealistic spirit Einstein has ever encountered. This was the Austrian socialist Friedrich Adler, who much later became *privatdozent* at Zurich, and appeared as a sort of saint in life and conduct. As a student he displayed the utmost zeal and patient devotion even during the most tedious lectures. For a long time Einstein believed that Adler was the only person who had really assimilated the tiresome matter of the course in astronomy.[48]

Perhaps the most remarkable document we have concerning Einstein's student years, his isolation, and his ostracism is a letter that Friedrich Adler sent to his father in 1908. It is so remarkable and has been so little known that I will quote at some length.

> . . . There is a man called Einstein who studied at the same time as I and with whom I attended several lectures together. Our development was fairly parallel, he married a student about the same [time] as I, had children, but had nobody who supported him, was starving for a time and was treated during his student years [the following word is illegible] rather contemptuously by his professors at the polytechnical institute; he was locked out of the library etc., he had no understanding of how to deal with people.
>
> Finally he found a position at the patent office in Bern and he continued to work on theoretical studies despite all miseries. . . .
>
> For the people [the professors] it is natural that on the one hand they

have pangs of conscience because of the way they treated him previously; on the other hand, the scandal is felt not only here, but also in Germany that such a man should be in the patent office.[49]

A Marxist in politics, a Machian in philosophy, a saint in his personal life, and a rebel against the passive political tactics of the older generation of social democrats, in Friedrich Adler were blended all the contemporary political currents. As Reiser tells:

Einstein learned a great deal about socialist teachings from his friend Adler, with whom he lived in the same house in Zurich. Indeed, were all people so conscientious, disciplined, and helpful as this Adler, the socialist paradise could certainly be realized. Adler was not only a socialist and physicist, but an enthusiastic philosopher. Like the young Lenin, he was a follower of the empirical-critical philosophy of Ernst Mach . . . Adler believed that natural science should come purely from experience . . .[50]

Adler's major writing on theoretical physics was published in 1905, the same year as Einstein's classical paper on the special theory of relativity. It won the approbation of Ernst Mach himself. Adler combined Machian concepts with a Marxist idiom; he wrote of the overcoming of "fetishism" and the achievement of a sound "Praxis." "The removal of metaphysics is always really a liberation from fetishism. Instead of 'Strivings,' 'Wishes,' 'Tendencies' of things, our predictions deal with phenomena," he affirmed. "An abundance of questions of science and practice are only to be solved through the calculation of energy quantities."[51]

Of all those who were trying at this time to combine Marxist and Machian ideas, Friedrich Adler alone could speak with scientific authority. His synthesis of the two became the principal philosophical tenet of what later became known as Austro-Marxism, and the prototype as well of the efforts of logical empiricists to assimilate historical materialism into their framework.

As a Marxist, Adler declared that the studies of Friedrich Engels, Marx's collaborator, were even after almost thirty years, "among the best and most elucidating work on natural science"; but then he went on to argue that these Marxist studies really found their natural fulfillment in Mach's work. To maintain this, Adler shed all the Hegelian metaphysics that still clung to Marxism. All that was meant by "dialectic," he said, was a historical-evolutionary conception of nature. Marx and Engels were the first to arrive at this standpoint with respect to the history of human society, "the first of the great discoveries in the field of an empirical history of evolution."[52] To think dialectically was to think in accordance with evolutionary conceptions: "all human knowledge, all science can only be understood as a dialectic-historical, as an evolutionary process." What then was Mach's contribution? "Mach had tried with conscious clarity to introduce the concept of evolution into physics." He had stressed that the next great advances in clarifying the foundations of physics must come from biology, more specifically the analysis of sensations. As Marx and Engels had aimed to eliminate metaphysical conceptions from the understanding of human history, so Ernst Mach had tried to eradicate them from natural science. Engels, moreover, had never quite completely mastered the sciences dialectically; therefore "a remnant of mechanical materialism" was found in his thought. Nevertheless, Engels too had rejected the fashionable mechanical materialism because he would not accept any theory founded on a hypothetical substance, in this case, matter. Moreover, said Adler, mechanical materialism was metaphysical in a second sense,—"it presents itself as a final solution and an eternal truth. It postulates an eternal being, through which processes occur, yet it does not understand its own process, its growth and decline, its history in time."

Adler very boldly associated Mach's critique of the kinetic theory of heat with Engels' polemic against mechanical materialism. They were examples of the same conception of scientific method. It was strange to find the world's leading socialists magazine, *Die Neue Zeit*, edited by Karl Kautsky, the esteemed inheritor of Engels' mantle, devoting several pages to the kinetic theory of gases. It required an unusual ideological sense to feel that the kinetic theory of gases was somehow an obstacle to proletarian success in waging the class struggle. But Adler's youthful ideology accomplished this synthesis. "The statement of mechanical materialism, that the manifestations of heat are manifestations of movement, naturally makes absolutely no sense, because that which we really know, the sensations of heat, are always given to us as something different from movements."[53] What then was the status of all the details of the kinetic theory of gases and the deduction of the gas and thermodynamic laws from the molecular hypothesis? Adler's reply was an engaging one:

The kinetic theory of gases, looked at closely, is really nothing else than a theory about the properties of systems of elastic spheres. As such it will retain its lasting value. On the other hand, to think that every body in the world constitutes such a system is an unfounded hypothesis, whose followers finally now seem to be decreasing.

So Clausius, Maxwell, and Boltzmann, it transpired, had been unconscious investigators of the behavior of systems of elastic balls; for them to think that they were investigating the phenomena of heat was an unfounded, illegitimate extrapolation: "When we make it clear to ourselves that the sensations of heat are given to us just as directly and immediately as movements, then it seems to us that an explanation of the one by means of the other is useless." Possibly such phenomena of heat and motion did coexist. Even so, the hypotheses and derivations of the phenomena of heat from those of a molecular realm still remained "superfluous complications." Adler wrote:

> The phenomenological physics which Mach sets as our goal is not only without contradictions, and from the standpoint of theory of knowledge, incontestable, but it is also considerably simpler (more in accordance with an economy of thinking) than a mechanistic physics.

Hitherto, said Adler, when one spoke of "historical materialism," one had in mind the views of Marx and Engels specifically on the history of mankind. With the advent of Ernst Mach's studies, that situation had changed:

> Now we can see that generally no other materialism than the historical is possible in any field. In all areas of science dialectic-historical materialism, or as we can also call it today, an empirical history of evolution, which was the basic conception of our masters, has achieved a victorious break-through.

There were Marxists like the respected historian Franz Mehring who felt that the "younger adherents" of Marxism were wasting time with a lot of "philosophical hair-splitting" about the meaning of materialism.[54] To them Friedrich Adler replied that their stand was equivalent to a "rejection of all theoretical work." For in the quarter of a century since Marx's death the natural sciences had made tremendous advances, which had great bearing for the formal development of the materialistic conception of history. The old way of approaching Marx through Hegel was obsolete, wasteful, uneconomical:

Yet however interesting the history of the origin of Marx's ideas is . . . it nevertheless seems to me that today a straighter path to his ideas is possible than through Hegel. Marx in his youth was a follower of Hegel; yet no one among all those thousands who today strive to grasp his teachings is a Hegelian.

Adler continued for several years in his writings in socialist journals to urge the union of Mach with Marx. On the occasion of Mach's seventieth birthday, he assimilated Mach's central significance to the cardinal Marxist tenet that change was the law of all things:

If we are today in a position to reply that the elimination of the idea of all absolutely unchangeable bodies from physics is possible, and therefore necessary, we owe thanks for this to the comprehensive critical labor, to the penetrating investigation of the entire science of physics, performed by Ernst Mach.[55]

When Adler wrote "What is the Use of Theory?" the amalgam of Mach with Marx produced unusual phrases of epistemological activism: "Theories are becoming constantly more important as tools in the class struggle . . . Proletarians of all lands, unite!"[56] Nonetheless, the Machian ingredient entailed the elimination from Marxism of the metaphysics of inevitability: "Does the theory enable us to know what will necessarily come? By no means. We know only what will happen in all probability."

Curiously, when Adler was old, and looked out upon a world which had confuted so much of his youthful, socialist, idealistic hopes, he could not recall that political discussion had played any part in his intimate friendship with Einstein. In 1957 he wrote:

My relationship to Einstein took place completely on the basis of physics. According to my memory, politics and even less the question of the war played no role whatsoever in our conversation. . . . From the time until 1911—when I went as a party secretary to Vienna—I don't have any recollections of conversations with Einstein which concern themselves with political questions, as I have already stated. I even have to admit that I considered him too naive in regard to these problems, in order to discuss them with him, while I had grown up with these questions.[57]

But Adler's memory, we must conclude, had thrown into oblivion the political discussions which according to Einstein's account, as set down by Anton Reiser, they did have. Adler's noted biographer and friend, Julius Braunthal, also writes:

There is for me not the slightest doubt that in those days they must have discussed for hours the thought of Marxism and Machism, for both of which Adler so ardently stood . . . But, to be true, I have unfortunately never touched this topic in my conversation with Adler.[58]

An ambiguity as to what is meant by "political" discussions may well account for an ambiguity in Adler's recollections. Einstein and Adler probably discussed the problems of Marxian theory; at the same time, however, they might rarely have exchanged views on tactics of socialist politics, a domain in which Adler regarded Einstein as uninformed.

Friedrich Adler indeed lived by the socialist ethic: "a straightforward, idealistic nature devoted with religious ardor to socialism and at the same time a level-headed scientist, he was a miracle in a brutal world."[59] In 1909 Adler's selfless intervention enabled Einstein to secure his first university post, a professorship of physics at the University of Zurich. A socialist majority at that time controlled both the town council and its board of education. Consequently, when the professorship became vacant, Adler, both a good socialist and an "excellent physicist," was the favored candidate. Upon hearing, however, that Einstein was available for the post, Adler withdrew his own candidacy:

He did not wish to compete with his fellow classmate. Adler would not think of depriving a friend of a position. But he never mentioned his renunciation . . . Albert saw this sacrifice as an additional confirmation of the portrait he had formed of the man's impeccable character.[60]

When eight years later Friedrich Adler was on trial for his life, Einstein recalled this self-sacrifice in an interview with the Berlin *Vossische Zeitung*: "Dr. Friedrich Adler relinquished with unselfish objectivity the much-longed-for professorship. Professor Einstein knows of many other examples of the unselfish objectivity of Adler, who, of course, always remained loyal to his principles."[61]

A common spirit of generational radical revolt moved the two friends, Friedrich Adler and Albert Einstein. In Adler, revolt finally took a political form within European socialism, and in Einstein it was directed against the basic assumptions of classical physics. Friedrich's father Victor, was, in the opinion of Friedrich Engels the ablest of the chiefs of the Second International. But Friedrich Adler, like Karl Liebknecht, son of a venerated German socialist leader, judged his father's party as having betrayed its principles. Both men felt that revolutionary socialism should be lived in one's personal life, and both regarded the socialist party's failure to oppose their countries' involvement in the world war as unpardonable treason to the working class. Friedrich Adler felt that his father whom he had idolized had betrayed his principles. As secretary of the Austrian party Friedrich had indeed been making the arrangements for an International Socialist Congress scheduled to meet in Vienna in August 1914; the war ended the Socialist International.[62]

When Friedrich Adler stood trial for murdering the prime minister, he made no personal defense, but dwelled on the betrayal by the party's elders: "When the party leaders have lost their revolutionary spirit, as they had in Austria, an individual act may revive this spirit in the party." He spoke of his own father's shameful weakness. Yet as he faced the court, he wondered whether in the light of the principle of relativity he could say that his act of assassination had been the more moral choice. Adler's was probably the only political trial in history in which a defendant has invoked the idea of relativity and the logic of the controversy between Copernicans and anti-Copernicans. "In a certain sense," said Adler, "both were right, and it depends upon whether one is standing on the earth or transfers oneself to the sun. Both views are logically possible." But given the relativist standpoint, how, Adler asked himself and his judges, could one say that acting in the interests of class struggle was more ethical than waging war on behalf of national struggle? Adler felt compelled to mitigate the relativist logic. "Even if both points of view were correct, that on the earth and that on the sun," he said, "both points of view are nevertheless not of equal value." Just as the relativist conception did not prevent natural scientists from continuing to adhere to the Copernican standpoint, so likewise, said Adler, could socialists remain confident that their standpoint of class struggle was most in accord with the advancement of humanity.

The trial was a strange medley of relativistic physics and relativistic ethics, combined with the longing for a Marxist moral absolutism and its counterpart in physical theory. With his life at stake, Adler insisted on the higher validity of the standpoint of the young militants. He assailed the "bureaucratic machine," which had imposed its supremacy on the labor movement, and the "lying spirit" that had entered the party. Cynicism, he said, had pervaded the party leadership; it was a "sin against the spirit." Stifled by both the socialist bureaucracy and the Imperial government, he had felt he could vindicate his commitment only by some deed of direct action, and killed the prime minister, a man whom, in fact, he personally respected. Adler said: "If you desire to comprehend my deed, and all that has led to it, an understanding of my revolt and my opposition to this sin that has smothered every vestige of manliness in Austria, must run, like a red line, through your considerations."[63] According to Julius Braunthal, "Adler's speech acted like a purge to the Austrian socialist movement, from which it emerged purified in spirit and in morale."[64] "To thousands of workers," another historian has written, "Adler became a hero who had avenged in part the physical and spiritual indignities which they had suffered for more than two years."[65]

Albert Einstein stood steadfastly by his old friend. He wrote to Adler on April 14, 1917, asking to be called to testify on his behalf as a character witness in the trial.[66] Then immediately after the trial proceedings, an interview with Einstein was published in the *Vossische Zeitung* (May 23, 1917) entitled "Friedrich Adler As a Physicist." Einstein spoke warmly of his friend's high moral character and discussed the studies in the physical theory of space and time with which Friedrich Adler occupied himself in prison. But privately, to Michele Besso, Einstein confessed his doubts concerning their friend. He felt it was morally incumbent on them to intercede on Adler's behalf. For Friedrich Adler was "a selfless, calm, industrious, conscientious man," but Einstein added, with a "sterile, Rabbinical head, rigid, without any sense of reality. Utterly selfless, with a strong thrust to self-flagellation, and even suicide. A strong martyr-nature." Einstein was prepared to say that Adler was a "conscientious thinker," seeking clarity, and with success, but he felt that Adler's writing on relativity was a long collection of subtleties "without the least value," so that Einstein was really embarrassed about having to give his opinion. Adler, he said, "rides the Machian nag to exhaustion."

It is noteworthy that Adler the physicist felt himself confronted by the same problem as Adler the political activist: How could he rescue some minimum of an absolutist standpoint from the dissolving criticism of the relativist logic? His solution was the same in both cases. In physics he proposed that "nature itself offers or suggests privileged coordinate systems upon which the natural phenomena, in particular, the movements of the weighable material are to be based." What Adler argued was that the center of gravity of the particular system whose movements are to be described defines the point of origin of its coordinate system. If one is concerned with movements on the earth, the coordinate origin is located by the earth's center of gravity; if one is concerned with planetary movements in the solar system, then the center of gravity of the solar system locates the coordinate origin. Adler then constructed a hierarchy of coordinate systems in accordance with the extent of their domain; the one with the higher order was characterized by more general laws of dynamics. Adler thus believed he had a natural, experiential criterion for affirming that the sun as a frame of reference was more privileged than the earth. Analogously he had argued in court that among the various social standpoints the Marxist ethic had a more privileged, wider, humanistic truth.

Friedrich Adler was condemned to death; his sentence, however, was commuted to imprisonment and the end of the war brought him amnesty. Both he and Einstein continued to share their socialist faith through many ordeals; both had, at the same time, misgivings concerning the methods of the Bolsheviks. Einstein regarded Lenin as one of mankind's moral-political pioneers: "I honor Lenin as a man who completely sacrificed himself and devoted all his energy to the realization of social justice. I do not consider his methods practical, but one thing is certain: men of his type are the guardians and restorers of humanity."[67] Years later, Einstein interceded with the finance minister of the German Republic, Rudolf Hilferding, to give asylum to Leon Trotsky.[68] Adler, the leading figure in the Labor and Socialist International, tried to preserve a critical friendship for the Soviet Union. The Soviet leaders, however, soon directed their abuse and hatred against him for his ethical, humanistic standpoint. Before World War I, during his seven years' residence in Vienna, Trotsky was an intimate friend of Adler, whom he called his "comrade in ideas and friend." In 1919, after the Bolshevik Revolution, Trotsky joined with Lenin in nominating Adler as the honorary secretary of the Communist International; they were disappointed when Adler declined. The Austrian Communist party similarly offered its leadership to Friedrich Adler, and then lamented that he "who had the talent to become the Lenin of Austria" chose instead to become its opponent. Adler was then made the target for an onslaught of criticism. Karl Radek, the Soviet publicist, wrote that Adler had revealed himself "as a tame petty bourgeois gone mad in 1916 during the impact of the war."[69] Twenty years later, with Trotsky in exile and Radek and his Bolshevik comrades purged, Friedrich Adler was still pursuing his lifelong aim of welding socialism to a higher ethic. In 1931, Stalin's Soviet regime in one of the first of its so-called Moscow trials—that of the Mensheviks—accused Friedrich Adler of having, together with other socialist leaders, conspired for "foreign imperialist" intervention into the Soviet Union. Adler rebutted the outlandish charges. During the later, more well-known Moscow trials, he denounced the judicially veiled crimes of Stalin's regime though still pleading "in favor of a peaceful and evolutionary development in the Soviet Union towards the establishment of the rights and liberties of the people."[70]

In exile Adler labored to preserve some of the idealism of the Austrian socialist movement. He saw Einstein only once during his years in American exile at the time of World War II. But the collapse of his youthful dreams rendered him shy even with an old friend.[71] After the Nazis were defeated in 1945, he prepared to return home. But the newly reconstituted Austrian Socialist party would not invite him back.[72] A new generation, which had little interest in Marx, Mach, or the socialist ethic, had arisen.

Ernst Mach: The Genesis of the Relativist Standpoint

When young Einstein was at Berne developing his conception of relativity, he used to watch how children developed their ideas of space and time. As his friend Solovine recalls, Einstein "employed with predilection the genetic method in the examination of fundamental notions. To clarify them, he used what he was able to observe among children."[73] The exploration of the unconscious origin of basic ideas, Einstein later said, helped liberate our thinking concerning them: "The attempt to become conscious of the empirical sources of these fundamental concepts should show to what extent we are actually bound to these concepts. In this way we become aware of our freedom, of which, in case of necessity, it is always a difficult matter to make sensible use."[74] Einstein's guide in these genetic inquiries was Ernst Mach. Mach was a scientist, wrote Einstein, who had "looked into the world with the curious eyes of a child."

The significance of such a spirit as Ernst Mach, according to Einstein, was that he had forged the instruments for the psychological emancipation of a generation of young scientists. "The fact is that Mach has had a great influence on our generation of scientists through his historical-critical writings in which he traces the development of the special sciences with so much devotion and where he pursues the path-breaking scientists to the inner recesses of their brain chambers." Even Mach's opponents, Einstein declared, "hardly know how much they have sucked in the Machian approach with their mother's milk (*wieviel von Machscher Betrachtungsweise sie sozusagen mit der Muttermilch eingesogen haben*)."[75]

What, then, was Einstein's view of Mach's contribution? He had with his genetic studies undermined the hold of traditional absolutes: "Notions which have proved useful in the ordering of things acquire such an authority over us that we forget their worldly origin, and accept them as irrevocable givens. They are then stamped as 'necessities of thought', 'a priori given', etc. . . ." It was no idle amusement when Mach taught scientists "to analyze the too familiar notions" and to show the conditions for their usefulness; "by doing so their excessive authority is broken," and they can, if necessary, be replaced by more useful concepts. In such ways, both directly and indirectly, he had been profoundly influenced, wrote Einstein, by his study of Mach's work on the history of mechanics.[76]

We therefore inquire into the genesis of Mach's conception of science because he was the teacher, precursor, and guide of Einstein's generation of 1905. Fortunately for our inquiry, Mach—as befitted a founder of genetic empiricism—was very interested in the study of his own unconscious, dreams, and early associations. Consequently, in the case of Mach, to our customary inquiry into sociological backgrounds, we can add one into the psychological origins of his thought. Mach would have been the first to welcome such an analysis.

In Mach's life we perceive the genuine pioneer, moved by strange emotions to depart from the familiar furrows of his contemporaries. He was a man who felt most at home on the frontiers, where the path from each novel idea had yet to be cleared, and where one never knew whether one was guided by insight or delusion.[77]

Like Einstein, the young Ernst Mach was a trial to his father and teachers; they despaired of his coming to anything. Born in February 1838, in a village in Moravia, Ernst was enrolled in 1847 in the beginning class of a Benedictine school in lower Austria. Ernst's father was struggling to better the family's circumstances; he had begun as a tutor in a nobleman's family but in 1840 acquired a farm. Thus Ernst grew up in the country, where happily his imagination could soar with windmills and experiments in atmospheric pressure, liberated from the constraints of Latin grammar. Consequently, "the good fathers found the boy very lacking in ability but allowed him to pass, advising his father, however, to have him learn some trade or business. . . ." The father, "greatly disappointed by his son's poor success," tried to teach Ernst himself. As a consequence, there were stormy scenes between father and son. The father railed at the backward Ernst. "Norse Brains," "Head of a Greenlander," were among the curious epithets he used for abuse. The son enjoyed the respite of working in the fields during the afternoons, "and from this experience he gained a lasting respect for that part of mankind who live by manual labor."[78]

Ernst Mach lived his formative adolescent years at a time when the revolutionary hopes of 1848 had foundered. Young Ernst always retained vivid memories of the revolution, its nobility, courage and ardor. He recalled student revolutionaries coming to his family's house, which was thirty kilometers from Vienna, and a professor and a Hungarian fugitive taking refuge there. Mach remained a child of the 1848 enthusiasm; a generation later he commented on the bloodthirstiness of the upper classes in 1870 (evidently referring to the Paris Commune).[79] In later years, toward the end of the century, Mach together with his friend Popper-Lynkeus, was associated with a Viennese Fabian Socialist group, the *Wiener Fabier Gesellschaft.*[80]

While Ernst was still in his boyhood, a political gloom settled over Europe; Schopenhauer became the favorite philosopher and men brooded over the null product of their restless wills. A "strong reactionary and clerical complexion" prevailed. Young Ernst wanted to leave the country and immigrate to the United States; to make this possible he wished to learn the trade of a cabinetmaker. His father gave him permission to do so, and for more than two years Ernst worked faithfully two days a week under the guidance of a skilled technician in a neighboring town. He always held this experience in "grateful remembrance," and found that "many an experience gained while thus working in wood proved very useful to him in his later vocation." He recalled in later years with pleasure "the agreeable feeling with which, when physically wearied in the evening, he would sit on the fragrant woodpile and at his leisure construct pictures of future machines, air-ships and the like. From this experience the thinker learned how much he owed to the laborer."[81] The carpentry shop was to the young Mach what the Berne patent office was to the young Einstein fifty years later.

At the age of fifteen, Ernst made another attempt to succeed at formal education. He entered the sixth class of a public Piarist *gymnasium* in Moravia. Compelled to participate in religious rituals against his own convictions, Mach acquired an abiding disgust for "church scent, crucifix, incense," and other such "unmentioned" paraphernalia.[82] He complained to his father of the tedious religious exercises but the father said they used to be much worse. And to complete the sorry picture, Ernst found he could not get along in the school "because he lacked the cleverness and cunning prevalent in schools . . ."[83] Persisting in his studies in spite of these difficulties, Ernst entered the university.

Einstein, has said that his difficulties and slowness in growing up psychologically accounted for his scientific curiosity in adolescence and manhood. Lateness in psychological development, or resistance to socialization can have its scientific compensation. In Mach's case, every one of his unusual emphases or contributions to scientific method was founded on some childhood experience, as he tells us, which left its impress on his subsequent scientific thinking.

Mach said that if he were writing a novel, its hero would be "a cockchafer, venturing forth in his fifth year . . ."[84] He told several times of how at the age of five he was affected by the experience of a windmill, how it was what we might call the "genetic event" of his philosophy. This event had a profound impression on him because it penetrated the deepest layers of his unconscious, evoking from it the controlling models and forms ("paradigms" in current usage) of his scientific thought.

> In my own case, for example—I remember this exactly—there was a turning point in my fifth year. Up to that time I represented to myself everything which I did not understand—a pianoforte, for instance—as simply a motley assemblage of the most wonderful things, to which I

ascribed the sound of the notes. That the pressed key struck the cord with the hammer did not occur to me. Then one day I saw a wind-mill, I saw how the cogs on the axle engaged with the cogs which drive the mill-stones, how one tooth pushed on the other; and, from that time on, it became quite clear to me that all is not connected with all.[85]

As Mach wrote in 1912 to the young American anthropologist Robert H. Lowie, this experience gave him insight into the nature of causal, that is, "functional thinking."[86] The traditional notion of "cause" was discredited in Mach's psyche, for to speak of "cause" meant assigning a primacy to one variable that was regarded as the basic, independent, activating factor. Thus even as a child, Mach rebelled against whatever smacked of such primacy or dominance. He tended to repress the causal (and sexual) significance of the pressed key and hammer in the piano; this partook too much, one gathers, of the suggestion of male dominance. But with the windmill it was otherwise; the interlocking cogs, the cogs engaged with cogs driving the millstones, here was a more egalitarian image, the protomodel of two classes of phenomena that were functionally dependent. Mach's notion was that one event could no longer be said to be the cause of another; all that could be said was that the two events were functionally interdependent. To be sure, for purposes of convenience and economy a given event could be regarded as the "independent variable," and another event as the "dependent variable," but to Mach's mind an equation showing x as a function of y could just as well be rewritten to make y a function of x.

To Mach at the age of five, toward the close of that period in childhood when the impulse for knowledge has been sharply awakened, especially by the problem of birth, the cogs meshing with cogs in the windmill were diagrammatic of relations in which a kind of equality was achieved. Each cog was as important as the other, and there was not that masculine dominance, that "kingship," about which the child Ernst had indeed protested to his mother. Mach began his very first book with a childhood reminiscence which stated the underlying psychological theme of his life's work:

He who calls to mind the time when he obtained his first view of the world from his mother's teaching will surely remember how upside-down and strange things then appeared to him. For instance, I recollect the fact that I found great difficulties in two phenomena especially. In the first place, I did not understand how people could like letting themselves be ruled by a king even for a minute. The second difficulty was that which Lessing so deliciously put into an epigram, which may be roughly rendered:

> *"One thing I've often thought is queer,"*
> *Said Jack to Ted, "the which is*
> *That wealthy folk upon our sphere*
> *Alone possess the riches."*

The many fruitless attempts of my mother to help me over these two problems have led her to form a very poor opinion of my intelligence.[87]

Such were what we might call Mach's "protoproblems"; they were his childhood perplexities about society. He could not understand by what right some men had authority over others, why some could rule others, why some were rich and others poor. To his mother who struggled to answer the young Ernst, no doubt the problem had its immediate relevance and challenge: Why should the father rule and the mother submit? These questions were fruitfully sublimated in Mach's challenge in later years to every methodological axiom in science that smacked of privilege and status for any given body or event in nature. Thus to Mach, pondering the meaning of "cause," it was clear that no event was privileged in status; all that nature presented was functional dependencies. It was arbitrary, authoritarian men who imposed a corresponding arbitrary status on some events as the causes of others.

The relativist conception of space and time was the outcome of Mach's deepest feelings. Evidently, it was not the product of an exercise in abstract methodology; it had the stamp of Mach's protest against a world of categories expressive of paternal dominance. "Poetical myths," wrote Mach, as "that of Time, the producer and devourer of all things," contravene the scientist's conception of ideas as useful economic devices.[88] The history of thought according to Mach, had hitherto been a history of mythologies; the mechanistic myth had replaced that of substances and forms which had previously superseded the animistic and demonological; in his own time, a dynamic mythology was overcoming the mechanistic. But now, Mach said, the time-mythology must be overthrown together with the mechanical. Time, the devourer of all things, was like a father-tyrant who would be eradicated by scientific criticism. "I must candidly confess that I hate the rubbish of history," wrote Mach.[89] Time was not the absolute existent which had enthralled, according to Mach, the intellects of Newton and Kant; to the natural inquirer, wrote Mach, "determinations of time are merely abbreviated statements of the dependence of one event upon another, and nothing more."[90]

If Time was the Devourer, Space was its ally, the Separator. As a child, Mach had always experienced difficulty in discerning the relationships of distance among objects; the fact of distance awoke some elemental resistance in him. He could not accommodate to the separations of objects or to the displacements of perspective. "I can remember distinctly that at three years of age all perspective drawings appeared to me as gross caricatures of objects," wrote Mach.[91] "In my earliest youth the shadows and lights on pictures appeared to me as spots void of meaning." When as a child, he drew the portrait of his family's pastor, it provoked his mother's anger, and the incident, said Mach, was "strongly impressed upon my memory"; he had shaded the pastor so much in black that his feelings toward the interloper on his mother's love were altogether too manifest.[92] His perception of distance was trauma laden, so that he feared his stereoscopy might fail him: "Once when I was walking on a dark night in a district with which I was unfamiliar, I was all the time afraid of running up against a large black object. This turned out to be a hill several kilometres distant. . . ."[93]

Again, the image of the mill returned to Mach's consciousness; it was linked to denials of the upwardness and downness of things, a challenge to the absolute status of direction: "I once dreamed vividly of a mill. The water flowed downwards in a sloping channel, away from the mill, and close by, in just such another channel, upwards to the mill. I was not at all disturbed by the contradiction." Mach, like Freud, held that "we frequently draw the elements of our dreams from the events of the preceding day," but we are, of course, totally unaware of what occurred on the day of his dream.[94] Nevertheless, the image of the female waters flowing both upward and downward in the neighborhood of the mill, that grinding symbol of sexuality with its interlocking cogs, is an eloquent protest against the spatial asymmetry that an arbitrary human perspective has imposed on a child's relations with his mother. Sometimes the protest against separation became overwhelming in Mach; at such a time he wrote in a mystical vein that the great contribution of the physical sciences would be "in making man disappear in the All, in annihilating him, so to speak. . . ."[95] In this one senses a self-destructive maternal mystic behind the relativist empiricist. Thus Mach wrote warmly of Buddhism, its "wonderful story," and how much it approached his view that the ego should be "nothing at all."[96] The eminent physicist Erwin Schrödinger could therefore say justly that the "idea of Mach" and his co-workers "comes as near to the orthodox doctrines as the Upanishads as it could possibly do without stating it *expressis verbis*."[97]

In conscious fantasy, Mach liked to play with the notion of an overturning of the space-time order, of a successful revolution against its devouring and separating. He wrote in such a vein:

> You see, physics grows gradually more and more terrible. The physicist will soon have it in his power to play the part of the famous lobster chained to the bottom of the Lake of Mobrin, whose direful mission, if ever liberated, the poet Kopisch humorously describes as that of a reversal of all the events of the world; the rafters of houses become trees again, cows calves, honey flowers, chicken eggs, and the poet's own poem flows back into his inkstand.[98]

Thus, with time's direction reversed, the quietude of return to the maternal source would achieve a final equilibrium. But it was especially in his dreams that Mach's unconscious expressed the disquietude with the world of ordinary space, and a longing for reversed relationships. He recounted:

> At a time when much engrossed with the subject of space-sensation, I dreamed of a walk in the woods. Suddenly I noticed the defective perspective displacement of the trees, and by this recognised that I was dreaming. The missing displacements, however, were immediately supplied. Again, while dreaming, I saw in my laboratory a beaker filled with water, in which a candle was serenely burning. "Where does it get its oxygen from?" I thought. "It is absorbed in the water," was the answer. "Where do the gases produced in the combustion go to?" The bubbles from the flame mounted upwards in the water, and I was satisfied.[99]

This last dream in particular plays boldly with the theme of a reversal of the laws of physics, and in a curious way is egalitarian in its motivation; it aims to eliminate privileged male status. We may venture to explicate its familiar symbols. In the beaker filled with water (the female container), a candle (the erect male organ), is "severely burning." Whence, however, asks Mach, does it get the oxygen it needs for burning? At this point, Mach's egalitarian feeling asserts itself; the oxygen is in the woman's fluids themselves. But then what happens with the products of the combustion, the intercourse? The "bubbles" (the semen), mount upward in the woman. Then, writes Mach, in the language of a lover, "I was satisfied." Thus the import of the dream is in making the woman man's equal, his equivalent, and a valid source of oxygen. The dream seems an eloquent testimony that in sexuality Ernst Mach achieved a liberation from fears and excessive discipline; the woman, enflaming him, giving him the oxygen to keep him burning, disenthralls him from the father-fear that menaced his potency.

Mach's very way of perceiving the ordinary world became relativized; his emotions recast his mode of perception. He would stand on a bridge, look fixedly at the water flowing beneath, and the water would seem to be in motion; but with "prolonged gazing," the water would assume the appearance of being at rest, while "the bridge, with the observer and his whole environment" would begin to move in the opposite direction: "the relative motion of the objects is in both cases the same. . . ."[100] A subjective "physiological reason" would be required, said Mach, to explain "why at one time one, and at another another," is felt to move. His relativist propensity was so strong that Mach got results in one such experiment that others did not; when William James and another investigator replicated Mach's experiment, they did not report similar relativistic findings.[101] But Mach's world fulfilled his relativistic demand; it became fluid without absolutes: "A short time ago, while making a steamboat excursion on the Elbe, I was astonished at getting the impression, just before landing, that the ship was standing still and that the whole landscape was moving towards it. . . ."[102] All of us have had a similar experience, but for Mach it became what the Danish philosopher Høffding called the "type-phenomenon" in terms of which he described all experience.

The relativized world, in which there was no privileged frame of reference, no absolute, was a universe in which one's longings for the dethronement of paternal, clerical, political, and sexual absolutes were projected. Like Einstein, Mach experienced a poignant rejection in childhood and youth; in Mach especially it evoked a tendency to solipsism, to the epistemology of loneliness. As emotional bonds were severed, other people, other selves, lost reality for him. In his "early youth," Mach told, he felt a "certain inclination towards solipsism," which he acknowledged he never entirely overcame. Henceforth, he could always empathize with the solipsist: "Having been through it in our early youth, I know this condition of mind well, and can easily understand it."[103] Mach felt himself separated as by a wall from the world of other things and other people's feelings. Then, in his eighteenth year, impelled by the aggressive emotions of adolescence, he felt himself carried away by a philosophical discovery: Only what was observed was real; there were no things in themselves. He needed no walls because the world was composed of elements that were all sensations. Mach describes this moment of philosophical experience:

Some two or three years later (after the age of fifteen) the superfluity of the role played by the "thing in itself" abruptly dawned upon me. Upon a bright summer day in the open air, the world with my ego suddenly appeared to me as one coherent mass of sensations, only more strongly coherent in the ego. Although the actual working out of this thought did not occur until a later period, yet the moment was decisive for my whole view—I had still to struggle long and hard before I was able to retain the new conception in my special subject. . . . At times, too, the traditional, instinctive views would arise with great power and place impediments in my way.[104]

This problem of the unreality of other persons and objects, Mach was aware, was emotional in its origin, the outcome of a psyche maimed in childhood. Mach described explicitly how the epistemological domain of the ego fluctuated with his emotions: "In conditions of despair, such as nervous people often endure, the ego contracts and shrinks. A wall seems to separate it from the world."[105] In later years, when he translated the contents of science into statements about sensations, he felt secure enough in so handling physical bodies, but an agony of insecurity would seize him when he dealt with other persons' feelings in a similar way. When we are dealing with bodies, he wrote,

we find ourselves in a thoroughly familiar province which is at every point accessible to our senses. When, however, we inquire after the sensations or feelings belonging to the body K'L'M' . . . , we no longer find these in the province of sense; we add them in thought. We have the feeling as if we were plunging into an abyss.[106]

The "wall" and the "abyss" were always evoked by Mach when he tried to imagine other people's sensations and feelings; they called forth in Mach a further childhood memory:

When I first came to Vienna from the country, as a boy of four or five years, and was taken by my father upon the walls of the city's fortifications, I was very much surprised to see people in the moat, and could not understand how, from my point of view, they could have got there; for the thought of another way of descent never occurred to me. . . .[107]

In later years, the perturbing question of how to explain human feelings by laws concerning physical entities again awoke the child's recollection of the wall. "The problem is not a problem," Mach argued:

A child looking over the walls of a city or of a fort into the moat below sees with astonishment living people in it, and not knowing of the portal which connects the wall with the moat, cannot understand how they could have got down from the high ramparts. So it with the notions of physics. We cannot climb up into the province of psychology by the ladder of our abstractions, but we can climb down into it.[108]

This curious memory of wall, moat, and connecting portal as holding the answer to the most baffling riddles was recurrent in Mach's consciousness: "I recall this feeling every time I occupy myself with the reflection of the text, and I frankly confess that this accidental experience of mine helped to confirm my opinion on this point, which I have now long held."[109]

Here then was the wall, the separation from the watery moat below: the unconscious problem was to find the portal to the world of human feeling. But associated with this solipsistic separation was a certain self-disparagement. His rejection in childhood by others left its residue in the guise of his partial rejection of himself. This hostility of others internalized into a partial self-hostility made him long to be rid of the burden of the ego. Mach reported how his image displeased himself:

Once, when a young man, I noticed in the street the profile of a face that was very displeasing and repulsive to me. I was not a little taken aback when a moment afterwards I found that it was my own face which, in passing by a shop where mirrors were sold, I had perceived reflected from two mirrors. . . .

Not long afterward, after a trying railway journey by night, when I was very tired, I got into an omnibus, just as another man appeared at the other end. "What a shabby pedagogue that is, that has just entered," thought I. It was myself; opposite me hung a large mirror. The physiognomy of my class, accordingly, was better known to me than my own.[110]

Thus, the young Ernst Mach regarded himself punitively, repressing his own features and the head which his father had ridiculed. Father-fear had come near extirpating his joy in self; he had built a defensive wall that seemed to lack a portal to the world of feelings. But then, to the naive, unsophisticated lad at the Piarist *gymnasium*, came the liberating powers of sexuality. This, said Mach, was the kind of innate impulse that the traditional empiricism with its single law of association could never reduce:

The most striking of these [innate impulses], because they make their appearance at a moment when the mental faculties and the power of observation are fully developed, are the first manifestations of the sexual impulse. I have been told by a perfectly trustworthy man, a person with a strong love of truth, that when he was a lad of sixteen, being quite innocent and inexperienced at the time, he saw a lady in a low-necked dress, and was startled to find that he was suddenly aware of a striking bodily change in his person; this change he took to be an illness, and consulted a colleague about it. The whole complex of entirely new sensations and feelings which were then suddenly revealed to him was colored by a strong additional element of fear.[111]

Evidently the burst of adolescent sexuality opened a portal to the world of feeling. Mach's elated ego could become coextensive and identical with the burgeoning world of nature: "when I say that the table, the tree, and so forth are my sensations, the statement, as contrasted with the mode of representation of the ordinary man, involves a real extension of my ego."[112] The awakening of sexuality brought an enhanced confidence, an ego expanded and coextensive with a world constituted by a "coherent mass of sensations." It was a world without separations, without the disrupting, sundering effect of absolute spatial and temporal differentiations, without hierarchy, dominance, or privileged variables, and without those hard, sundering atoms prized by many physicists. It was a world in which the ego merged with its maternal fluence. For Mach the answer to solipsism was to dissolve the self in the mass of fluent sensations. He suggested the emotional, nonlogical basis for his perspective when he wrote of its moral and spiritual significance:

The ego must be given up. It is partly the perception of this fact, partly the fear of it, that has given rise to the many extravagances of pessimism and optimism, and to numerous religious, ascetic, and philosophical absurdities. In the long run we shall not be able to close our eyes to this simple truth, which is the immediate outcome of psychological analysis. We shall then no longer place so high a value upon the ego , . . . We shall then be willing to arrive at a freer and more enlightened view of life, which shall preclude the disregard of other egos and the overestimation of our own.[113]

Phenomenalism, Mach felt, fulfilled a moral need; it led to an ethical ideal which was neither ascetic nor that of the "overweening Nietzschean 'superman.' "[114]

The connection made by Mach between his ethics and his phenomenalism is hardly convincing. It has as little cogency as the similar argument that would link ethical relativism to physical relativism. For a phenomenalist might without logical inconsistency be either an ascetic, Nietzschean, or whatsoever; that collection of sensory elements known as the person can be characterized without contradiction as moved by either selfish, altruistic, aggressive, or self-destructive instincts. But to Mach, for psychological reasons, viewing the world as a "coherent mass of sensations" coextensive with his ego brought an emotional liberation. The pervasive paternal threat was gone; the thing-in-itself, which was sometimes the repressive metaphysical agent and at others the inaccessible realm of separated selves leaving one in isolation, was eradicated. Then, even death seemed for Mach to lose its dominion: "death viewed as a liberation from individuality may even become a pleasant thought."[115]

From this longing for a world without separations and freed from masculine dominance, there also issued a most unusual doctrine which later estranged Mach from most of the greatest physicists of his time. Mach nourished a tremendous emotional aversion to any theory that proposed the existence of atoms. The atomic hypothesis, in his view, was a "mechanical mythology"; to regard "molecules and atoms" as "realities behind phenomena," he wrote, was to substitute "a mechanical mythology for the old animistic or metaphysical scheme."[116] As physical science matures, Mach wrote, "[it] will give up its mosaic play with stones and will seek out the boundaries and forms of the bed in which the living stream of phenomena flows."[117]

Emotions of the most intense kind were aroused among scientists in the debate provoked by Mach's antiatomism. When the illustrious Ludwig Boltzmann, professor of theoretical physics at Vienna, committed suicide in 1906, the virulence of fanatical Machians was regarded as having contributed to his act of self-destruction.[118] Methodological fanatics had invoked the authority of the names of Mach and Ostwald. Then, Max Planck, the founder of quantum theory, excoriated Mach as a "false prophet." Refusing to give ground, Mach replied that "the physicists are on the surest road to becoming a church." He remained steadfast against the atomists:

> I simply answer: "If belief in reality of atoms is so essential to you, I hereby abandon the physicist's manner of thought. I will be no regular physicist. I will renounce all scientific recognition; in short the communion of the faithful I will decline with best thanks. For dearer to me is freedom of thought."[119]

Clearly the hypothesis of atoms aroused an intense feeling in Ernst Mach, a degree of emotional aversion altogether disproportionate and out of place in a scientific discussion of the question. The atomists evoked an association with his father's churchmen, and he passionately invoked the spirit of liberty against the atoms which seemed to have become symbols of his father's domination. He called the atoms "stones," the classical Biblical metaphor for testicles. The world without stones that he sought, in which all phenomena flowed into a living stream, was a reality depaternalized and as a whole liberated from the menacing father.[120] In short, a reality deatomized was a projection, we might infer, in which the father himself was unmanned; Ernst Mach was a son who had projected into physical theory a most elemental, mythological revolt against a father. Scientific objectivity, according to Mach, was a frame of mind in which one could envisage the most horrendous events with a spirit of neutrality. One could even talk about cannibalism without becoming morally indignant. Only thus, he said, could the "true inquirer" preserve the "sublime glow of freedom" of the scientific ethnographer, the spirit of his own five-year-old boy who cried out, wrote Mach: " 'We are cannibals to the animals!' "[121] The motive power in scientific research of the sublimated intellectual parricide echoed in Mach's depiction of scientific objectivity and was expended mostly against the hypothesis of atoms. As Philipp Frank wrote, Mach's antagonism to atoms was a "historically and individually conditioned one."[122]

In his dreams, certainly, Ernst Mach felt himself menaced by some masculine power: "We dream of a man who rushes at us and shoots . . . ," wrote Mach.[123] He himself had a fantasy of tearing the flesh from some human: Was this a strange symbol of reprisal against the man who rushed at him, and fired? "Less often, I see in the evening before falling asleep, various human figures, which alter without the action of my will. On a single occasion I attempted successfully to change a human face into a fleshless skull." As he rid the world of atoms and stones, so did Mach take another man's flesh.

Let us be cautious in placing excessive weight on a single, random memory of a dream or traumatic incident, or in attaching to any single incident a bearing of crucial evidence. Yet, when we have made such allowances, it is nonetheless true that a consistent pattern seems to pervade both Mach's intellectual and emotional life, and in keeping with his own genetic method, we have tried to define that pattern as simply as possible. It was from such a complex of emotions that Mach's antipathy to atoms issued. When he expounded his sensationalist standpoint and declared: "The molecular physics of to-day certainly does not meet this requirement," Mach was affirming a requirement that was nonlogical in its basis.[124]

Curiously, as we have seen, Mach knew that the standpoint that burst upon him was not the "instinctive" one, and he recorded that within himself he experienced great resistance to the view he adopted. Clearly, the inception of that view had to do much more with a drama of conflicting emotions than one of logical ideas. "What inquirer," wrote Mach, "does not know that the hardest battle in the transformation of ideas is fought with himself?"[125] Mach knew the power of unconscious ideas. He once, for instance, discussed an obsessive symptom of his: Why, when he had "had a handshake from a damp, perspiring hand," and was prevented from washing, he asked, did "I preserve a sense of discomfort, the cause for which I sometimes forget . . . ?[126] He believed that "ideas once formed, even if they are no longer in consciousness, nevertheless maintain their existence." And such ideas, formed in his childhood, and "no longer in consciousness," seem to have been the underlying determinants of Mach's distinctive requirements of scientific method. According to Freud's disciple, Sandor Ferenczi, Mach's "hypersensitiveness" with respect to bodily moisture sprang ultimately from a defense against certain "sexual ideas and memories," and such persons "are wont to shy at the intellectual contact with sexual things." Perhaps indeed Mach's antiatomism and anticausalism were at bottom an intellectual defense against certain sexual fears. We need not follow

why not?

the details of Ferenczi's conjecture. We have seen, however, that in Mach's unconscious there prevailed a strong emotional desire to demasculinize and equalize the conception of reality, to rid it of separations, and achieve the unity of maternal fluence. And it is noteworthy that in exploring the role of unconscious ideas in scientific discovery, Mach found in 1896 that the phenomena he described were related to the "phenomena which Breuer and Freud described recently in their book on hysteria."[127]

Those who met Mach—from William James in 1882 to Einstein, in Mach's old age—were struck by the simplicity and generosity of his spirit. James, a visitor in Prague, heard Mach give a "beautiful physical" lecture, "one of the most artistic lectures ever heard." He gave Mach his visiting card; then, as James tells: "Mach came to my hotel and I spent four hours walking and supping with him at his club, an unforgettable conversation. I don't think anyone ever gave me so strong an impression of pure intellectual genius. He apparently has read everything and thought about everything, and has an absolute simplicity of manner and winningness of smile when his face lights up, that are charming."[128] James at this time was an unknown American assistant professor of philosophy. The two men later wrote to each other and exchanged writings. Mach can indeed be regarded as a European forebear of James's pragmatism and radical empiricism.[129] At the same time that English intuitionists and idealists were deriding James's forthright, unpretentious philosophizing, Mach was welcoming his bold essay "The Sentiment of Rationality" (1884) as a "fine paper."[130] Hegelians everywhere were appalled by James's irreverent handling of their master, but Mach in 1909 found that only James enabled him finally to understand the German dialectician.[131] Academic logicians and metaphysicians alike were appalled by the venturesome psychological method that James used in *The Varieties of Religious Experience*, but the free, untrammeled Ernst Mach wrote him: "Your fine and remarkable book . . . has gripped me powerfully. Religious inspiration is certainly very similar to the scientific inspiration which one feels when new problems first present themselves in a form which is as yet not wholly clear. There is an as yet unmeasured depth into which one is gazing."[132] This sense of an "unmeasured depth" later eroded among Mach's philosophical disciples, but with a scientist's modesty, Mach reiterated it as ultimate truth.

The Appeal of Mach's Philosophy to the New Revolutionary Generation

The revolutionary student culture of what we might call the "generation of 1905" was a blend of the epistemology of Ernst Mach and the social standpoint of Karl Marx. At this time, young Marxists throughout Europe were very much attracted to Mach's philosophy. Marxism was the most scientific theory of sociology and economics, they thought, but it lacked a philosophical outlook toward the world. Mach's theory of knowledge seemed to fulfill this need. Friedrich Adler advocated a synthesis of Marxist and Machian ideas and led this trend in Austria and Switzerland. Among Russians, its most noted exponents were Alexander Bogdanov, Anatoly Lunacharsky, and Nikolay Valentinov.[133] Bogdanov was a founder of Bolshevism and was, together with Lenin, its recognized leader from 1904 to 1907.[134] Lunacharsky, later the first Soviet commissar of culture, studied at the University of Zurich under Professor Richard Avenarius whose so-called empirio-criticism was similar to Mach's philosophy.[135] In 1902 Leon Trotsky traveled from Zurich to London to see Lenin. Of this time he later recalled: "In philosophy, we had been much impressed by Bogdanov's book which combined Marxism with the theory of knowledge put forward by Mach and Avenarius. Lenin also thought at the time, that Bogdanov's theories were right."[136]

The Russian Machians held that Mach's philosophy was not only consistent with but was in the spirit of Marx. Mach, they wrote, had linked the development of scientific knowledge to the processes of social labor. Truth, according to the Machians, was not an eternal, unconditional objectivity; rather truths were tools, "the organizing form of human experience," an ideological form that was transformed by the development of its social basis. According to Bogdanov, such concepts as cause, space, time, and atoms, did not correspond to objective realities but were social forms for organizing experience. It followed from the Marxian-Machian conception that a scientific revolution would go hand in hand with a social revolution; for corresponding to the new technology and the emerging novel social forms, there would also be a new scientific truth, a new social form for organizing experience.[137]

What attracted Marxists to Mach was, first, the fact that his theory of science sounded like a corollary of Marx's historical materialism and labor theory of value. "The first real beginnings of science," wrote Mach, "appear in society, particularly in the manual arts, where the necessity for the communication of experience arises. . . ."[138] Science, like capital, was a kind of congealed labor; natural laws were economical summations of experience, labor-saving devices, that enabled the labor of other men to be substituted for one's own. Mach acknowledged that the principles of political economy were those of his theory of science. Occam's Razor, the so-called principle of simplicity in scientific method, was no longer to be founded on a metaphysical belief in the simplicity of the universe; rather it was a principle of economy expressive of man's biological aim to do things with the least expenditure of energy.

Second, the Marxist felt a kinship with Mach's critique of absolute space and time; it was all of a piece with Marx's critique of metaphysics in historical and economic theory. If Marx expelled unobservable, idealistic principles from social science, Mach did the same with space and time; "poetical myths" concerning time, Mach wrote, were of no import to the scientist, for whom the "determinations of time are merely abbreviated statements of the dependence of one event upon another, and nothing more."[139] The relativistic standpoint was rapidly winning adherents at this time. In 1904, in the sixth edition of his *History of Mechanics*, Mach announced proudly that "the number of decided relativists who deny the barely intelligible hypothesis of absolute space and time is growing rapidly and soon there will not be one prominent partisan of the contrary opinion."[140] Einstein in later years rendered homage to Mach's influence:

> It was Ernst Mach who, in *History of Mechanics*, shook this dogmatic faith; the book exercised a profound influence upon me in this regard while I was a student. I see Mach's greatness in his incorruptible skepticism and independence; in my younger days, however, Mach's epistemological position also influenced me very greatly, a position which today appears to me essentially untenable.[141]

Mach's radical critique of absolute space and time and his rejection of them as unobservable entities found enthusiastic disciples among the radical student circles. When Mach died in 1916, Einstein wrote in tribute: "Apparently Mach would have arrived at the theory of relativity, if at the time when his mind still had the freshness of youth, the question of the velocity of light had already engaged the attention of the physicists."[142]

As a second-year student at the Federal Polytechnic Einstein became intensely absorbed in the problem of the measurement of the earth's velocity in absolute space (or the ether) and the nature of light. He sacrificed mathematical seminars to spend his time in the physical laboratory. "I worked most of the time in the physical laboratory, fascinated by the direct contact with experience," wrote Einstein, and his chosen problem was, remarkably, the same as that of Michelson and Morley.[143] As Reiser tells us:

> He encountered at once, in his second year of college, the problem of light, ether and the earth's movement. This problem never left him. He wanted to construct an apparatus which would accurately measure the earth's movement against the ether. That his intention was that of other important theorists, Einstein did not yet know. He was at that time unacquainted with the positive contributions, of . . . Hendrik Lorentz . . . and of Michelson. He wanted to proceed quite empirically, to suit his scientific feeling of the time, and believed that an apparatus such as he sought would lead him to the solution of a problem, whose far-reaching perspectives he already sensed.[144]

On this same experimental problem that A. A. Michelson and Einstein had faced independently, there supervened in Einstein's case the influence of Ernst Mach. As a second-year student he was still an adherent of the absolute space. But in the latter Zurich years the Machian influence, added to the knowledge probably garnered at the time that Michelson and Morley had already indicated that his experimental effort would have a null result, must have moved Einstein from physical absolutism to physical relativism.

NO

Members of the Marxian-Machian Zurich circle (three of whom constituted the "Olympia Academy") were relativists in history, economics, and physics, and had the youthful revolutionary perspective that sustained Einstein's burst of scientific originality. That Einstein regarded himself as engaged in the work of a revolutionary in science is clear from his correspondence at the time. After a preliminary paragraph of comradely scolding addressed to his fellow Olympian Conrad Habicht for not having written ("you cold-blooded old whale, you dry bookish creature"), Einstein confided that he was engaged in four works: the first "deals with the radiation and energy characteristics of light and is very revolutionary," wrote Einstein; the fourth was "still a mere concept," dealing with electrodynamics "by the use of a modification of the theory of space and time."[145] This work, Einstein continued, had a "purely kinematic part" that would undoubtedly interest Habicht. A few months later, Einstein wrote Habicht that the relativity principle was leading him to the "remarkable" consequence that mass is a direct measure of energy: "This thought is amusing and infectious but I cannot possibly know whether the Good Lord does not laugh at it and has led me up the garden path."[146] A year later, in 1906, Albert Einstein, still in his mid-twenties, wondered whether his revolutionary years were already over. To his faithful friend Maurice Solovine, he wrote on May 3, 1906 telling sadly that he was not then obtaining many scientific results: "Soon I will arrive at the stationary and sterile age where one laments over the revolutionary mentality of the young.[147]

Meanwhile, Friedrich Adler's efforts to amalgamate Mach with Marx were arousing considerable interest in the world socialist movement. In 1908 Adler published his German translation of Pierre Duhem's *La Théorie Physique* to which Mach himself wrote an appreciative introduction. Duhem's standpoint was closely allied, apart from his theological views, to that of Mach.[148] In the same year Adler also published an article in the official Austrian socialist journal *Der Kampf* celebrating Mach's seventieth birthday; it attracted so much attention that it was republished in the United States in the burgeoning socialist movement's principal intellectual organ, *The International Socialist Review.*

The Machian philosophy at this time occasioned the most bitter polemics within the Russian socialist movement. Lenin, at first sympathetic, was intellectually, a nineteenth-century Russian materialist and soon turned furiously on those who were experimenting with Machian ideas. He wrote a vitriolic book directed against the Russian Marxist-Machians. "The philosophy of Mach, the scientist, is to science what the kiss of Judas is to Christ," he said. He was relatively gentler with Adler whom he ridiculed as the "good-natured Adler (Fritz)" who, yielding to the strange doctrine that physical theory doesn't deal with real objects, comforted his Marxist conscience with the reflection that, after all, "Engels did not know Mach as yet."[149] To Lenin the Machian standpoint raised the specter of solipsism, the annihilation of the external physical world through its reduction to a collection of sensations. Actually Einstein never seemed, even in his most Machian phase, to have followed the doctrine through to such an extreme; for him the belief in the reality of the external world was consistent with the wish to eliminate from physical theory such unobservables as absolute space and time.[150] Lenin, by this time called the "old man," was suspicious and hostile to this novel theory which was attracting young Marxists; his use of the Christ-Judas metaphor (himself, indeed, the betrayed Jesus) indicates the depth to which his emotions were stirred. To the great damage of later Soviet science, Lenin never grasped the positive significance of Mach's ideas.[151]

In later years Friedrich Adler raised his voice against the terrorist methods used by the Bolsheviks during the Civil War; Leon Trotsky thereupon drew a brilliantly mordant portrait of his late friend in the pamphlet, *Terrorism and Communism.* Trotsky averred that Adler's empirical skepticism had finally corroded his Marxist revolutionary spirit: "Friedrich Adler is the most balanced representative of the Austro-Marxian type. . . . Friedrich Adler allowed his ironical scepticism finally to destroy the revolutionary foundations of his world outlook." According to Trotsky, Adler's generational revolt against his father had finally failed:

> The temperament inherited from his father more than once drove him into opposition to the school created by his father. At certain moments Friedrich Adler might seem the very revolutionary negation of the Austrian school. In reality, he was and remains its necessary coping-stone . . . Friedrich Adler is a sceptic from head to foot; he does not believe in the masses. . . . At the time when Karl Liebknecht . . . went out to the Potsdamerplatz to call the oppressed masses to the open struggle, Friedrich Adler went into a bourgeois restaurant to assassinate there the Austrian Premier. By his solitary shot, Friedrich Adler vainly attempted to put an end to his own scepticism. After that hysterical strain, he fell into still more complete frustration.[152]

In truth, Adler's skepticism concerning the "masses" was shared by Einstein; in their later lives it filled them both with melancholy. Generational rebellion and empirical skepticism could build a new system of the physical world far more easily than it could fashion a new social system.

The Olympia Academy: A Scientific Countercommunity

Apart from Friedrich Adler, Einstein was encouraged in his most difficult days of material hardship and intellectual gestation by a remarkable group of young men, most of them somewhat adrift in the academic world and looking for jobs after their studies. Their personal characters, to Einstein's good fortune, were much like his own. "Whoever had such good fortune," wrote Carl Seelig, "as was accorded to . . . [this] writer . . . , to meet some of Einstein's oldest and most trusted friends, would have acquired the impression that they, like himself, were examples of an unselfish comradeship, genuineness and simplicity."[153] In Berne, where after two mostly lost years, Einstein began work on June 23, 1902 as a technical assistant, third class, in the Swiss patent office, the friends formed a circle which they called the "Olympia Academy."[154] The Olympia Academy was among the most fruitful of countercommunities in the history of science; it was comparable to Freud's psychoanalytic circle and the English "Invisible College" in the seventeenth century. As Seelig writes: "The role played by 'Olympia,' as they christened their private academy, during Einstein's first years in Berne, cannot be underestimated in its influence and its intellectual importance."[155] Its three members were Conrad Habicht, Maurice Solovine, and Einstein, but like all academies, it had what might be called its "corresponding fellows." The group usually met in Einstein's room where Turkish coffee was brewing, and read Mach, Poincaré, Mill's *Logic*, Hume. Einstein acted as treasurer and organizer, dunning them for their dues and scolding them playfully for their absences from sessions. These godfathers of the theory of relativity were a cosmopolitan group, including, beside Maurice Solovine and Conrad Habicht, his brother Paul Habicht, Michele Angelo Besso, and in Zurich, Marcel Grossmann. They were, above all, young generational rebels, in revolt against the science of the establishment. In 1953, toward the end of his life, Einstein wrote to Maurice Solovine:

To the Immortal Olympia Academy

In your brief and active existence you took a childlike joy in all that was clear and intelligent. Your members created you in order to make merry over your eminent, old and puffed up sisters. How much they thereby hit the truth, I have learned to appreciate fully through painstaking observations over the years.

All of us three members have at least proven ourselves to be durable. Even if they are somewhat crumbling, still something of your bright and life-giving light shines on our life's lonely path; for you have not become old and grown up like a lettuce grown into a stalk.

To you is offered my loyalty and devotion until the last very learned gasp! The member now only corresponding,

A. E.[156]

Maurice Solovine was an impecunious Rumanian Jew of philosophical bent who was auditing lectures on physics; he was one of that tribe of revolutionary seekers, hungry for scientific learning, who abounded in the foreign student colony. One day in Berne in 1902 he chanced to read a newspaper advertisement offering private lessons in physics for three francs an hour by Albert Einstein, a former student of the Federal Polytechnic School. Solovine went to Einstein's room and the two young men got to discussing philosophy. Einstein confided that when he was younger he had been attracted to philosophy, but its vague and arbitrary character had turned him away. They talked for two hours, and for another half-hour in the street. Einstein enjoyed their discussion so much that he would have no nonsense about payment of fees; the mutual discussion of problems, he said, was more interesting and they agreed to keep up their intellectual companionship.[157] "Solo" (as Einstein affectionately called Solovine) took part in the socialist May Day demonstration; Einstein "was struck by my participation," said Solovine in 1957, more than a half-century later. "Einstein was a great liberal, a very enlightened spirit with very high opinions, a great liberal of the epoch of 1848."[158] Einstein was worried about Solovine's impractical, bohemian ways and expressed his concern to Conrad Habicht: "Solo spends much time giving lessons; he still cannot bring himself to sit for his examinations. I am very sorry for him, for he leads a miserable existence. He looks very harassed. But I do not think it is possible to make him lead a more bearable, orderly life."[159]

When Einstein visited France in 1922 for the first time after the end of the World War, he enjoyed a reunion with Solovine. Together with the physicist Langevin and the astronomer Nordmann, they went on April 9, to visit the graves of the French and German soldiers buried side by side. They walked through land with deadened trees. "We ought to bring all the students of Germany to this place," said Einstein, "all the students of the world so that they can see how ugly war is." Langevin and Nordmann had had some misgivings about how Frenchmen would react to a visitor from Germany. But at luncheon, two French officers and a lady recognized Einstein: "When we got up from the table they all three rose without saying a word, all moving together, and bowed low and respectfully to the great physicist."[160] Evidently French people remembered that Einstein had been among the handful of professors at Berlin who had refused to sign the pro-German *Manifesto of the Ninety-Three Intellectuals*, and had with two others signed instead the countermanifesto, *Appeal to Europeans*, against Prussian militarism.

Michele Angelo Besso was evidently the first to whom Einstein expounded his theory of relativity; "I could not have found a better sounding-board in the whole of Europe," said Einstein. Besso came from a distinguished Italian-Jewish family and was a graduate of the University of Rome; he was six years older than Einstein. He came to the Federal Polytechnic School to study engineering, and evidently in 1897 urged the study of Mach's *Mechanics* on Einstein. Two years after Albert got his post at the Swiss patent office, he suggested that Besso also apply for a post as a technical examiner. Besso followed Einstein's suggestion and got the job. The two friends could then discuss social questions, philosophy, and physics, both on and off the job. Einstein was hoping to turn the patent office into Olympia territory. He wrote to Conrad Habicht in 1905 expressing the hope that "perhaps we may manage to smuggle you in among the patent boys. You would soon recover your old vigor."[161] Michele Besso was clearly a man who was thinking his way through fundamental challenges of ideas and life. He was to venture far from his family's ways. His uncle Marco Besso had fought with Garibaldi at Rome in 1867. His aunt Amalia Besso, a well-known painter, founded in 1917 the first fascist organization in Italy, antedating by two years the advent of Mussolini. It was an association of women called *Fascio Femminile*. Amalia wanted to raise the morale of the Italian people in the war against Germany. Later she became secretary of the women's group in Mussolini's Fascist party and supervisor of the elementary schools in Rome. She worked hard for laws to safeguard the welfare of children. This was, of course, during the time when, as Guido Bedarida wrote, "the Italian Jews [little more than forty thousand individuals] . . . followed the birth and development of the Fascist movement with interest and sympathy, without nourishing the least apprehension for possible anti-semitic developments."[162] But finding congenial the social tradition of the reformers, Zwingli and Calvin, Michele Besso joined the Swiss Reformed Church,[163] whose pastors of the Reformed Church played a major part in labor and democratic activities in Zurich.[164] Michele Besso and Einstein continued an extensive correspondence throughout their lives; their letters ranged over political and religious questions, Jewish traditions, the state of Israel, and Christian religious instruction. More than two hundred letters were exchanged from 1903 to 1955.[165] At the end of his 1905 paper on relativity Einstein gave thanks to Besso, his "friend and colleague" who had rendered "loyal assistance" in the elaboration of the questions and to whom he was indebted for "several valuable suggestions."

Toward the end of his life, Michele Besso wrote his philosophical creed in a short article entitled "Experimental Metaphysics." Its eccentric, expansive prose pleaded still for a liberal standpoint that sought in every domain of reality its appropriate experimental metaphysics. Thus cultivated, it would, like the branch of a tree, grow to the limits of validity. In a plurality of standpoints; there would be a "dialecticization" of theories of reality. The editors of the journal *Dialectica* prefaced the essay with a brief tribute to its author.

> Some meritorious persons show themselves in full view, others remain as if hidden underground, and flower only here and there for everybody to see. How much gratitude deserve those who "help others to think." M. Besso is one of such. He always knew better than anyone else how to listen and understand , . . . and to challenge if it were necessary. The thought which is trying to take form has always found in him the test which clarifies, the echo which amplifies.[166]

Conrad Habicht, who together with Solovine, witnessed Einstein's marriage in January 1903 to Mileva Maric, was a gentle spirit and a close friend of both husband and wife. In 1904 he became a teacher of mathematics and physics at the Protestant Educational Institute in the small town of Schiers. His brother Paul operated the Turkish coffee machine at the meetings of Olympia. Together with Einstein, the Habicht brothers labored to perfect a highly sensitive voltmeter, for which they duly secured a patent;

but the instrument was not a commercial success. Later, Paul Habicht became a manufacturer of electrical instruments.

Such was the Olympia circle and its associates: the two "patent boys," Einstein and Besso, Solovine, the Rumanian "eternal student," and the brothers Habicht, a future village teacher and a would-be inventor and instrument technician. This modest circle of friends never anticipated how Olympian its impact on the future of science would be.

This Zurich-Berne countercommunity drew foreign student wives into its circle. It thus became even more a radical students' international. In 1903 Adler married a Russian girl, Katya Germanischskaya, who was studying physics at the University of Zurich. Her fellow students called her admiringly "the second Sonya Kovalevski," after the famed Russian woman mathematician.[167] Einstein's wife Mileva, a taciturn girl from a Serbian peasant family, four years older than Albert, withdrawn and afflicted with a limp, was a student of mathematics and physics, working to achieve a teacher's certificate. She had radical political views, and evidently something of the Slavic students' back-to-the-people spirit. Before the two marriages, Adler and Einstein together with their two Slavic future wives used to attend the lectures on analytical mechanics given by the great Professor Hermann Minkowski.[168] But Einstein was regrettably unfaithful in his attendance so that a few years later Minkowski marveled at his achievement: "For me it came as a tremendous surprise, for in his student days Einstein had been a lazy dog. He never bothered about mathematics at all."[169]

The Zurich-Berne circle was predominantly composed of students of Jewish origin—Einstein, Adler, Besso, Solovine. One might, therefore, be led in explaining the creativity and freedom of this group, to assign a primary role to their Jewish origin. In a remarkable essay published in 1919, the year in which Einstein's fame rose to unprecedented proportions, Thorstein Veblen undertook to account for the fact that Jews were "particularly among the vanguard, the pioneers, the uneasy guild of pathfinders and iconoclasts, in science, scholarship, and institutional change and growth." Veblen argued that it was "renegade Jews," gifted youths who had escaped from the archaic, traditional Jewish cultural environment, who became scientific leaders. They came to science, he said, with a "skeptical animus"; moreover, though they were estranged from their own Jewish culture, they could not accept the Gentile "conventional verities." Thus, the Jewish scientist became "a disturber of the intellectual peace . . . an intellectual wayfaring man, a wanderer in the intellectual no-man's land. . . ." The young Jew "is a skeptic by force of circumstances over which he has no control."[170]

There is much truth in Veblen's theory; it reads almost like a sketch of Einstein's personal history. What it omits, however, is the significance of such countercultural, radical student circles as Olympia. Where such circles did not exist, or where they failed to attract Jewish students, youths merged without revolt into the surrounding national cultures. Henri Bergson as a student at the Ecole Normale saw himself a continuator of the French philosophical tradition; Durkheim and Lévy-Bruhl followed the traditions of Comtist positivism. In Italy, young Jews such as Padoa, Enriques, and Levi-Civita in no way constituted an alienated, marginal group. It was the distinctive, revolutionary student culture of Zurich and Berne, the revolutionary international of young intellectuals, that provided the cultural soil in which a young Einstein with his unique personal traits could find the necessary supporting nutriments. Without such a group, he himself felt that he would have perished intellectually. Such a group, with its Marxian and Machian cultural vectors, standing physically outside the scientific establishment, was found only in Zurich and Berne.

Countercultures and Isoemotional Lines

When the Zurich-Berne student circle was attracted to Mach's critique of absolute space and time, they were moved strongly apart from scientific arguments, by sociological, emotional, nonlogical factors. For the most powerful logicians of the time regarded Mach's arguments as utterly unconvincing. Charles Peirce, the founder of pragmatism, thought that Mach simply had a metaphysical aversion to absolute space and time which had nothing to do with science:

> His [Mach's] metaphysics tells him that there is no such thing as absolute space and time, and consequently no such thing as absolute motion. The laws of motion must be revised in such a way that they shall not predict that result of Foucault's experiment which they did successfully predict. . . . Is this not making fact bend to the theory?[171]

The young Bertrand Russell in 1903 likewise found Mach's argument against the absolute rotation of the earth "very curious" and one that fell under analysis to the ground. It seemed to Russell that an intellectual experiment could be performed that would contravene Mach's use of the fixed stars as points of reference: one could imagine that the fixed stars were removed, yet the phenomena constituting experimental evidence for the earth's absolute rotation would be unaffected.[172] Distinguished physicists today still find Mach's arguments unpersuasive; the use of the fixed stars as a universal frame of reference, writes Henry Margenau, for instance, would involve one in such "operationally absurd" concepts as action at a distance and absolute simultaneity.[173]

Logically, a plausible enough case could have been built against Mach's standpoint. However, from a sociological viewpoint, it can be seen that the emotions of the young Zurich rebels moved them to dethrone absolutes. In addition, Einstein had at hand the result of the Michelson-Morley experiment which could buttress revolt far more effectively than would mere insistence on Mach's philosophical critique. Extremely precise experiments and measurements with light-beams had failed to disclose any evidence of the earth's motion in an absolute space. What Mach had tried to do with philosophical arguments (which were never convincing to the nonrebellious) Einstein simply took as the empirically grounded starting-point for the principle of the constant velocity of light; the consequential corollaries of a relativist space and time were boldly accepted. To be sure, the Zurich-Berne revolutionary culture circle had prepared Einstein emotionally and intellectually to regard the physical facts in this unprecedented way.

The question arises: How were the political-philosophical ideas of the Zurich-Berne student culture circle related in a causal sense to the bold reasoning that went into the formulation of the special theory of relativity? Here we are in the most difficult domain of the sociology of science—the tracing of a specific causal line between a political-philosophical standpoint in a given cultural circle and an achievement in physical theory.

Werner Heisenberg has written that "the spirit of a time is probably a fact as objective as any fact in natural science,"[174] and the eminent psychologist E. G. Boring, defining the *Zeitgeist* as "the sum total of social interaction as it is common to a particular period and a particular locale," has affirmed that it sometimes helps scientific progress, but at other times hinders it.[175] Yet the notion of a

Zeitgeist is too vague to enable us to isolate specific, intellectual causal relations. We shall attempt to define the specific influences of a particular culture circle which, in a very definite sense, deviated from the modal culture of Western Europe.

With regard to European bourgeois culture, the Zurich-Berne circle of student socialists was in some respects what we might call a "counterculture." It was, in other words, a countercommunity wherein the alienated and the rejected found support. When his friend Marcel Grossmann died in 1936, Einstein wrote to the widow:

> I remember our student days. He the irreproachable student, I myself, unorderly and a dreamer. He, on good terms with the teachers and understanding everything, I a pariah, discontent and little loved. . . . Then the end of our studies—I was suddenly abandoned by everyone, standing at a loss on the threshold of life.[176]

There is such a thing, Einstein once wrote, as "the mentality characteristic of a particular generation. . . ."[177] The counterculture of the Zurich-Berne revolutionary students was (as the translator conveys Einstein's words) a "pariah culture."[178] It stood opposed to the official culture of the established scientific community. When Einstein, according to Seelig, submitted his work on relativity as a thesis to the University of Zurich, it was declared by the faculty to be "inadequate." As one professor recalled: "It was more or less clearly rejected by most of the contemporary physicists." The professor of experimental physics returned Einstein's study, saying: "I can't understand a word of what you've written here."[179]

In its extremes, a revolutionary counterculture is not what is often called a "subculture," for the defining characteristic of a subculture is that it remains compatible with the values of the culture as a whole or its dominant segment; a subculture merely appends more complex, distinctive traits which are expressive of the particular group from which it arises.[180] Thus, a subculture does not seek to disrupt the cultural equilibrium; it is one component in the cultural plurality which is relatively "orchestrated" (as Horace Kallen uses the term) with the others. A pure revolutionary culture cannot be a subculture; if it becomes one it has ceased to be revolutionary, and is reformist within the framework of the total cultural system's values. The Zurich-Berne group stood midway in the continuum between subculture and counterculture. In short, our task is to explicate the common, underlying emotional base of a group of intellectuals who were indeed partially estranged from but not enemies of the dominant European culture. For this reason, the *Zeitgeist*, the notion of an overall "climate of opinion," is too encompassing to provide an adequate explanation. Instead, we shall introduce the concept of the *isoemotional line* to illumine the social roots of Einstein's theory of relativity.

What is an *isoemotional line* in a given cultural universe? It consists of a class of isoemotional theories or ideas; one idea is isoemotional with another or with any cultural manifestation when it is an expression, reflection, outcome or projection of the same emotion. When a given section of the intellectual class shares and is moved by a common underlying emotion, it becomes an intellectual community; its intellectual products, which are expressive of a common emotional decision-base, constitute an isoemotional line. A variety of isoemotional lines crisscross the cultural universe of any given society in the way in which isobaric lines (linking places with the same barometric pressure) and isothermal lines (linking places with the same temperature) partition a meteorological field. A group of intellectuals alienated from their surrounding society in some common emotional respect, at odds with the dominant ethos of the *Zeitgeist*, may at a given time evolve a class of isoemotional ideas which may extend, for instance, from economic relativism to moral and physical relativism. Our inquiry therefore leads to the questions: With what class of ideas in its cultural universe was Einstein's theory of relativity isoemotional? What socioemotional standpoint was basic to the formative background of the theory of relativity?

The Sublimation of Revolutionary Emotion into Physical Theory

Einstein has been described in contrast with Max Planck, the founder of quantum theory, as a "revolutionary genius."[181] Einstein was consciously trying to develop a new foundation for physical science; his intent was revolutionary. Whereas Planck was a reluctant revolutionist, unwilling but compelled by the sheer weight

Heisenberg !

of experimental facts to break with the traditional mode of thought, Einstein derived a positive joy from the overturning of categories; as he said many years later, he was playing a good joke.[182] Einstein shared this outlook with the Zurich-Berne revolutionary circle. What modes of thought, however, were the common expression of their isoemotional standpoint? Or, in other words, what isomorphism in theoretical standpoint, would have been founded on an isoemotional outlook shared by scientific and political revolutionists?

A clue is afforded us by an unusual biographical fact: Albert Einstein's favorite contemporary writer was Thorstein Veblen. This unusual sympathy may have had an intellectual significance not unlike the rapport which both Charles Darwin and Alfred Russel Wallace experienced with Thomas R. Malthus' writings on population. Thorstein Veblen, wrote Einstein, was the only "other contemporary scientific writer," apart from Bertrand Russell, to whom he owed "innumerable happy hours."[183] In an article entitled "Why Socialism?" Einstein indicated which of Veblen's ideas he found especially congenial. He was drawn to Veblen's historical relativism, to the notion that every stage of social evolution has its own specific sociological laws, that the laws of economics were not universal for all systems but were always relative to some given particular system. Thus, according to Veblen, the economic laws of a socialist society would be qualitatively unlike those of a capitalist society; the economic laws of our present predatory order were entirely relative to the existing economic foundation, but these laws were not universal necessities. Given a socialist economic basis, an entirely different set of laws as yet unknown would prevail. Einstein wrote:

> But historic tradition is, so to speak, of yesterday. Nowhere have we really overcome what Thorstein Veblen called "the predatory phase" of human development. The observable economic facts belong to that phase and even such laws as we can derive from them are not applicable to other phases. Since the real purpose of socialism is precisely to overcome and advance beyond the predatory phase of human development, economic science in its present state can throw little light on the socialist society of the future.[184]

What Einstein found significant in Veblen's thought was identical with Karl Marx's description of his own dialectic method. It was a topic that we may be sure was discussed in the Olympia circle because Friedrich Adler himself was engaged in an empiricist evaluation of Engels' philosophy.[185] Veblen, like Marx, had argued that economists of the right wing tended to confer upon the economic laws of business enterprise an absolutist, privileged status. Veblen called upon his fellow thinkers to "penetrate behind the barrier of conventional finality"; then they would perceive how transient was the system of business enterprise. For economic systems are not economic absolutes. Thus Veblen was proposing a more genetic and relativistic approach to economic law. Veblen's teacher, Richard T. Ely, described the intellectual standpoint of the young radicals at the end of the nineteenth century: "The most fundamental things in our minds were, on the one hand, the idea of evolution, and on the other hand, the idea of relativity. These two ideas meant far more than the debate on methodology."[186] Veblen had indeed drawn from Marx this historical-relativist conception that no economic law was to be regarded as valid for all places and times.[187] And Einstein, giving expression to his socialist emotions and faith, used its idiom and formulations.

Curiously, the relativistic ideas of Mach were from the outset associated with a milieu of emerging social and ethical revolutionary notions. Ernst Mach drew his emotional sustenance from a Viennese group whose leading personality was Josef Popper-Lynkeus, a man whose character and combination of interests were not unlike those of Friedrich Adler. "When forty years ago," wrote Mach in 1912, "I first expressed the ideas explained in this book, they found small sympathy, and indeed were often contradicted. Only a few friends, especially Josef Popper, the engineer, were actively interested in these thoughts and encouraged the author."[188] Popper-Lynkeus, a man of Jewish birth endowed with a tremendous imagination, unusual force of character, originality, and social vision, earned his living for twenty-five years in a variety of humble occupations from railway clerk to boiler-engineer; anti-Semitism closed to him the possibility of an academic career.[189] His powers of insight were such that Freud rendered him rare homage, describing him not only as a precursor, but as one who came as near to being a man "wholly without evil and falseness and devoid of all repressions" as he had ever heard of.[190] Popper-Lynkeus, an advocate of Voltairean enlightenment, was also a socialist reformer and philosopher, the inventor and advocate of bold schemes for national service and the abolition of poverty. When Einstein came to Vienna in 1921 after World War I, he called on two men: one was Josef Breuer, the co-worker of both Mach and Freud; the other was

Popper-Lynkeus, "the nearest friend of Ernst Mach" who at the age of eighty, and confined to a sofa, was still concerned with scientific and social revolution, with relativity and plans for reconstructing the social system.[191] As a final act of piety to his emotional-intellectual forebear, Einstein in 1954 wrote an introduction to the first publication of Popper-Lynkeus' social writings in the United States: "Popper-Lynkeus was a prophetic and saintly person, and at the same time a thoroughly modern man."[192] It was in this tradition that in 1954, during a period when he felt intellectuals in American would do best to once more withdraw from the establishment into countercommunities, Einstein declared:

If I were a young man again and had to decide how to make a living, I would not try to become a scientist or scholar or teacher. I would rather choose to be a plumber or a peddler, in the hope of finding that modest degree of independence still available under present circumstances.[193]

We are now in a position to set forth our hypothesis concerning Einstein's theory of relativity and its relationships to the political theories of the Zurich-Berne culture circle. There is "such a thing as the spirit of a given time, the mentality characteristic of a particular generation," wrote Einstein in a discussion of the generations of German students.[194] Imagine then the youthful genius Einstein in the Zurich setting of a radical student group in which the revolutionary ideas of Marx commingle with those of Mach. Einstein imbibes a notion of the relativity of social laws to transient social systems; the laws of contemporary society are in reality the expression of bourgeois relations, and are not immutable absolutes. He and Fritz Adler and others, we may surmise, would even argue whether bourgeois and socialist observers could describe a common social world, or whether the description of social events varied with the social standpoint of the observer; for this too was an issue which Austro-Marxist philosophers debated endlessly. Thus the mind of the young "revolutionary genius" in physics would transform, sublimate, and finally express the Marxist-Machian revolutionary emotion and vision. The emotions that gave rise to sociological relativity might then seek to express themselves in a physical relativity; transposed and projected upon the study of the physical world, they would issue in an overthrow of absolute space and time, and in a conception of the relativity of length and time measurements to the observer's state of motion. The revolutionary-minded young physicist would stand ready to challenge the notion of absolute simultaneity, the basic axiom of the Newtonian conception, and to affirm that two events which, viewed from one system of coordinates, are simultaneous, are not simultaneous when described from another coordinate system which is moving relatively to the first.[195] The young scientific revolutionist would be prepared to regard the ordinary Newtonian laws of mass, energy, and motion as descriptions valid only within certain spatio-temporal limits, that is, for bodies moving at ordinary speeds relative to the observer. Thus a common isoemotional line would join the revolutionist in politics with the revolutionist in physics.

The word "relativity" and the expression "the principle of relativity" became emotive symbols of the new generational mode of thought, symbols for the isoemotional line of generational rebellion. The first philosophical article in the United States on the subject, published in 1912, said "the slogan of the new school" was "the Principle of Relativity," and that "the man who started this movement and was the first to formulate it in concise language and to base it upon close argument was Professor Einstein. . . ."[196] This socioemotional significance of the word "relativity" enables us, moreover, to explain the curious naming of Einstein's theory. Theoretical physicists have observed that "the principle of relativity" was in a sense a misnomer for the logical content of Einstein's theory. Einstein defined it as a principle of invariance: "The laws by which the states of physical systems undergo change are not affected, whether these changes of state be referred to one or the other of two systems of coordinates in uniform translatory motion."[197] In particular, in conformity to the principle, a law of nature was not truly such unless it was invariant with respect to different coordinate-systems in accordance with the Lorentz transformations. It was repeatedly pointed out that Einstein was indeed aiming at a higher absolutism, a more general objectivity of natural law, rather than at relativism or subjectivity. The expression "theory of relativity," wrote the eminent physicist Arnold Sommerfeld, was a "widely misunderstood and not very fortunate name." "Not the *relativizing* of the perceptions of length and duration are the chief point for him, but the independence of natural laws, particularly those of electrodynamics and optics, of the standpoint of the observer," was Sommerfeld's judgment.[198] In 1928 Einstein himself indicated that "principle of covariance" would have been a more accurate description than "relativity."[199] Why, then, did Einstein choose to give his principle of invariance the name "principle of relativity"? Here the influence of the Zurich-Berne culture circle under which the young Einstein was working was, we venture to say, the crucial factor.

The logical content of the principle of relativity was indeed an absolutist one, a statement of a principle of invariance.[200] Given the requirement, however, of the conformity of laws of nature to the Lorentz transformation, and the principle of the constancy of the velocity of light, there followed the remarkable consequences of the relativized status of time intervals and spatial distances; to maintain the Lorentz invariance, the measurements of temporal and spatial segments would vary with the relative uniform velocity of the observer. The startling relativist consequence—rather than the absolutist postulate—was what affiliated Einstein's theory emotionally with the relativist school. He had vindicated Machian relativity with consequences that went far beyond anything that Mach had conceived in his critique. It was from this emotional identification with the relativist school that Einstein's theory took its name. Curiously, Marx too, in his exposition of the dialectic method, had stressed the relativity of social laws to the technological foundation; but this relativist consequence had also been derived from an underlying, invariant, materialistic conception of history. It was the aspect of the relativity of things, however, that stirred the young generation of Marxist-Machians.

The vocabulary a person chooses to express his concepts is generally indicative of the isoemotional culture circle of which he feels himself and his work to be a part. The young, rebellious thinker of the Zurich-Berne circle would thus have been drawn to use the word "relativity" even as today's rebellious thinker of the younger generation feels impelled to use the word "alienation" in conveying the import of his researches. The logical content of Einstein's paper had, as Sommerfeld writes, "absolutely nothing to do with ethical relativism, with the 'Beyond Good and Evil.'" Nor had it anything to do logically with the revolt against economic or sociological absolutes. Yet the terminological similarity expressed a common emotional standpoint; it conveyed that the scientific inquirer approached the problems of interpreting the physical facts with emotional longings which he shared with his friends and associates, social, political, and ethical rebels. Together their activities constituted an isoemotional cultural line.[201] From the purely logical standpoint, there was no isomorphism between ethical and physical relativity. For in the case of ethical relativity, the word "relativity" signifies not only that there is no absolute good valid for all men, but that each man's good is incommensurable with that of others—it cannot even be said that their respective conceptions are "equally valid." In the case of physical relativity, however, each observer's frame of reference is equally valid with those of other observers, and there exist laws of transformation that correlate the different descriptions of an objective reality, so that finally invariant laws, valid for all observers, can be stated. But the emotive force of the word "relativity," the emotive symbol of those in rebellion against the absolutes of the establishment in all fields, was the controlling factor in the choice of language.[202] Einstein, indeed, in his later years commented on the bond by which the standpoints of ethical and physical relativists were united:

> But what is the origin of such ethical axioms? Are they arbitrary? Are they based on authority . . . ? For pure logic all axioms are arbitrary, including the axioms of ethics. But they are by no means arbitrary from a psychological and genetic point of view. . . . Ethical axioms are found and tested not very differently from the axioms of science.[203]

What Einstein wrote of ethical axioms was the same as what Helmholtz, his great precursor, had written of geometrical axioms.

Einstein as a Generational Revolutionist

We tend to forget the extent to which Einstein's paper in 1905 on the "restricted" theory of relativity was a document of generational rebellion. Its style, as C. P. Snow has observed, was something unwonted in technical journals, with a strange, almost poetic freedom "as though he had reached the conclusions by pure thought, unaided," and indeed with "very little mathematics."[204] "There are no references; no authorities are quoted," writes Leopold Infeld, a later collaborator of Einstein, "and the few footnotes are of an explanatory character. The style is simple, and a great part of this article can be followed without advanced technical knowledge."[205] As the philosopher of science, Morris R. Cohen, once said: "Like so many of the very young men who have revolutionized physics in our own day, he has not been embarrassed by too much learning about the past, or what the Germans call the literature of the subject."[206] David Hilbert, regarded by many as the greatest mathematician of his time, once told a mathematicians' meeting: "Do you know why Einstein said the most original and profound things about space and time that have been said in our generation? Because he had learned nothing about all the philosophy and mathematics of time and space."[207]

In short, what the young Einstein brought to the scientific facts was a desire to theorize about them in a new way, more in keeping with the emotions of the Zurich-Berne revolutionary student circle than with the official scientific establishment. His categories and thought patterns of analysis were not wholly shaped by either sensory observations or purely rational considerations. Einstein admired what he regarded as Kant's "really significant philosophical achievement," namely, his showing that our dominant concepts "can not be deduced by means of a logical process from the empirically given (a fact which several empiricists recognize, it is true, but seem always again to forget)." Kant, however, said Einstein, was wrong when he asserted that the concepts we use are "the necessary premises of every kind of thinking."[208] Our modes of thought, then, are not structural necessities of the human mind; the dominant intellectual patterns of a given era are sometimes transcended. Such was the influence of the Zurich-Berne revolutionary countercommunity on the categories and principles with which the young Einstein approached the physical data.

We would in no way denigrate the primary significance of the experimental facts that pose the problems for theoretical explanation, and the experimental verifications that help us decide among rival theories. Beyond the questions of logic and confirmation, however, there is the sociological problem of what motivated the young Einstein to seek a certain kind of scientific explanation. Einstein himself indicated that the relativist mode of thought was not derived from the empirical facts by a *solely* logical process. Our inquiry therefore is not so much concerned with the nonlogical determinants of Einstein's mode of thought as, more accurately speaking, with the extra- or socio-logical determinants. A nonlogical drive finally counterposes itself against logical or experimental considerations; an extralogical motive may control the choice of a preferred hypothesis, but it will not legislate the choice against the weight of evidence.

In no way does our inquiry deny the significance of Einstein's individual traits as an inquirer or his distinctive psychological characteristics. The uniqueness of the individual persists in its recalcitrance to sociological determinism. Einstein, for instance, often told Infeld that from the time he was fifteen or sixteen years old, "he puzzled over the question: what will happen if a man tries to catch a light ray? For years he thought about this very problem. Its solution led to relativity theory." Not only did he think about the man running after the light ray; he was also haunted by the image of "the man closed in a falling elevator." If the image of the man running after the light ray led to the special theory of relativity, "the picture of the man in a falling elevator led to general relativity theory."[209] Einstein once tried to explain "how it happened," as he said, "that I in particular discovered the relativity theory." The answer seemed to him "to lie in the following circumstance."

> The normal adult never bothers his head about space-time problems. Everything there is to be thought about it, in his opinion, has already been done in early childhood. I, on the contrary, developed so slowly that I only began to wonder about space and time when I was already grown up. In consequence, I probed deeper into the problem than an ordinary child would have done.[210]

It would be an intriguing venture to unravel, if this could be done, the psychological factors that caused ordinary concepts of space and time to take on a problematic character in the consciousness of the young adult Einstein. We have, however, not entered upon such psychological inquiries for which the biographical facts and psychological knowledge may not be available. The formative sociological influences on Einstein's thought can be traced in a more verifiable fashion.

"From the point of view of logic," wrote Einstein, "all concepts, even those which are closest to experience, are . . . freely chosen conventions. . . ."[211] From another point of view, however, the choice among alternative concepts has a sociological ingredient, to be explained as far as possible, sociologically: such an explanation is particularly relevant with respect to the origin of a hypothesis which falls outside the domain of equiplausible admissible hypotheses consistent with the dominant, established scientific culture.[212]

Finally, we must observe too that, apart from the theory of relativity, Einstein's bold and innovative use of a non-Euclidean geometry in his cosmology was also allied to the mode of thought of his

Zurich-Berne circle. His use of a Riemannian metric in the development of the general theory of relativity and the notion of a "curvature" in space as a consequence of his law of gravitation, were dramatically novel ideas for physicists. Strangely enough, non-Euclidean geometries had much occupied the thoughts of Einstein's English and French contemporaries. Whitehead studied them closely, Russell gave lectures on the subject in 1896 at Johns Hopkins University, and Henri Poincaré wrote the classical philosophical discussion of their significance.[213] It was Poincaré's often-quoted conclusion that carried the endorsement of scientific opinion at the beginning of the twentieth century:

> One geometry cannot be more true than another; it can only be more convenient. Now, Euclidean geometry is, and will remain, the most convenient: 1st, because it is the simplest, and it is not only so because of our mental habits or because of the kind of direct intuition that we have of Euclidean space; it is the simplest in itself, just as a polynomial of the first degree is simpler than a polynomial of the second degree; 2nd, because it sufficiently agrees with the properties of natural solids, those bodies which we can compare and measure by means of our senses.[214]

To Einstein, on the other hand, with his revolutionary emotion, the non-Euclidean geometries were transformed from an exercise in the mathematical imagination that was irrelevant to physics, into an instrument for forging a novel cosmology. There can be little doubt that within the Zurich circle the apostle of non-Euclidean geometry was Einstein's close friend, Marcel Grossmann, the loyal spirit who persuaded his father to intercede to secure for Albert his first regular employment at the patent office. Geometry, particularly non-Euclidean geometry, was Marcel's best-loved subject during and after his student years. His dissertation in 1902, "Ueber die metrische Eigenschaften kollinear Gebilde (On the Metrical Properties of Collinear Structures)" was followed by the publication in 1903 of an article in non-Euclidean geometry, "Die Konstruktion des geradlinigen Dreiecks der nichteuklidischen Geometrie aus dem 3 Winkeln (The Construction of the Rectilinear Triangle of Non-Euclidean Geometry from the 3 Angles)."[215] That same year, Marcel also published a study, "Die fundamentalen Konstruktionen der nichteuklidischen Geometrie," (The Fundamental Constructions of Non-Euclidean Geometry) intended evidently to awaken teachers of mathematics to an interest in the subject, as well as a further article, "Metrische Eigenschaften reziproker Bündel (Metrical

Properties of a Reciprocal Bundle)."[216] He was concerned too with the pedagogical problems involved in presenting the foundations of geometry, and published a series of essays on such themes beginning in 1909 with his "Ueber den Aufbau der Geometric (On the Construction of Geometry)."[217] Grossmann took part in the International Congress of Mathematicians at Cambridge in 1912 to present a paper, "Die Zentralprojektion in der Absoluten Geometrie (The Central Projection in Absolute Geometry)." In his later years he devoted himself especially to the problems of the reform of secondary and higher education in Switzerland; these interests led to a series of articles in Swiss educational journals and newspapers.

Einstein felt a kinship between his own aims and those of Marcel Grossmann. He wrote to him from Berne and the patent office in 1904:

> There is an extraordinary similarity between us. We, too, are expecting a youngster next month. You, too, will get a work from me which I sent a week ago to Wiedemann's *Annalen.* You are dealing with geometry without the parallel axiom, while I deal with the atomic induction theory without the kinetic hypothesis.[218]

Einstein dedicated "to my friend Dr. Marcel Grossmann," his doctoral thesis, "A New Definition of Molecular Dimensions."[219] In 1912 the friends were reunited when Einstein, thanks once more to efforts initiated by Grossmann, was appointed to the chair in theoretical physics at the Federal Polytechnic.[220]

Meanwhile, however, Einstein had begun to feel the need for more advanced, powerful mathematical methods in generalizing the theory of relativity. He had at first been skeptical of the need for the formalism of a four-dimensional geometry introduced by Hermann Minkowski in 1908. At this time the eminent mathematician David Hilbert remarked: "Every boy in the streets of our mathematical Göttingen understands more about four-dimensional geometry than Einstein. Yet, despite that, Einstein did the work and not the mathematicians."[221] From Prague Einstein now appealed for help to his friend Marcel Grossman: "Help me, Marcel, or I'll go crazy."[222] Marcel came; then Einstein removed to Zurich once more and they became collaborators. Marcel guided Albert through such works as Christoffel's *Absolute Differential Geometry.*[223] When Einstein's first work on general relativity appeared in 1913 under the title *Entwurf einer Verallgemeinerten Relativitätstheorie und Eine Theorie der Gravitation*, the first part, the physical one, was by

Einstein while the second part, the mathematical, was by Grossmann. Marcel also published that year in the quarterly of the Natural Research Society of Zurich an article entitled "Mathematische Begriffsbildungen zur Gravitationstheorie (Mathematical Concept Formations of Gravitational Theory)" as a companion piece to Einstein's "Physikalische Grundlagen einer Gravitationstheorie." There was utter confidence and affection between the two friends. Thus, when the physicist Arnold Sommerfeld expressed his misgivings that Einstein's generous acknowledgment to Marcel Grossmann might lead the latter to claim the distinction of a codiscoverer of general relativity, Einstein replied in July 1915, simply that "Grossmann will never make the claim of being a codiscoverer. He only helped me in orienting me in the mathematical literature, but did not help materially with the results."[224]

Yet surely, it is significant that it was a fellow member of the Olympia circle who persuaded Einstein that non-Euclidean geometry and the absolute differential calculus were powerful new instruments for physical theory. When Marcel died after much suffering endured stoically for many years, Einstein wrote with emotion of their close friendship, of Marcel's "magnificent strength of mind," of his help when "at the end of our studies—I was suddenly abandoned by everyone, standing at a loss on the threshold of life," and of the days of "our feverish work together on the formalism of general relativity."[225] Thus the two friends of the Zurich-Berne student circle created a new conception that wedded speculative mathematics with physical reality—a non-Euclidean world.

Social and Ideological Contrast of the Zurich-Berne Circle with the Cantabrigians and Parisians

We have seen that the emotional standpoint of the Zurich-Berne revolutionary student circle provided the supporting social environment, motivation, and modes of thought for the conception of the theory of relativity. Yet it must be noted that there also existed at this time other European circles composed of highly gifted young scientists. Can we explain, at least partially, the fact that the theory of relativity was not discovered among the latter? Did the scientific communities of England and France lack certain ingredients that were present in the Zurich-Berne countercommunity? Would their modes of thought and emotion have impeded or retarded the kind of emotive insight that led to Einstein's formulation? To answer these questions we now direct our attention to the defining characteristics of other leading European scientific communities of Einstein's time.

The Cantabrigians

The emotional standpoint of young English scientists at the outset of the twentieth century was quite the opposite of that of Einstein, Friedrich Adler, and the Olympia group. Here were two isoemotional lines that did not intersect. There was a brilliant circle in Cambridge, England's scientific center; foremost were Bertrand Russell, A. N. Whitehead, and a bit later John Maynard Keynes. These young men found their prophet in G. E. Moore, who exercised a remarkable emotional and intellectual influence over his friends. Moore seemed to the Cantabrigians to have "suddenly removed from our eyes an obscuring accumulation of scales, cobwebs, and curtains, revealing for the first time to us, so it seemed, the nature of truth and reality, of good and evil and character and conduct. . . ." Common sense had supplanted Hegel.[226]

Yet Moore was a philosopher of absolutes; he believed in a pure, non-sensory intuition of "good," a direct insight into a Platonic absolute, valid for all mankind. Of this, he persuaded his friends by the intensity of his conviction. Moore also exercised his powers on behalf of an absolute theory of space and time. When Bertrand Russell published an essay in the British philosophical organ, *Mind*, entitled "Is Position in Time and Space Absolute or Relative?" he acknowledged Moore: "The logical opinions which follow are in the main due to Mr. G. E. Moore, to whom I owe also my first perception of the difficulties in the relational theory of space and time."[227] The influence of Moore's *Principia Ethica*, published shortly afterward, was (as Keynes later described it) "not only overwhelming," but "exciting, exhilarating, the beginning of a renaissance, the opening of a new heaven on a new earth, we were the forerunners of a new dispensation. . . ."[228] Its effect, Keynes wrote, dominated and perhaps still dominates, everything else. We were at an age when our beliefs influenced our behaviour, a characteristic of the young which it is easy for the middle-aged to forget. . . . It is those habits of feeling, influencing the majority of us, which make this Club a collectivity and separate us from the rest.

The emphasis was on "platonic," "timeless" states of mind, "not associated with action or achievement or with consequences." Pleasures were regarded askance. There was a curious caste snobbishness in this group, very unlike the impoverished, outcast Olympians: "The divine resided within a closed circle," it was a religion of "neo-platonism," wrote Keynes. They were far from the Machian relativists:

> Indeed, we combined a dogmatic treatment as to the nature of experience with a method of handling it which was extravagantly scholastic. Russell's *Principles of Mathematics* came out in the same year [1903] as *Principia Ethica*; and the former, in spirit, furnished a method for handling the material provided by the latter.

The year 1903 was the *annus mirabilis* for Cambridge philosophy, wrote Leonard Woolf; the two works of Russell and Moore were like two new Tables of Law revealed from above. "It was a pure, sweeter air by far than Freud cum Marx," wrote Keynes, and it "served to protect the whole lot of us from the final *reductio ad absurdum* of Benthamism known as Marxism."[229]

In emotional standpoint, the Bloomsbury-Cambridge culture circle was thus diametrically opposed to the Olympia circle in almost every respect. The Cantabrigians were Platonist, absolutist, anti-Marxist, upper class conscious, concerned with personal relations, and identified with the establishment. The Zurich-Berne friends were Machian, relativist, Marxist, democratically class conscious, concerned with socialist movements, and alienated from the establishment. The Cantabrigians were almost all Englishmen; those of Zurich-Berne were a cosmopolitan youth from all parts of Central and Eastern Europe.

While the Olympians and their friends were boldly questioning absolute space and time, their young Cantabrigian contemporary Russell wrote with assurance that the arguments against absolute time were based on an "antiquated logic" and capable of "an easy and simple refutation"; he defined "event" and "simultaneity" in an absolutist sense. The experimental argument for absolute rotation, which Mach tried to interpret in a relativist fashion, sufficed, said Russell, in the absence of other logical grounds, "to turn the scale in favor of absolute position."[230] Russell found it a favorable aspect of the absolutist theory that, unlike the relational one, it defined spatial relations which held "eternally" and "timelessly." This was an isoemotional line far from that which took Marxian-Machian delight in historical change.

A. N. Whitehead, later Russell's collaborator on *Principia Mathematica*, was at this time engaged in the research which led to his memoir of 1905 "On Mathematical Concepts of the Material World." But with Whitehead too, as his most scholarly American student writes, a Platonic attitude was "almost second nature." Though he inclined to a Leibnizian theory of space, Whitehead, treated time entirely from the absolutist point of view, referring to the young Russell's article for support. "It seems most unlikely," wrote Whitehead, "that existing physicists would, in general, gain any advantage from deserting familiar habits of thought." Five years later in 1910, in an article for the *Encyclopaedia Britannica*, Whitehead wrote that as far as the relational and absolute theories of space were concerned, "no decisive argument for either view has at present been elaborated."[231]

We may take these noted Cantabrigians as expressive of the dominant emotional perspective among the ablest and boldest of their young contemporaries.Their rebellion was essentially confined to traditional boundaries. Like his predecessor Henry Sidgwick, Moore was an ethical absolute intuitionist.

The official English scientific establishment likewise differed most unusually from its continental counterpart.[232] It remained friendly to religion and had felt the void which threatened the human spirit with the encroachments of science. There was an agnostic sorrow in Britain at the end of the nineteenth century;[233] several of its most noted scientific figures were drawn to spiritualism and psychical research in the hope of bringing good tidings. Three Fellows of Trinity College, Cambridge, were among the founders of the Society for Psychical Research; among the physicists who served as the society's presidents were Sir William Crookes, Oliver Lodge, and Lord Rayleigh. The mathematical logician De Morgan was among their predecessors in psychical inquiry.[234] Such a temper of mind was evidently unknown at the Federal Polytechnic School at Zurich. It abounded in ideologies, Marxisms of all kinds, anarchism, Zionism, and populism; but adherence to spiritualism would have constituted a candidacy for eccentricity or worse.[235] The Zurichers waged war against metaphysical entities. Not even their professors would have undertaken to multiply the latter. Nowhere among English scientific or academic communities was there a circle similar in aim and composition, to the Zurich-Berne countercommunity. Given the nature of Cambridge University, for instance, the cosmopolitan assemblage of impecunious student revolutionists of all nations and both sexes could never have taken place.

The Parisians

In 1904 at the International Congress of Arts and Science at St. Louis, the famed French mathematician Henri Poincaré used (in its physical sense) the expression "the Principle of Relativity." Perhaps he was the first to do so. There are those who would therefore assign the priority of the conception of the theory of relativity to him.[236] Such was not the feeling however among the younger generation of students of physics in the early twentieth century. To them Einstein was the leader in the relativist reconstruction. Max Born recalls:

> I then [1907] returned to my home city Breslau. I was working at that time on a relativistic problem, which was an offspring of [Hermann] Minkowski's seminar, and talked about it to my friends. One of them, Stanislaus Loria, a young Pole, directed my attention to Einstein's articles, and thus I read them. Although I was quite familiar with the relativistic idea and the Lorentz transformations, Einstein's reasoning was a revelation to me.[237]

There was a generational audacity in Einstein's paper that made it a virtual generational manifesto:

> But for me—and many others—the exciting feature of this paper was not so much its simplicity and completeness, but the audacity of challenging Isaac Newton's established philosophy, the traditional concepts of space and time. That distinguishes Einstein's work from his predecessors and gives us the right to speak of Einstein's theory of relativity, in spite of Whittaker's different opinion.[238]
>
> . . . For my contemporaries, Einstein's theory was new and revolutionary, an effort was needed to assimilate it. Not everybody was willing or able to do so. Thus the period after Einstein's discovery was full of controversy, sometimes of bitter strife.[239]

There was a basic difference in the status that the principle of relativity had for Einstein as compared to that which it possessed for his precursors Lorentz and Poincaré. For Lorentz, the respected Dutch physicist, the principle of relativity was a *derived* law; one might believe that compensatory electronic phenomena made it impossible for an observer to determine the velocity of his absolute motion in space. Thus, physicists could speak of a "conspiracy of nature" to make the determinations of absolute motion impossible. Poincaré often used the word "relative" with respect to our human knowledge (*dans notre monde relatif toute certitude est mensonge*), but the word always had for him the connotations that the Comtist school attached to their tenet of the "relativity of knowledge."[240] Auguste Comte did not deny that an absolute reality existed, but he maintained that all our knowledge was always relative to our human status as perceivers: "We acknowledge the search for *Causes* to be beyond our reach, and limit ourselves to the knowledge of *Laws*."[241] "Invariable relativity" was an ineradicable attribute of our knowledge, "repudiating all absolute principles"; "positive" included "relative" in its signification. It was "not so absurd" to believe that because of the intervening action of physical causes we might be constrained, conceded Poincaré, to accept a law of relativity.[242] Lorentz continued to hope that some nonelectronic procedure might still be contrived so that the notion of absolute time would be preserved and the means provided to measure it independently of the observer's frame of reference.

It was for Einstein, however, that the principle of relativity became a foundation, an ultimate, *underived* fact of the physical world.[243] One began with the principle of relativity, one tried in no way to derive it as a corollary of intervening physical variables. It was a basic principle; one followed its consequences *usque ad finem*, even when they entailed a radical reconstruction of the notions of space and time. Within the intellectual standpoints of Poincaré and Lorentz, there still hovered agnostic, absolutist notions concerning the significance of "relativity"; absolute motion might exist but we could only know relative motions. For Einstein, there were no such absolutist, agnostic residues; the principle of invariance, of relativity, was intrinsic to the nature of physical reality itself. Thus the principle of relativity became synonymous with a new world outlook.

Unlike Einstein, Poincaré did not have the spirit of the revolutionist in physics. The eminent physicist Paul Langevin was in the

United States with Poincaré for a week after the 1904 St. Louis lecture at which Poincaré had used the expression "the principle of relativity." Langevin recalled the "passionate interest" with which Poincaré was following "all the phases of the revolution" that was taking place in fundamental scientific conceptions. But Poincaré "saw with some anxiety (*avec un peu d'inquiétude*) the undermining of the old edifice of Newtonian dynamics which he had recently crowned with his admirable works. . . ."[244] By contrast, Einstein had no such anxiety, emotional attachment, or investment, in the classical physical theory and enjoyed seeing it undermined. Indeed, Einstein's impression in 1911 was that Poincaré, despite his ingenious constructions, had an inadequate conception of the situation in contemporary physics.[245]

Poincaré's philosophy, we might say, expressed the spirit of the Third French Republic, the "republic of professors." He believed in democracy, was anticlerical and antimilitarist, but he also believed in the primacy and indispensability of the intellectual elite. In this sense, he was a man of the establishment. He had been born into a family of parliamentarians, civil servants, and soldiers. His own scientific career and education reflected the merger in France between the scientific and administrative elites. A graduate of the Ecole Polytechnique, Poincaré became a mining engineer in the state service and later was employed in the railway administration.[246] He acquired a reputation for cool courage when he went down into a smoking mine where an explosion had just claimed sixteen victims.[247] He spoke proudly to his fellow alumni of the Ecole Polytechnique of their achievements; in their ranks, he said, were

> the dean of our generals, several ministers, scientific engineers, the director of a great company, those who have conquered our empire and those who have organized it. . . . Our comrades have distinguished themselves in War, the Finances, in the Administration, in the Public Works; on the cabinet as on the field of battle, in the colonies, on board ships, on the sea and on the land, and even under land.[248]

He felt that if scientists participated in politics, it should be in all parties, even "on the other side of the handle" (*dû côté du manche*), because, said Poincaré in 1904, science needs money, and people in power must never be able to say: "Science, that's the enemy."[249]

As a man of the scientific elite, Poincaré was finally drawn into the Dreyfus case by the utter violation of scientific method on the part of the military. He intervened when Bertillon, the reputed government handwriting "expert," invoked the calculus of probabilities to prove that Dreyfus was the author of alleged documents. Poincaré wrote caustically that since Laplace and Condorcet, such efforts had been scientifically worthless and devoid of common sense; if Dreyfus were to be condemned, it would have to be on other grounds. In the final trial, on behalf of the court's commission of three scientists, Poincaré edited the technical report against the army's evidence. He found the task tedious because the questions were so elementary.[250] But he wrote the memorable words of the report analyzing the "prejudice," the "absolute lack of critical capacity and scientific thought," and the "taste for the ridiculous" of the military intelligence experts; they had applied the calculus of probability to matters to which it was inapplicable. They had applied "bad logic to false documents."[251]

French scientists at this time were highly involved in political-intellectual circles. Léon Blum, France's first socialist premier, recalled, for example, his meetings with the physicist Jean Perrin. When they first met they were both under twenty.

> I met him again eight or ten years later at the time of the [Dreyfus] Affair in the little circle of "intellectuals," which the apostolate of Lucien Herr had converted to socialism and which grouped itself around the Bellais-Péguy bookstore. Jean Perrin was one of this number as was Paul Langevin.[252]

Poincaré was the exemplar of the Voltairean republican. He warned against conformity: "Uniformity is death," he said, and he pleaded for tolerance of those rebellious souls who adopt new ideas which their teachers feel are the negation of morality.[253] Yet he was keenly aware that he himself had grown up in a generation that had rejected revolutionary, Utopian aspiration. He spoke of a predecessor generation, that of Sully-Prudhomme, as having allowed itself to be seduced by humanitarian Utopias; its sense of reality was restored by the debacle of the Franco-Prussian War and the horrors of the siege of Paris. "Ah, then, how he renounces his previous errors, and with what *élan* he writes his poem on *Repentance*; how he loves France, and those who died for her:

> *If all men are my brothers*
> *What henceforth are those to me.*[254]

To Poincaré the significance of his generation's experience, was that it had revealed the illusions of internationalism, revolution, and Utopian hope.

Poincaré was an Anatole France in science, classical in model, severe, skeptical, and without illusions. Like Anatole France, when he reflected on the impact of scientific determinism on morals, he held at the end: "Instinct is stronger than all metaphysics. . . ." In 1912, in a pragmatic spirit, Poincaré said that whatever one's philosophy, it would always be the same morality one would teach, that without which the nations would perish.[255] He spoke in praise of the relativist temper of mind, but it was the relativism of the Voltairean skeptic. He praised Berthelot: "he knew that we can know nothing but the relative," a creed Berthelot had derived from his boyhood friend, the skeptical Hebraic scholar, Ernest Renan.[256] And Poincaré warned against confounding the love of truth with the desire for certitude, saying, as we have seen, "far from that, in our relative world all certitude is a lie."[257]

The Comtist agnosticism of Poincaré, impressed his leading English reviewer, Arthur Schuster. As he read Poincaré's book, Schuster wrote:

> We feel ourselves all the time walking along a precipice, at the bottom of which is written, "You shall never know anything of the real construction of the material universe!" The author seems to enjoy taking us very near the edge of that pit, and when he whispers into our ears, "what you cannot know is not worth knowing," we feel as if he intended to throw us over.[258]

However, when he sought for a political analogy to explain the status of laws of nature, the spirit of the French *république des professeurs* and the role of the administrative intellectual was evoked: "our laws are therefore like those of an absolute monarch, who is wise and consults his Council of State."[259]

This classical Voltairean skeptic and the men of his scientific circle, linked closely to the scientific establishment of the Third Republic, were hardly motivated by the kind of revolutionary longing that filled the young Zurich-Berne revolutionists whose alienated countercommunity provided a natural setting for the intended relativist revolution.

It was Mach above all who was self-consciously the leader of the aggressive school of relativists, an ideological group as it were, and Einstein inherited the mantle. Poincaré was not, in this sense, a "relativist." Thus, when Mach reviewed the spread of relativism among European and American scientists, he did not include Poincaré among them:

> The view that "absolute motion" is a conception which is devoid of content and cannot be used in science struck almost everybody as strange thirty year ago, but at the present time it is supported by many and worthy investigators. Some "relativists" are: Stallo, J. Thomson, Ludwig Lange, Love, Kleinpeter, J. G. MacGregor, Mansion, Petzoldt, Pearson.

The number of relativists was growing, said Mach, and he expected his view to become ascendant. But Poincaré was not numbered by Mach among the "relativist" school.[260]

One might indeed venture the hypothesis that precisely because France was a "republic of professors" and the ablest talent of the Ecole Polytechnique was promptly absorbed into the military and engineering cadres, it was less likely that a very fundamental break with the received principles would take place there.[261] A "scientific revolution" finds its most fertile soil in a countercommunity. Where administrative responsibilities soon beckon the young scientist, his energies are less available for sublimation in radical research curiosity. The very openness of the French administration to scientific talent was perhaps more important for explaining its scientific conservatism than other factors which have usually been emphasized, its centralized control, the dominance of Paris, the lack of competition among French universities, the hegemony of professors over their assistants, and the gerontocracy in the Academy of Sciences.

Hermann Minkowski: Architectural Formalist of the Theory of Relativity

In 1908 the mathematician Hermann Minkowski, carried through the second stage of the revolution that Einstein had begun. Minkowski began a celebrated address that was frankly "radical" in its intent with words that have become classic and are often quoted:

The views of space and time which I wish to lay before you have sprung from the soil of experimental physics, and therein lies their strength. They are radical. Henceforth space by itself, and time by itself, are doomed to fade away into mere shadows, and only a kind of union of the two will preserve an independent reality.[262]

Minkowski demonstrated that within a four-dimensional, space-time continuum, the interval between any two events was an invariant, that is, its numerical value remained the same throughout all the different possible measurements that could be made of the distances and time lapses between the two events, by all their possible observers moving relatively to each other with uniform velocities. Minkowski said, therefore, that "the word *relativity-postulate* for the requirement of an invariance . . . seems to me very feeble." He replaced the terminology of relativity with that of a new, higher absolutism:

Since the postulate comes to mean that only the four-dimensional world in space and time is given by phenomena, but that the projection in space and time may still be undertaken with a certain degree of freedom, I prefer to call it the postulate of the absolute world (or briefly, the world-postulate).[263]

It was Minkowski who created the haunting image of the whole universe as composed of world-lines in which the substantial points in their everlasting career moved along the parameter of time, and in which physical laws were defined as the reciprocal relations between the world-lines.

Minskowski remains a shadowy figure. He was, of course, one of Einstein's professors of mathematics at the Federal Polytechnic School, an excellent teacher, said Einstein, though according to Frank, "not a very good lecturer." However, as we have noted, Einstein happened at that time to have lost interest in pure mathematics, believing that only the most primitive mathematics would be required to state the fundamental laws of physics. Thus Einstein was not much interested in Minkowski's classes, and to his later chagrin, did not secure "a sound mathematical education."[264]

Hermann Minkowski came from the world of Eastern European ghetto Jewry. He was born in 1864 in the village of Alexotan, in Lithuania, a backward *shtetl* without libraries or schools worthy of the name. On Saturday afternoons, the old men would gather at the synagogue to discuss politics, or (in later years), the most recent "letter from America."[265] Hermann spent his childhood years in such surroundings. Then, to escape the pogroms the family moved to Germany.[266] In their new environment, the three Minkowski sons met the problem of anti-Semitism in diverse ways. Max, who was a merchant, became the spiritual guide of the Russian-Jewish students at the University of Königsberg and chairman of the Maskilim, the local society of modern Hebraists. Oskar Minkowski, a physician later known for his research on diabetes, renounced Judaism in order to advance his professional status. Hermann, the third brother, was (according to the noted Zionist writer, Shmarya Levin) "so important that the government had to sacrifice its anti-Semitism."[267] When Minkowski gave his celebrated address in 1908, he was professor of mathematics at the University of Göttingen.

Göttingen was, as Harald Bohr tells,

> a provincial town, calm and peaceful . . . the richest scientific life flourished there. A spirit of genuine international brotherhood of a rare intensity reigned there among the many young mathematicians who came from nearly all over the world to make a pilgrimage to Göttingen, bound together as they were by their common interest in and love for their science.

Here, among "an elite of young mathematicians," Minkowski felt truly at home. And although Göttingen was a center for other sciences as well, "mathematicians were by far the majority among the young scientists because of the especially great traditions for mathematics. . . ." It seemed to Harald Bohr "as if the town were populated almost entirely by mathematicians." An anecdote about Minkowski reflected the character of Göttingen life. Walking down the street one day, Minkowski saw a young man, unknown to him, buried deep in thought, and looking worried. Thereupon, "Minkowski went up to him, gave him a pat on the back and said encouragingly: 'It is sure to converge.' "[268] And indeed the young man was a brooding mathematician. It was in this setting that Minkowski conceived his great space-time generalization. The greatest mathematics has frequently flourished in provincial centers, not industrial ones.

Something of that ferment of alienation and rebellion, which was so marked in the Zurich-Berne student circle, had also entered the experience of Minkowski. Upon his death in 1909, his colleague and warm friend David Hilbert wrote a guarded obituary in a mathematical journal of Imperial Germany; he told of Minkowski's lively concern for politics and the theater, and his unswerving optimism, his conviction that the good and right would finally triumph.[269] It was inferred that Minkowski felt that the German social order was far from having realized that triumph of the good; evidently Minkowski stood aside from the German nationalistic culture of his place and time. Evidently too, Minkowski had a love for the idiom of Spinoza's metaphysics, so typical of the first generation of sons to emerge from the ghetto. The son of the ghetto, in whom arise the values of protest and "justice hunger" (to use Meyer Liben's phrase) invariably finds a prototype in Spinoza. Thus Minkowski as a radical physical theorist wrote of the "substance" of the world, and then, using the Leibnizian language, of "the idea of a pre-established harmony between pure mathematics and physics."[270] The postulate of the absolute world had, above all, a Spinozist note. One wonders whether a Spinozist emotion in Minkowski made congenial a mathematical theory in which space and time would fade into "mere shadows" as they do in Spinoza's substance. It was a Spinozist note, moreover, that dominated Einstein's emotions during the latter part of his life, profoundly shaping the character of his thought during his last thirty-five years.

The Subsidence of the Relativist Spirit: Einstein as a Generational Conservative

By the end of World War I, Einstein's relativist mood began to subside. His thinking was no longer isoemotional with revolutionary trends. His longing was for harmony, indeed, for a realization of God's mind in nature. A shift took place in his middle years which we might describe as from Hume to Spinoza. His references to Spinoza became more numerous than and finally supplanted those of empiricist critics. Leopold Infeld noted that in the 1930s Einstein's religious beliefs had an "affinity to those of Spinoza's." Another collaborator, Cornelius Lanczos, observed that Spinoza's "cosmic philosophy was so near to Einstein's heart. . . ." In 1937 when Niels Bohr continued at a personal meeting the long-standing disagreement he and Einstein had had with respect to determinism, their discussion was transformed into "a humorous contest concerning which side Spinoza would have taken if he had lived to see the development of our days. . . .[271] Asked by a rabbi whether he believed in God, Einstein responded: "I believe in Spinoza's God, who reveals himself in the harmony of all being, not in a God who concerns himself with the fate and actions of men." This belief was no decorative, ceremonial creed but one that emotively expressed Einstein's working faith as a scientist. As Arnold Sommerfeld wrote:

> It would have been impossible for Einstein to give the rabbi a more pointed reply or one which came closer to his innermost convictions. Many a time, when a new theory appeared to him arbitrary or forced, he remarked: "God doesn't do anything like that." I have often felt and occasionally also stated that Einstein stands in particularly intimate relation to the God of Spinoza.

This God provided Einstein with a kind of cosmic regulative principle for the choice of hypotheses: He entailed the "inner consistency and the logical simplicity of the laws of nature." Or as Einstein put it in the sentence that he had engraved in his reception room: "God is slick but he ain't mean."[272]

In Einstein's last years when this Spinozist mood prevailed the rationalistic criterion for the truth of propositions tended to supersede the experimental. Thus he wrote to Max Born in May 1952:

> Even if the deflection of light, the perihelial movement or line shift were unknown, the gravitation equations would still be convincing because they avoid the inertial system (the phantom which affects everything but is not affected).[273]

Einstein also wrote that he found himself drawn to "Buddha and Spinoza as religious personalities." His Spinozist faith, however, was more than a personal admiration or even religious ethic; it became a regulative principle for discovery of the laws of nature.[274] As a child, Einstein had longed for a religious existence, and resented his father's "ironic and unfriendly talk about dogmatic ritual," for Albert's father prided himself that Jewish rites were not observed in his house. Albert, however, "grieved over the fact that the Jewish dietary laws were neglected." He gave expression to his religious longings in songs that were filled with a Spinozistic pantheism identifying God and Nature.[275] This search for a Spinozistic order of nature was evidently part of Einstein's first generational rebellion against (and partial reconciliation with) his father's arid materialism.

From Spinoza Einstein also derived a fortitude against the "feeling of despondency and isolation" which was produced "by the frightful events of these times." The bold, optimistic, revolutionary hopes of 1905 were gone: "The confidence in the sure and constant progress of mankind that inspired people in the nineteenth century has given way to a crippling disillusionment." Spinoza's belief in the values of a determinist outlook commended itself to Einstein in this crisis of disenchantment:

> The spiritual situation with which Spinoza had to cope peculiarly resembles our own. The reason for this is that he was utterly convinced of the causal dependence of all phenomena, at a time when the success accompanying the efforts to achieve a knowledge of the causal relationship of natural phenomena was still quite modest. . . . He had no doubt that our notion of possessing a free will was an illusion resulting from our ignorance of the causes operative within us. In the study of this causal relationship he saw a remedy for fear, hate and bitterness, the only remedy to which a genuinely spiritual man can have recourse.[276]

The indeterminacies of quantum theory therefore awoke deep anxieties in Einstein. They seemed to conjure a disorderly world in which the Jew would find himself deprived of his birthright. Einstein fell back on the Biblical metaphors that were closest to him. He wrote to Max Born on December 4, 1926: "Quantum mechanics demands serious attention. But an inner voice tells me that this is still not the true Jacob. The theory accomplishes a lot, but it scarcely brings us closer to the secret of the Old One. In any case, I am convinced that He does not play dice."[277] The young quantum theorists were indeed like Jacob stealing the birthright of physics. But to Einstein, however, they remained children of Esau. And identifying himself with the blind father Isaac, he felt that they too were deceiving him, that they were not truly what they purported to be. Sharing a kinship, however, with the Hebrew Jacob and not with the Edomite Esau, Einstein used a conventional metaphor that reversed the characters of the stories. For the blind Isaac cried: "The voice is the voice of Jacob, but the hands are the hands of Esau . . ."[278]

Underlying Einstein's tenacious resistance to the trend of contemporary younger physicists toward an indeterminist mode of thought was his emotional conviction that the foundations of wisdom were being undermined; the principle of indeterminacy seemed to him a child of contemporary nihilism. With the beginning of the German Revolution of 1918, Einstein, like so many others, had experienced the exhilaration of the coming of a bright new order, but he was soon brought face to face with the evil that was latent in idealism itself. As early as November 1918, Einstein began to lose faith in the revolution and the revolutionary students in Berlin. On one memorable occasion, Einstein, Max Born, and the psychologist Max Wertheimer had tried to intervene with the revolutionary student leaders on behalf of the preservation of academic freedom; they were not received warmly by the revolutionary committee. We owe to Max Born a vivid account of this incident during the revolution:

In November of that year, "the German chiefs of staff suddenly capitulated, and revolution broke out all over Germany." Born was lying ill in bed, when the telephone rang. It was Einstein. He said that the students at the university, following the example of the newly organized workers' and soldiers' councils (German "soviets"), were forming a students' council. The student soviet, emulating the "seizure of power," had begun its revolution by deposing the rector of the university, and imprisoning him, together with other dignitaries. Einstein was asked to intercede with the student extremists. Since he was known as a "leftist," it was believed he would be able to exert an influence on the student activists, and negotiate with the students' council for the release of the academic prisoners of the class struggle; then a rational order might be instituted. The students' council was at the time in session at the Reichstag. Einstein asked Born to join with him as a delegation, and together with Max Wertheimer, the Gestalt psychologist, they proceeded by foot and tram to the revolutionary Reichstag center.

Masses of people were encamped around the building; cordons of red-banded, heavily armed revolutionaries were on patrol. But fortunately, someone recognized Einstein, and paths were opened for them.

Inside the Reichstag building, the chairman of the student soviet greeted the professional committee courteously, and requested that they wait until a deliberation on the new university statutes was completed. The professors waited patiently, listening. Then finally, when the deliberation was over, the chairman said: "Before I take up your request, Professor Einstein, may I be permitted to ask what you think about our new regulations for the students?" Einstein thought for a few minutes, then said: "I have always believed that the most valuable thing in the German universities is their academic freedom; no one tells the lecturer what to teach, and the student chooses which lectures to attend, with little supervision or control. It seems to me, however, that your new statutes do away with all that, and replace it with strict regulations." A downcast silence came over the ambitious, young student leaders. They showed little inclination to consider Einstein's request, and pronounced themselves as lacking in jurisdiction. They directed him to take the matter to the new government at the Wilhelmstrasse, and gave the professorial delegation the necessary pass.

At the palace of the Reichschancellor, the old order had not quite given way to the new. Imperial lackeys stood at the stairs and hallways, but running up and down them were shabby men with briefcases, socialist parliamentarians, and representatives from the workers' and soldiers' councils. The main hall was filled with excited people, talking loudly. Again Einstein was recognized, and was taken to see the new President of the Republic, the chief of the Social Democratic party, the former saddler, Friedrich Ebert. He impressed upon them that with the survival of the Reich at stake, he could not deal with smaller matters like the incarcerated rector. Nonetheless, he gave them a written note to the responsible minister, and the affair was concluded.

"We left the palace of the Reichschancellor," writes Born, "in the highest of spirits, with the feeling of having taken part in a historic event, and hoping indeed that the time of Prussian arrogance was finished, that it was all over with the Junkers, the hegemony of the aristocrats, the cliques of officials, and the military, that now the German democracy was victorious."[279]

However, the German student movement became increasingly hostile to Einstein. They placed obstacles to his public lectures at the University of Berlin. According to David Baumgardt in "Looking Back on a German University Career" (1965), Einstein was scheduled to give a course in the winter of 1919 or 1920, in the Auditorium Maximum, the largest in the university. The leadership of the student body felt that Einstein's fame was mainly the product of Jewish propaganda. They complained that much coal and electricity would be required for using the lecture hall, and their president suggested that Einstein hire his own hall in town since most of his listeners, in any case, would be nonstudents. Einstein replied: "Yes, my dear sir, leave the decision about this to me." He gave the lectures at the scheduled place and time, but they were evidently the last he ever gave at the University of Berlin.

The next generation saw revolutionary hopes turn to nullity. For a number of years Einstein retained his hope that perhaps the socialist revolution would build a new society in which men would rise to rationality. He lent his name to left-wing committees and societies. When the League of the Friends of the International Workers' Aid (*Bund der Freunde der Internationalen Arbeiterhilfe*) was organized on March 16, 1923, Einstein was made its honorary president and co-sponsor together with such well-known left-wing artists as Kathe Kollwitz and George Grosz. The instigator of the group was Willy Muenzenberg, the famed, adroit, and indefatigable designer of many communist fellow-traveling groups, and years earlier in Zurich the secretary of the Swiss Socialist Youth Movement. In June 1923, Einstein became a charter member of the Society of the Friends of New Russia (*Gesellschaft der Freunde des neuen Russland*), and in 1927, honorary president, together with Henri Barbusse, of the League Against Imperialism.[280]

But his faith in the revolutionary student youth ebbed steadily as he observed academic freedom vanishing under the blows of German student activists. In 1931 he wrote that "today it is no longer the college students and university professors who embody the hope and ideals of men." The following year he asserted that "the conduct of the students . . . is one of the saddest symptoms of our time. . . ."[281]

The experiences of Nazi Germany and Stalinist Bolshevism made Einstein look back in later years upon the naiveté of his hopes in 1918. He wrote to Born in 1944:

> Do you still recall how a little less than twenty-five years ago we took a tram together to the Reichstag, convinced that we could really help make honest democrats out of those fellows there? How naive we were as men forty years old. I cannot help but laugh when I think about it. We both did not see how much more remains in the spinal cord than in the cerebrum, and how much more firmly it is entrenched there. I must think about that now, if the tragic mistakes of that time are not to repeat themselves. We must not allow ourselves to be surprised if the scientists (the great majority of them) are no exception, and when they do turn out to be different, it is not because of their intellectual abilities, but rather because of their stature as human beings.[282]

Disillusioned with the consequences of political and economic revolution, Einstein was similarly postrevolutionary in his mood as a scientist. His positivist friend Philipp Frank was surprised around 1932 to discover that Einstein did not share his positivist opinions and refused to accept "the new fashion" that was arising in physics. Frank protested: "But the fashion you speak of was invented by you in 1905?" Einstein rejoined: "A good joke should not be repeated too often." Einstein the Spinozist said: "I shall never believe that God plays dice with the world." It was all a bit unfathomable to the stanch logical positivist Frank. His invaluable biography of Einstein never even mentioned Spinoza, as if ill at ease with a metaphysical interloper; Einstein's thought was clearly no longer isoemotional with positivism.

The social causes with which Einstein became identified, furthermore, showed a different pattern in later years. He no longer cosponsored organizations for Soviet friendship and workers' aid. He became more concerned typically with homeless Jews, wandering from Poland, drifters into Central Europe, members of the "exiled sections of the Russian bourgeoisie." Was he moved by the memory of his own father, the unsuccessful maker of electrical items, wandering uncertainly from country to country? Einstein became so active on behalf of his luckless fellow Jews that Freud commented in February 1929: "How often do I not envy Einstein the youth and energy which enable him to support so many causes with such vigor." By contrast Freud saw himself as "old, feeble and tired."[283]

The transition from the relativism of the Zurich-Berne revolutionary circle to the Spinozist absolutism of the postrevolutionary era was more than a drama of the aging of man. It prefigured something deep in the sociology of ideas—the workings of the cycle from relativism to absolutism. A generation later, with the end of World War II, a revolt against relativities—ethical, sociological, and anthropological—took place. Whether this outlook will inspire fruitful perspectives in physical theory, whether novel experimental facts will prove themselves malleable in such an emotional direction, are still open questions.

The sympathy that Einstein felt for those intellectuals rebelling against pomposity and position, however, never diminished; in spirit he remained a comrade of the Olympia Academy all his life. The warmth and encouragement he showed to Upton Sinclair in that doughty reformer's many causes was an outstanding example of this trait. Sinclair, a man much derided by the academic establishment and the bourgeois public, published a book, *Mental Radio* (1930), in which he reported on his wife's presumably successful experiments with telepathy.[284] Einstein found the book "highly worthy," unlike Russell who refused to read it on the grounds that the whole subject was "humbug." Einstein agreed to write the preface for a

German edition of the book. "I am convinced," he wrote, "that it deserves the most earnest attention, not only of the laiety, but also of the specialists in psychology." Though its results went "beyond what an investigator of nature considers to be thinkable," wrote Einstein, nevertheless he felt that "so conscientious an observer and writer as Upton Sinclair" merited a hearing; perhaps, suggested Einstein, here was evidence of "some unknown hypnotic influence." Einstein hailed Sinclair's muckraking, pomp-deflating activities: "To the most beautiful joys of my life belongs to me your wicked tongue!" he wrote him in 1932. When Sinclair was told in 1933 that his social protest was "undignified," Einstein comforted him in verse which he inscribed on a photograph and brought personally to Sinclair's home:

To whom does the dirtiest pot not matter?
Who knocks the world on its hollow tooth?
Who suspects the now and swears by the morrow?
Who never troubles about "undignified"?
Sinclair is the valiant man.
If anyone, then I can attest it.

He welcomed Sinclair's candidacy for the governorship of California in 1934, and thought his epic plan to create "a quasi isolated economic state within the state would be really worth while to be undertaken." Twenty years later Einstein felt that Sinclair had like himself finally concluded that the "cause of evil" was "not in a particular group, but in the nature of man itself. One must already be tolerably old, before this becomes clear to one."[285]

Einstein took pride and pleasure in such friendships. He was more at home with the Sinclairs of the world than with any other group. For this reason he took perverse delight in introducing Sinclair to the neighbor who had been avoiding him, Dr. Robert A. Millikan, a chieftain of the scientific establishment: "Dr. Millikan, I want you to meet my friend Upton Sinclair."[286] Such was the spirit of the counterculture from which had issued the theory of relativity.[287]

Even his feelings toward women were touched with emotions of social protest. Einstein's second wife said that "out of pity he was attracted to women who did physical work." Therefore she assured Vera Weizmann, wife of Chaim Weizmann, when they were traveling together in 1921 on behalf of the World Zionist Organization that "she did not mind her husband's flirting with [Vera] as 'intellectual women did not attract him.' " Vera found Einstein "young, gay, and flirtatious."[288]

Like his aging friend Friedrich Adler, Einstein finally relinquished the hopes he had entertained after the Bolshevik Revolution for an era in which the human spirit would be liberated. It seemed that a socialist, planned economy could also plan the extirpation of human liberties; if God was not mean enough to play dice with physical entities, evidently His order reduced human idealists to pathetic players laboring under great odds in games of social chance. Einstein fell back on his own mockery and his own prayer to the gods. He wrote some political verses evidently in 1952.

The Wisdom of Dialectical Materialism

Through sweat and pain beyond compare
To arrive at a small grain of truth?
A fool he who thus miserably toils and moils
We simply ordain through the Party line.

And those who dare to doubt
Will have their skulls knocked in speedily.
Indeed, thus one educates, as never before,
Dissenting spirits to conformity.

Epitaph for the Marx-Engels Institute

In the realm of truth-seekers there is no human authority. He who attempts to play the ruler there will run afoul of the laughter of the gods.[289]

As Einstein grew older, he brooded on the conflict of generations among scientists and wondered whether his friend Paul Ehrenfest had chosen the rational way out in committing suicide. He often thought of and discussed Ehrenfest's action, which he regarded in good part as the outcome of a generational cleavage. It was not the usual conflict of generations, however, wrote Einstein's friend Antonina Vallentin; such conflict was exacerbated in Einstein's own case by his intellectual isolation late in life. Einstein's was the drama of a man who, "in spite of his years persisted in continuing on his own way, which was becoming more and more deserted; for most of his friends, and all the young men around him, declared that the way led nowhere."[290] Einstein, the revolutionist of 1905, felt estranged from the revolution of indeterminacy that had commenced in the disorder of the postwar world. He wrote of Ehrenfest in 1934 in words that were remarkably applicable to his own isolation during the last twenty-five years of his life:

In the last few years, the situation was aggravated by the strangely turbulent development which theoretical physics has recently undergone. To learn and to teach things that one cannot fully accept in one's heart is always a difficult matter, doubly difficult for a mind of fanatical honesty, a mind to which clarity means everything. Added to this was the increasing difficulty of adaptation to new thoughts which always confronts the man past fifty. I do not know how many readers of these lines will be capable of fully grasping that tragedy. Yet it was this that primarily occasioned his escape from life.[291]

This inevitable conflict of two generations of thought was one, said Einstein, "which no university professor over fifty can escape in one way or another." As Antonina Vallentin wrote: "He said this with resignation, for it was also his own conflict."[292]

This tragedy of generational estrangement is probably felt most sharply by those whose youthful thought patterns were dominated by emotions of generational revolt. The revolutionists of a preceding generation are usually the most alienated from the revolution of the next; a permanent revolutionist is an eccentricity of human nature. Max Planck, the innovator of quantum theory, once wrote: "A new scientific truth does not triumph by convincing its opponents and making them see the light, but rather because its opponents die, and a new generation grows up that is familiar with it."[293] Whatever sociological truth this generational relativism might have, it was a standpoint repugnant to Einstein, the lover of objective truth, international, interracial, intergenerational. Yet the story of his work is part of the history of the grandeur and defeat of the generation of 1905.

Will the Zurich-Berne countercommunity that nurtured Einstein's scientific creativity have its counterparts in the present world of science, so opulent in opportunities for young practitioners? Will the fact that scientific research is becoming both increasingly well rewarded and bureaucratized render obsolete the self-formation of such groups as the Olympia Academy? Will the unusual, individualistic insight tend likewise to be submerged or even ostracized through the unconscious pressures exerted by a highly organized scientific community?

In his later years, Einstein's thoughts returned often to the happy days at Zurich and the Berne patent office when the Olympia Academy followed its own impulses without regard for the scientific establishment. He used to advise young scientists to work at a "shoemaker's job," withdrawn from both business and academic pressures.[294] He wondered how an equivalent for his patent-clerk period could be contrived for fledgling scientists in a modern industrial society. Might the young, wayward scientific theorist be allowed by society to spend his years of creative isolation as a lighthouse-keeper? Einstein said in 1933:

> When I was living in solitude in the country, I noticed how the monotony of a quiet life stimulates the creative mind. There are certain occupations, even in modern society, which entail living in isolation and do not require great physical or intellectual effort. Such occupations as the service of lighthouses and lightships come to mind. Would it not be possible to place young people who wish to think about scientific problems, especially of a mathematical or philosophical nature, in such occupations?[295]

Einstein thought that such isolated employments were better for the young scientist than scholarship grants, for he felt the young scientist would then be free from the pressure "to arrive as soon as possible at definite conclusions," a pressure Einstein felt to be "harmful to the student of pure science." He urged the young inquirer to enter some practical profession in which he could earn his livelihood, provided that it left him sufficient time and energy for his scientific work. As a child, Albert had been fond of the Jewish sayings in the *Pirke Aboth* (The Ethics of the Fathers). The elucidations of them which he heard as a boy from his rabbi in Munich "were of such profound religious significance that Einstein could never forget them."[296] One of them said: "Beautiful is the study of the Torah when joined with ordinary occupation."[297] Einstein remained faithful to this teaching.

The history of the Zurich-Berne student circle provides much material for reflection in our own vexed time of generational strife. Generational rebellion is a powerful motivating force in the progress of science. It seems, however, to be a much more uncertain force in determining the direction of social change. In science, generational rebellion widens the imagination, deepens the intuition, proposes challenging hypotheses, provokes laborious calculation and patient observation; but additionally, science has the common discipline and criteria of scientific method, which bind into a joint enterprise both intergenerational and intersubjective, the diverse, opposed hypotheses and emotional standpoints. In politics such common discipline and criteria are less easily defined, and more easily rejected; generational rebellion then, for all its idealism, can be transmuted into a self-destructive force. Einstein's energies of generational revolt were sublimated into abiding works of theoretical genius; those, on the other hand, of his friend Friedrich Adler, whose avowed aim in life was to raise "the moral standards of socialism," went into blind alleys of revolutionary fervor which hardly left their intended impress on either the social system or an evolving human nature.[298] The generational revolutionist in human affairs runs risks of delusion and self-destruction, and yet, as we have seen, his actions share a common source with the achievements of his comrade in science. By a strange fate Einstein's work became the abiding monument of the generation of 1905 and his achievement was, in a sociological sense, the most revolutionary in its consequences.

Notes

1. Peter Kropotkin, *The Conquest of Bread* (1906; New York, 1926), p. 103.

2. Philipp G. Frank, ed., *The Validation of Scientific Theories* (New York, 1961), p. 22.

3. Carl Seelig, *Albert Einstein: A Documentary Biography*, trans. Mervyn Savill (London, 1955), p. 51.

4. Anton Reiser, *Albert Einstein: A Biographical Portrait* (New York, 1930), pp. 48–51. Einstein attested in a prefatory note that the "facts of the book" were "duly accurate." Miss Helene Dukas, who was Einstein's secretary for many years, informed the author that "Anton Reiser" was the pseudonym of Rudolf Kayser, the husband of Einstein's stepdaughter, Ilse. (Interview, August 30, 1968, Princeton, New Jersey.)

5. General Inquiry Official, *Guide to Zürich* (Zurich, 1903), pp. 2, 178. According to Henry Demarest Lloyd, Zurich's population in 1900 was 150,700. Cf. *The Swiss Democracy*: H. D. Lloyd, *The Study of a Sovereign People* ed. John A. Hobson (London, 1908), p. 5.

6. Jan Marinus Meijer, *Knowledge and Revolution: The Russian Colony in Zürich, 1870–1873* (Assen, 1955), pp. 1, 47ff.

7. Stepniak, (pseud.) Kravchinsky, Sergyei Mikhailovich, *Underground Russia: Revolutionary Profiles and Sketches from Life* (New York, 1883), p. 19–20.

8. Seelig, *Einstein: A Documentary Biography*, p. 95. Philipp Frank and his wife often told the author during the summers of 1940 and 1941 in Cambridge, Massachusetts, that Einstein used to say that Germany had not deserved such a woman as Rosa Luxemburg.

9. J. P. Nettl, *Rosa Luxemburg* (London, 1966), vol. 1, pp. 60, 65.

10. Katharine Anthony, "Alexandra Kollontay: The World's One Woman Ambassadress," *North American Review*, vol. 230 (1930), p. 278. Isabel de Palencia, *Alexandra Kollontay: Ambassadress from Russia* (New York, 1947), pp. 33–36.

Alexandra Kollontai, *The Autobiography of a Sexually Emancipated Communist Woman*, trans. Salvator Attanasio (New York, 1971), pp. 11, 12.

11. Florence Kelley, "My Novitiate," *The Survey*, vol. 58 (April 1927), pp. 33–35.

12. Kropotkin, *Memoirs of a Revolutionist* (Boston, 1899), p. 269.

13. Eduard Bernstein, *My Years of Exile: Reminiscences of a Socialist*, trans. Bernard Miall (New York, 1921), pp. 110–111. The Swiss socialist, Fritz Brupbacher, who in 1901 married a Russian revolutionary student, wrote a vivid description of the hectic ideological life of the Russian student factions, "all the nuances, all the controversies of the different Russian political tendencies . . ." (*Socialisme et Liberté*, trans. Jean Paul Samson [Neuchatel, 1955], pp. 204–205).

14. Anatoly Vasilievich Lunacharsky, *Revolutionary Silhouettes*, trans. Michael Glenny (London, 1967), p. 87.

15. Julius Braunthal, *History of the International: 1864–1914*, trans. Henry Collins and Kenneth Mitchell (London, 1966), p. 326.

16. Chaim Weizmann, *Trial and Error: The Autobiography of Chaim Weizmann* (New York, 1949), p. 50.

17. Isaac Gruenbaum, *The History of Zionism*, pt. 2 (Tel Aviv, n.d.), pp. 16, 27, 37; *Letters and Papers of Chaim Weizmann*, ed. Leonard Stein and Gedalia Yogev (London, 1968), vol. 1, p. 158.

18. Ibid., pp. 79–80.

19. Chaim Weizmann, *Trial and Error: The Autobiography of Chaim Weizmann* (New York, 1949), pp. 50–51.

20. *Letters and Papers of Weizmann*, pp. 209–210.

21. Ibid., p. 201.

22. Ibid., p. 278.

23. Ibid., p. 167.

24. *The Hebrew University of Jerusalem* (Jerusalem, 1957), pp. 2–6.

25. Gaudens Megaro, *Mussolini in the Making* (London, 1938), p. 56 ff.

26. Margharita Sarfatti, *The Life of Benito Mussolini*, trans. Frederic Whyte (New York, 1925), pp. 104, 112.

27. Laura Fermi, *Mussolini* (Chicago, 1961), p. 36.

28. Warren Lerner, *Karl Radek: The Last Internationalist* (Stanford, 1970), p. 8. Max Nomad, *Dreamers, Dynamiters and Demagogues* (New York, 1964), pp. 35–36. "Zurich was in those years the only city with a German-speaking population where anarchists could present their views without having to think of the Damocles sword of possible persecution that always threatened their comrades in Germany and Austria," wrote Max Nomad (p. 15).

29. N. K. Krupskaya, *Reminiscences of Lenin*, trans. Bernard Isaacs (Moscow, 1959), pp. 175, 311, 322, 326, 327, 316; *On Education: Selected Articles and Speeches*, trans. C. P. Ivanov-Mumjiev (Moscow, 1957), pp. 77–80.

30. Lloyd, *The Swiss Democracy*, pp. 1, 201, 205, 207.

31. Ibid., p. 196.

32. Ibid., pp. 176–177, 188–189; William Harbutt Dawson, *Social Switzerland: Studies of Present-Day Social Movements and Legislation in the Swiss Republic* (London, 1897), p. vii.

33. Lloyd, *The Swiss Democracy*, pp. 7, 206, 249.

34. Ibid., p. 211.

35. Ibid., p. 257.

36. Ibid., p. 256.

37. Dawson, *Social Switzerland*, p. vii.

38. G. D. H. Cole, *A History of Socialist Thought*, vol. 3, pt. 2, *The Second International, 1889–1914* (London, 1956), 608 ff.

39. Sigmund Freud, *On the History of the Psychoanalytic Movement, Standard Edition of the Complete Psychological Works*, ed. James Strachey, vol. 14 (London, 1957), p. 27.

40. C. G. Jung, *Memories, Dreams, Reflections*, trans. Richard and Clara Winston, Vintage ed. (New York, 1965), pp. 111–117.

41. Franz Alexander, Samuel Eisenstein, and Martin Grotjahn, eds., *Psychoanalytic Pioneers* (New York, 1966), pp. 169 ff.

42. Yves Duplessis, *Surrealism*, trans. Paul Capon (New York, 1950), p. 15. Maurice Nadeau, *The History of Surrealism*, trans. Richard Howard (New York, 1965), pp. 56–57, 61–62.

43. Philipp G. Frank, *Einstein: His Life and Times*, trans. George Rosen (New York, 1947), p. 20.

44. Seelig, *Einstein: A Documentary Biography*, p. 95.

45. Julius Braunthal, *Victor und Friedrich Adler: Zwei Generationen Arbeiterbewegung* (Vienna, 1965), p. 178.

46. Ibid., p. 179.

47. Ibid., pp. 182–183.

48. Reiser, *Einstein: A Biographical Portrait*, pp. 50–51.

49. Braunthal, *Victor und Friedrich Adler*, p. 196.

50. Reiser, *Einstein: A Biographical Portrait*, pp 76–77. Also cf. David Reichinstein, *Albert Einstein: A Picture of His Life and His Conception of the World*, trans. M. Juers and D. Sigmund (Prague, 1934), p. 53. Reiser's account evidently blurs two periods of Einstein's friendship with Adler: that period in which they were both students, and a later one when Einstein returned to Zurich in 1909 to teach at the university. They lived in the same house during this later period.

51. Ernst Mach, *History and Root of the Principle of the Conservation of Energy*, trans. P. E. B. Jourdain (Chicago, 1911), p. 94. Friedrich Wolfgang Adler, *Bemerkungen über die Metaphysik in der Ostwald' schen Energetik* (Leipzig, 1905), pp. 38, 47. A copy inscribed to Karl Kautsky is in the Rare Documents Collection of the University of Washington at Seattle. It was offprinted from the original in *Vierteljahrsschrift für Wissenschaftliche Philosophie und Soziologie*, vol. 29 (1905), pp. 287–333. Also cf. Fritz Brupbacher, *60 Jahre Ketzer: Selbstbiographie von Fritz Brupbacher* (Zurich, 1935), p. 152.

52. Friedrich Adler, "Friedrich Engels und die Naturwissenschaft," *Neue Zeit*, Jahrg. 25, no. 18 (1906–1907), pp. 620–638.

53. Friedrich Adler, "Der 'Machismus' und die Materialistische Geschichtsauffassung," *Neue Zeit* Jahrg. 28, Bd. 1, no. 19 (February 4, 1910), pp. 671–682. Max Planck himself replied to Friedrich Adler's article on Ernst Mach (1909). He complimented Adler for not writing in the highly personal, biased style of Ernst Mach, but thought Adler made Mach's principle of economy vague by including in it the independent component of a search for a stable world-picture. He then argued that both Mach and Adler had an imperfect understanding of the principles of thermodynamics. Adler, he felt, failed to realize that the content of Clausius's second law actually did involve anthropomorphic notions. Planck evidently felt that the strict Machian program of de-anthropomorphization would destroy those portions of thermodynamical theory in which words like energy-reservoir or machine appear. Max Planck, "Zur Machschen Theorie der physikalischen Erkenntnis, Eine Erwederung," *Physikalische Zeitschrift*, vol. 11 (1910), pp. 1186 ff.

54. Franz Mehring, "Eine Antwort an Friedrich Adler," *Neue Zeit*, Jahrg. 28, vol. 1 (1910), p. 682, reprinted in *Zur Geschichte der Philosophie* (Berlin, 1931), pp. 239–248.

55. Friedrich Adler, "The Discovery of the World Elements (In Honor of Ernst Mach's Seventieth Birthday)," *International Socialist Review*, vol. 8, no. 10, (April 1908), p. 579. The original essay, "Die Entdeckung der Weltelemente," was published in *Der Kampf*, Band 1, (February 1908), pp. 231–240. Ernest Untermann, the translator of Adler's article, and himself a foremost American socialist theoretician, wrote to the editor:

> You will note the similarity between the views of Avenarius, Mach, and Dietzgen. In fact Mach has endorsed the views of Joseph Dietzgen, and Avenarius is regarded by scientists like Adler as a counterpart of the proletarian philosopher (p. 638).

Dietzgen was a Marxist workingman, self-taught in philosophy. Marx felt that his writing, though at times confused, contained "much that is excellent and—as the independent product of a working-man—admirable." Cf. Karl Marx, *Letters to Dr.*

Kugelmann (New York, 1934), p. 80. V. I. Lenin, *Materialism and Empirio-Criticism: Critical Notes Concerning a Reactionary Philosophy*, trans. David Kvitko (New York, 1927), pp. 204 ff.

56. Friedrich Adler, "What Is the Use of Theory?," trans. William E. Bohn, *International Socialist Review*, vol. 9, no. 5 (November 1908), p. 339. The original essay, "Wozu Brauchen wir Theorien?," appeared in *Der Kampf*, Band 2 (March 1909), pp. 256–263. Karl Kautsky, the chief expositor of Marxism in the Socialist International, wrote that Adler's combination of Marx and Mach was dominant among "the younger generation of Marxists." See *Die Materialistische Geschichtsauffassung*, (Berlin, 1927), Vol. 1, p. 15.

57. Letter of Friedrich Adler to Dr. Otto Nathan, March 26, 1957 (translated). Einstein visited Victor Adler in 1911. Many years later, on July 12, 1930, he recalled the character of the socialist statesman:

> The evening that I spent as a friend of Fritz Adler with Viktor Adler in 1911, will be a lifelong beautiful memory for me. This man combined a gracious understanding for human weaknesses with the fire of an active friend of humanity. This fatherly sympathetic understanding for all human beings and conditions was felt by all; all political parties gave him their trust and sympathy. . . . One can hardly imagine this today. In old Vienna with its patina, a tolerance was alive which has unfortunately practically vanished in Austria, because of the troubles of the period. If present-day Austria had a leader who would like Viktor Adler combine a superior intelligence with a high sense of justice, then the serious mistakes made by the powers at Versailles would not have produced the low level, at least in cultural matters, which has now regrettably begun (from the Archives of Albert Einstein, July 12, 1930, this author's translation).

58. Letter of Julius Braunthal to the author, September 24, 1968. Friedrich Adler introduced his future bride to "the small circle of friends who discussed half nights long on questions of natural science, philosophy, and socialism" (Braunthal, *Victor und Friedrich Adler*, p. 190).

59. Reiser, *Einstein: A Biographical Portrait*, p. 73. The Swiss socialist Fritz Brupbacher felt that Adler's political incapacity joined to his purity of character made him a tool for others. Cf. Fritz Brupbacher, *60 Jahre Ketzer: Selbstbiographie von Fritz Brupbacher*, p. 141.

60. Reiser, *Einstein: A Biographical Portrait*, pp. 72–73.

61. Kurt Joel, "Friedrich Adler als Physiker: Eine Unterredung mit A. Einstein," *Vossische Zeitung* (May 23, 1917). Paul Arthur Schilpp, ed., *Albert Einstein: Philosopher-Scientist*, (Evanston, 1949), p. 706.

62. Braunthal, *History of the International: 1864–1914*, p. 348; Ronald Florence, *Fritz: The Story of a Political Assassin*, (New York, 1971), pp. 133, 136–138, 147.

63. Friedrich Adler, *J'Accuse: An Address in Court* (New York, 1917), pp. 16, 17, 20–21, 24, 28, 34. The greater part of Adler's speech was also printed in "J'Accuse!" Friedrich Adler's Address in Court, Pt. 1," *The Class Struggle*, vol. 1, no. 2 (July-August 1917), pp. 102–114; "Pt. 2," vol. 1, no. 3 (September–October 1917), pp. 63–71.

Max Nomad, who knew Friedrich Adler in Zurich, writes that his act was directed against the government and "symbolically against his own father who did not oppose those tendencies" (*Dreamers, Dynamiters and Demagogues: Reminiscences by Max Nomad*, p. 38).

64. Braunthal, *In Search of the Millennium* (London, 1945), pp. 195, 198.

65. Charles A. Gulick, *Austria from Habsburg to Hitler* (Berkeley, 1948), vol. 1, p. 38. Also cf. Adolf Sturmthal, *The Tragedy of European Labor, 1918–1939* (New York, 1951), pp. 51–52.

66. Seelig, *Einstein: A Documentary Biography*, p. 97. Victor Adler, *Briefwechsel mit August Bebel und Karl Kautsky*, comp. Friedrich Adler (Vienna, 1954), p. 633. Adler's trial in Vienna was held on May 18–19, 1917. Cf. Peter Michelmore, *Einstein:*

Profile of the Man (New York, 1962), p. 70. Albert Einstein, Michele Besso, *Correspondance 1903–1955*, tr. Pierre Speziali (Paris, 1972), pp. 105–108.

67. Reiser, *Einstein: A Biographical Portrait*, p. 14.

68. Einstein wrote Hilferding humorously: "If, however, the Herr Minister will not allow the sick Leon Trotzki to enter, and give him asylum, then . . . if he weren't a Minister I would grab him by his ear. . . ." (Letter in possession of Stephen Lehmann, Berkeley, California.) The letter was sent by Einstein together with some enclosed verse on March 14, 1929, in response to a message of congratulations from Hilferding on Einstein's fiftieth birthday. Trotsky at this time sought in vain for a grant of asylum from the German government. Cf. Leon Trotsky, *My Life*, trans. Max Eastman (New York, 1930), pp. 568–572.

69. Isaac Deutscher, *The Prophet Armed: Trotsky 1879–1921*, Vintage ed. (New York, 1954) vol. 1, pp. 185–186. *International Communism in the Era of Lenin*, ed. Helmut Gruber (New York, 1967), pp. 203, 267. Gulick, *Austria from Habsburg to Hitler*, p. 72.

70. Friedrich Adler et al., *The Moscow Trial and the Labour and Socialist International* (London, 1931), pp. 7–22. Friedrich Adler, *The Witchcraft Trial in Moscow* (New York, 1937), p. 31. Also cf. Friedrich Adler, *The Anglo-Russian Report: A Criticism of the Report of the British Trade Union Delegation to Russia from the Point of View of International Socialism*, trans. H. J. Stenning (London, 1925).

71. "It is true, as Adler told me, that he had visited Einstein in Princeton only once during his stay in the United States. I am inclined to explain this puzzle by this extreme shyness and still more perhaps by his feelings of frustration about the failure of the international socialist movement to which he had devoted the greater part of his life's work" (Letter of Julius Braunthal to the author, September 24, 1968).

72. Joseph Buttinger, *In the Twilight of Socialism: A History of the Revolutionary Socialists of Austria* (New York, 1953), p. 541.

73. Einstein, *Lettres à Maurice Solovine* (Paris, 1956), p. vii.

74. Einstein, *Relativity: The Special and General Theory*, p. 141.

75. Einstein, "Ernst Mach," *Physikalische Zeitschrift*, vol. 17 (April, 1916), pp. 101–104.

76. Ibid.

77. Richard von Mises, "Ernst Mach und die Empiristische Wissenschaftsauffassung," in *Selected Papers of Richard von Mises*, ed. P. Frank et al. (Providence, 1964), vol. 2, p. 495.

78. Paul Carus, "Professor Mach and his Work," *The Monist*, 21 (1911), pp. 21–25.

79. Robert Bouvier, *La Pensée d'Ernst Mach: Essai de Biographie Intellectuelle et de Critique* (Paris, 1923), pp. 233–225.

80. Werner J. Cahnman, "Adolf Fischhof and his Jewish Followers," *Publications of the Leo Baeck Institute of Jews from Germany*, Year Book IV (London, 1959), p. 124.

81. Ibid.

82. Letter of Ernst Mach to Joseph Popper-Lynkeus, August 5, 1878, in Einstein Archives, Princeton, New Jersey.

83. Carus, "Professor Mach and his Work," pp. 21–25.

84. Ernst Mach, *Popular Scientific Lectures*, 4th ed. trans. Thomas J. McCormack (Chicago, 1910), p. 86.

85. Mach, *History and Root of the Principle of Conservation of Energy*, pp. 64–65.

86. Robert H. Lowie, "Letter from Ernst Mach to Robert H. Lowie," *Isis*, vol. 37 (1947), p. 66.

87. Mach, *History and Root of the Principle of Conservation of Energy*, p. 15.

88. Mach, *Popular Scientific Lectures*, p. 204.

89. Ibid., p. 85.

90. Ibid., p. 204.

91. Ibid., p. 76.

92. Ibid., pp. 81–82.

93. Mach, *The Analysis of Sensations and the Relation of the Physical to the Psychical*, trans. Sydney Waterlow from the 5th German ed. (Chicago, 1914), p. 136.
94. Ibid., p. 254.
95. Mach, *Popular Scientific Lectures*, p. 88.
96. Mach, *The Analysis of Sensations*, p. 356.
97. Erwin Schrödinger, *My view of the World*, trans. Cecily Hastings (Cambridge, England, 1964), p. 237.
98. Mach, *Popular Scientific Lectures*, pp. 60–61.
99. Mach, *The Analysis of Sensations*, p. 254.
100. Mach, *The Analysis of Sensations*, pp. 142–143.
101. Ibid., p. 144.
102. Ibid., p. 142.
103. Ibid., p. 358.
104. Ibid., p. 30.
105. Ibid., p. 13.
106. Ibid.
107. Ibid., p. 15.
108. Mach, *Popular Scientific Lectures*, p. 208.
109. Mach, *The Analysis of Sensations*, p. 15.
110. Ibid., p. 4. Freud discussed Mach's dreams in his exploratory article, "The Uncanny" (1919); they provided striking examples of the double (*der doppelgänger*) issuing from the recesses of some repressed complex (Sigmund Freud, *Collected Papers*, trans. Joan Riviere [London, 1925], vol. 4, p. 402).
111. Ibid., pp. 239–240.
112. Ibid., p. 13.
113. Ibid., pp. 24–25.
114. Ibid.
115. Ibid., p. 5.
116. Mach, *Erkenntnis und Irrtum*, cited in Federigo Enriques, *The Historical Development of Logic*, trans. J. C. Rosenthal (New York, 1929), p. 226.
117. Mach, *Popular Scientific Lectures*, p. 207.
118. Thus Peter Kapitsa writes:

> Boltzmann committed suicide in 1906 for no other reason that that the basic, bold idea which he had had used as the premise of his work in the kinetic theory of matter . . . was neither understood nor recognized. The period's leading scientists . . . did not at all wish to recognize the atomistic theory . . ." (*On Life and Science*, trans. Albert Parry [New York, 1968], pp. 142–143).

Cornelius Lanczos, a collaborator of Einstein on problems of general relativity, writes: "Boltzmann was driven to despondency by the vicious attacks to which he was subjected on account of his atomistic views . . ." (*Judaism and Science* [Leeds, 1970], p. 12).

119. Cited in Carus, "Professor Mach and his Work," p. 33.
120. Mach, *Popular Scientific Lectures*, p. 207.
121. Mach, *The Analysis of Sensations*, p. 40.
122. Philipp Frank, *Between Physics and Philosophy* (Cambridge, 1941), p. 41.
123. Mach, *The Analysis of Sensations*, pp. 96, 204, 253.
124. Ibid., p. 30.
125. Mach, *Popular Scientific Lectures*, pp. 63–64.
126. Cited in Sandor Ferenczi, *Further Contributions to the Theory and Technique of Psycho-Analysis*, trans. J. I. Suttie, et al. (New York, 1927), pp. 394–396, from Ernst Mach, *Die Principien der Wärmelehre* (Leipzig, 1896), pp. 441–442.
127. Mach, *Die Principien der Wärmelehre*, p. 442.
128. *The Letters of William James*, ed. Henry James (New York, 1920), vol. 1, pp. 211–212. Ralph Barton Perry, *The Thought and Character of William James* (Boston, 1935), pp. 60, 65.
129. Perry describes Mach as "an important forerunner of pragmatism," and observes that "James felt a kinship with Mach" (Ibid., vol. 2, pp. 389, 579).
130. Ibid., vol. 1, p. 588.

131. Ibid., vol. 1, p. 594.

132. Ibid., vol. 2, p. 341.

133. Cf. Nikolay Valentinov (N. V. Volsky), *Encounters with Lenin*, trans. Paul Rosta and Brian Pearce (London, 1968), pp. 156, 188, 226, 258.

134. S. V. Utechin, "Philosophy and Society: Alexander Bogdanov," in *Revisionism: Essays on the History of Marxist Ideas*, ed. Leopold Labedz (New York, 1962), p. 117.

135. Lunacharsky, *Revolutionary Silhouettes*, introd. Isaac Deutscher, pp. 10–12. Reichinstein, *Albert Einstein: A Picture of His Life and His Conception of the World*, pp. 178–179.

136. Trotsky, *My Life*, pp. 144; *Lenin* (New York, 1925; Capricorn ed., 1962), p. 9.

137. For a general account of Bogdanov's thought, see Julius F. Hecker, *Russian Sociology: A Contribution to the History of Sociological Thought and Theory* (London, 1934), pp. 279–296. Bogdanov, notes Hecker, "insists that the economic approach to the study of social evolution is essentially psychological. . . ." Nicolas Berdyaev recalls amusingly that Bogdanov regarded his idealist tendencies as "symptoms of psychic abnormality," and used to ask him frequently how had he slept, how did he feel in the morning, and so on. Bogdanov himself died as a consequence of experiments on himself with transfusions of blood (Berdyaev, *Dream and Reality: An Essay in Autobiography*, trans. Katharine Lampert [New York, 1962], pp. 130–131). Also cf. Gustav A. Wetter, *Dialectical Materialism: A Historical and Systematic Survey of Philosophy in the Soviet Union*, trans. Peter Heath (New York, 1958), p. 95. There are translated excerpts from Bogdanov's *Empiriomonism*, 2d ed. (St. Petersburg, 1905), in Robert V. Daniels, *A Documentary History of Communism* (New York, 1962), vol. 1, pp. 49–50. S. V. Utechin, "Philosophy and Society: Bogdanov," p. 144. Nikolai Bukharin, the talented coworker of Lenin and the Bolsheviks' most gifted theorist, later executed at Stalin's behest, accepted "absolutely" Bogdanov's theory of religion, and referred to his "ingenious attempt to dispense with philosophy." Bukharin's use of the concepts of equilibrium to clarify Marx's dialectic method clearly owes much to Bogdanov (Bukharin, *Historical Materialism: A System of Sociology* [New York, 1925], pp. 75, 83, 171).

138. Mach, "The Economical Nature of Physical Inquiry" (1882), in *Popular Scientific Lectures*, pp. 191, 192, 193, 196.

139. Ibid., p. 204.

140. Cf. A. V. Vasiliev, *Space, Time, Motion: An Historical Introduction to the General Theory of Relativity*, trans. H. M. Lucas and C. P. Sanger (London, 1924), pp. 90–91.

141. Schilpp, *Albert Einstein: Philosopher-Scientist*, p. 21.

142. Cf. Vasiliev, *Space, Time, Motion*, pp. 206–207.

143. Schilpp, ed., *Albert Einstein*, p. 15.

144. Reiser, *Albert Einstein*, p. 52.

145. Seelig, *Einstein: A Documentary Biography*, pp. 74–76.

146. Ibid.

147. Albert Einstein, *Lettrès a Maurice Solovine*, trans. Maurice Solovine (Paris, 1956), p. 5.

148. Pierre Duhem, *The Aim and Structure of Physical Theory*, trans. P. P. Wiener (Princeton, 1954), pp. 21, 39. Otto Neurath, *Le Développement du Cercle de Vienne et l'Avenir de l'Empirisme Logique* (Paris, 1935), pp. 34, 40. Carlton Berenda Weinberg, *Mach's Empirio-Pragmatism in Physical Science* (New York, 1937), p. 110.

149. V. I. Lenin, *Materialism and Empirio-Criticism: Critical Notes Concerning a Reactionary Philosophy*, pp. 32, 267, 42, 302. Friedrich Adler, on reading Lenin's book, wrote to his father in 1909:

> He [Lenin] only proves thoroughly that he doesn't know what it's all about. Therein the book is a great achievement in patient sitting, for in about a year he went through and criticized a whole literature of which he previously had not the slightest idea . . ." (Braunthal, *Victor und Friedrich Adler*, pp. 198–199.)

150. Insofar as Mach held to a view of science "which led him to reject the existence of atoms," Einstein regarded him as a "deplorable philosopher," as he said in 1922. Cf. Emile Meyerson, *La Déduction Relativiste* (Paris, 1925), pp. 61–62.

151. Some scholars now argue that the "official ideology" of dialectical materialism has not significantly impeded the development of Soviet science. Their arguments, however, hardly withstand the evidence of such firsthand experience as that narrated most notably by the physicist George Gamow. He tells how the official ideologists, exercising their "philosophical dictatorship," enforced a ban on Einstein's theory of relativity "on the ground that it denied a world ether, 'the existence of which follows directly from the philosophy of dialectical materialism.' " Gamow, together with four other students and assistants, including Lev Landau, inscribed a "joking teleletter" to the chief physical ideologist:

> Being inspired by your article on the light-ether, we are enthusiastically pushing forward to prove its material existence. Old Albert is an idealistic idiot!
> We call for your leadership in the search for caloric, phlogiston, and electric fluids.

Four of the signers were tried by their institutes in a "condemnation session" and punished; a jury of machine shop workers found them guilty of anti-revolutionary activity. Landau was dismissed from his teaching post. Two graduate students were deprived of their stipends and had to leave town. As compared to the harsh measures that were imposed against geneticists, those directed against physicists were "mostly of nuisance value." When Gamow, however, undertook to lecture upon Heisenberg's principle of uncertainty, a dialectical-materialist ideologist interrupted him and dismissed the audience. The following week, Gamow received strict instructions never to speak ("at least in public") about the subject. Later the pressure was relaxed.
Gamow finally fled the Soviet Union in 1933: "The increasing pressure of the dialectical-materialist philosophy was too strong, and I did not want to be sent to a concentration camp in Siberia because of my views about the world ether, the quantum-mechanical uncertainty principle, or chromosomic heredity—which could have happened in due course."
Several years later Lev Landau was imprisoned as a "Trotskyist spy" (*My World Line: An Informal Autobiography* [New York, 1970], pp. 93–99, 120).
Max Born wrote Einstein on August 12, 1929 of the case of Georg Rumer who left Russia "because relativists are badly treated there (truly!). The theory of relativity is thought to contradict the official 'materialist' philosophy and, as I have already been told by Joffé, its adherents are persecuted." Subsequently Rumer was condemned to many years in a "terrible" Arctic camp. After Stalin's death, he was appointed head of the Institute of Physics in Novosibirsk (*The Born-Einstein Letters*, trans, Irene Born [London, 1971], pp. 101–102).

152. Leon Trotsky, *Terrorism and Communism: A Reply to Karl Kautsky* (reprint ed., Ann Arbor, 1961), pp. 180–181. Cf. *The Letters of Lenin*, trans. Elizabeth Hill and Doris Mudie (New York, 1937), p. 402.

153. Seelig, *Einstein: A Documentary Biography*, p. 42.

154. Unable to obtain regular employment after his graduation in the fall of 1900, Einstein was "forced to lead the life of a vagabond, to give himself up to chance, to taste all the uncertainties and constraints of a roaming life . . ." (Reiser, *Albert Einstein*, p. 60). He worked briefly as a calculator in the Federal Observatory, then for two months between May 15 and July 15, 1901, he was a substitute teacher at a technical school in Winterthur. Through an advertisement, he became tutor to two boys in Schaffhausen but was dismissed from this post because the master saw in his influence "the beginning of a widespread revolution . . ." (Ibid., p. 62). During "an entire year" that followed his examinations, Einstein found (in his words) "any consideration of any scientific problems distasteful to me . . ." (Schilpp, *Albert Einstein*, p. 17). To this consequence of academic strain were added the anxieties of the jobless. Einstein "felt himself jobless in hard times" (Reiser, *Albert Einstein*, p. 61).

155. Ibid., p. 61.

156. Einstein, *Lettres à Maurice Solovine*, p. 125.

157. Ibid., p. vi.

158. Dr. Otto Nathan, Interview of Maurice Solovine, Paris, August 19, 1957.

159. Seelig, *Einstein: A Documentary Biography*, p. 75.

160. Charles Nordmann, "Einstein Visits France: Impressions of the Battlefields on a Great Mind" trans. from *L'Illustration*, *The American Hebrew*, June 9, 1922, pp. 111, 118.

161. Seelig, *Einstein: A Documentary Biography*, pp. 33, 71, 75.

162. Cf. Guido Bedarida, "The Jews under Mussolini," *The Reflex*, vol. 1, no. 4 (October 1927), p. 56. During the first years of fascist rule, Jewish youth was "by and large, entirely swallowed up in the world of Fascist ideals." Fireside Discussion Groups, Anti-Defamation League, B'nai Brith, no. 24, *The Jews of Italy* (Chicago, 1939), p. 7. "Indeed, Jews figured prominently even among the 'Fascists of the First Hour.'" Angelica Balabanoff, "Anti-Semitism in Italy," *Socialist Review*, vol. 6, no. 8 (September–October, 1938), p. 18. Michael A. Ledeen, "Italian Jews and Fascism, *Judaism*, vol. 18 (Summer 1969), pp. 277–298. "Amalia Besso," "Davide Besso," "Marco Besso," *The Universal Jewish Encyclopedia*, (New York, 1940), vol. 2, pp. 248–249.

> Marco and Davide Besso were uncles of my father. Davide lived with a third uncle, Benjamin, an engineer, the husband of Amalia Besso. My father spent his last year of studies for the baccalaureate at the home of my uncle Benjamin. (Letter of Vero Besso to the author, October 4, 1968).

163. "The information given by Philipp Frank is inaccurate. My father joined the Swiss Reformed Church in which the social tradition of the reformers Zwingli and Calvin is preserved rather than the more ecclesiastical one of Luther." Letter of V. Besso to the author, October 4, 1968. Frank had written that Besso had converted to Catholicism. Cf. Philipp Frank, *Einstein: Sein Leben und seine Zeit* (Munich, 1949), p. 90.

164. Eduard Bernstein, *My Years of Exile: Reminiscences of a Socialist*, p. 102.

165. Letter of Pierre Speziali to this author, October 4, 1968.

166. M. Besso, "Metaphysique Expérimentale," *Dialectica*, vol. 2, no. 1 (1948), pp. 44–45.

167. Author's interview with Mrs. Sofie Lazarsfeld, New York, December 9, 1972.

168. Seelig, *Einstein: A Documentary Biography*, pp. 38, 96. Philipp Frank writes of Mileva: "She was a free-thinker and progressive in her ideas, like most of the Serbian students" (*Einstein: His Life and Times*, trans. George Rosen [New York, 1947], p. 23). While Seelig reports that Mileva had "no outstanding mathematical talent," Michelmore tells that she was "as good at mathematics" as the very gifted Marcel Grossmann. Michelmore also writes that Mileva "came from a wealthy family —her father was a judge" (*Einstein: Profile of the Man*, p. 35).

169. Seelig, *Einstein: A Documentary Biography*, p. 28.

170. Thorstein Veblen, "The Intellectual Pre-eminence of Jews in Modern Europe," *Political Science Quarterly*, vol. 34 (1919), pp. 33–42.

171. C. S. Peirce, "Mach's Science of Mechanics," *The Nation*, vol. 57 (1893), p. 252.

172. Bertrand Russell, *The Principles of Mathematics* (Cambridge, 1903), pp. 492–493. "Mach's contemporaries regarded his ideas on absolute motion as so many impractical proposals of an over-critical mind. And it must be admitted that to a certain extent they were right." See also Tobias Dantzig, *Henri Poincaré: Critic of Crisis* (New York, 1954), p. 101.

173. Schilpp, *Albert Einstein: Philosopher-Scientist*, pp. 251–252.

174. Werner Heisenberg, *Physics and Philosophy: The Revolution in Modern Science* (London, 1959), p. 97.

175. Frank, *The Validation of Scientific Theories*, p. 188.

176. Seelig, *Einstein: A Documentary Biography*, pp. 207–208.

177. Otto Nathan and Heinz Norden, eds., *Einstein on Peace* (New York, 1960), p. 133.

178. Carl Seelig, *Albert Einstein: Eine Dokumentarische Biographie*, rev. ed.

(Zurich, 1960), p. 246. Einstein described himself in his years after graduation: "Ich abseits und unbefriedigt, wenig beliebt . . . ich plötzlich von allen verlassen, ratlos vor dem Leben stehend" (I apart and dissatisfied, little liked . . . I suddenly deserted by everybody, standing helpless before life). When Marcel's intervention brought him a job, "Es war eine Art Lebensrettung, ohne die ich wohl zwar nicht gestorben, aber geistig verkümmert wäre" (It was a kind of saving of my life, for though without it I would not have died, I would however have been spiritually crippled).

179. Seelig, *Einstein: A Documentary Biography*, p. 88. Nonetheless, as Anton Reiser tells: "On the strength of his work entitled 'A New Determination of Molecular Dimensions,' Albert Einstein received his degree of doctor of philosophy at Zurich University, from Professor Kleiner . . ." (*Einstein: A Biographical Portrait*, p. 69). The degree was granted in 1905. Schilpp, *Albert Einstein*, vol. 2, p. 694.

180. Guenther Roth, *The Social Democrats in Imperial Germany: A Study in Working-Class Isolation and National Integration* (Ottowa, 1963), p. 159. Milton Yinger, "Contraculture and Subculture," *American Sociological Review*, vol. 25 (1960), pp. 625–635.

181. Heisenberg, *Physics and Philosophy*, p. 35.

182. For a number of years," wrote Planck, he tried "to fit the elementary quantum of action somehow into the classical theory . . . , and they cost me a great deal of effort. Many of my colleagues saw in this something bordering on a tragedy" (*Scientific Autobiography and Other Papers*, trans. Frank Gaynor [New York, 1949], pp. 44–45). Werner Heisenberg writes: "Planck, who was conservative in his whole outlook, did not like this consequence at all, but he published his quantum hypothesis in December of 1900." Planck tried persistently, but without success, to reconcile the hypothesis of discrete energy quanta with the traditional physical theory, but "finally convinced himself that there was no way of escaping from this conclusion" (*Physics and Philosophy*, p. 35). Also cf. Frank, *Einstein: His Life and Times*, p. 216. Max Born said of Planck: "He was a revolutionary not by character but through his willingness to acknowledge the power of evidence" (Earl of Birkenhead, *The Professor and the Prime Minister: The Official Life of Professor F. A. Lindemann, Viscount Cherwell* [Boston, 1962], p. 95). See also Max Born, "Max Karl Ernst Ludwig Planck 1858–1947," *Obituary Notices of Fellows of the Royal Society, 1948–1949*, vol. 6 (London), p. 167. Professor Edward Teller writes: "While Planck was a conservative and Einstein radical, the circle of Bohr had in my opinion a liberal philosophy. I am using the word liberal in an older and to me, admirable sense" (Letter to the author, September 17, 1968).

183. Einstein, "Remarks on Bertrand Russell's Theory of Knowledge," in *The Philosophy of Bertrand Russell*, ed. A. Schilpp (Evanston, 1944), p. 279. It is not clear when Einstein began to read Veblen. Miss Helene Dukas, Einstein's secretary, thinks that possibly he was introduced to Veblen's writings by his colleague at Princeton, Professor Oswald Veblen, an eminent mathematician and nephew of Thorstein Veblen. In that case, Einstein's reading of Veblen would have begun only after his coming to America (Interview, August 30, 1968, Princeton, New Jersey).

184. Albert Einstein, "Why Socialism?" *Monthly Review*, vol. 1, no. 1 (May 1949), p. 9. Einstein's essay was made available to this magazine through the good offices of his friend and later executor, Dr. Otto Nathan. Dr. Nathan later parted company with the magazine which evolved in a pro-Maoist direction with an anti-Zionist standpoint. See Otto Nathan, "Letter to the Editor," *Israel Horizons* vol. 16, no. 2 (February 1968), p. 24.

185. Braunthal, *In Search of the Millennium*, p. 139.

186. Joseph Dorfman, *Thorstein Veblen and His America* (New York, 1934), pp. 203, 162. Richard T. Ely, *Ground under Our Feet: An Autobiography* (New York, 1938), p. 154.

187. Karl Marx, *The Poverty of Philosophy* (New York, n.d.), p. 102. Thorstein Veblen, *The Place of Science in Modern Civilization* (New York, 1919), p. 61. Karl Korsch, *Karl Marx* (London, 1938), p. 24 f.

188. Ernst Mach, *The Science of Mechanics: A Critical and Historical Account of Its Development*, 6th ed. rev. through the 9th German ed., trans. Thomas J. McCormack (Lasalle, 1960), p. xxvi.

189. Fritz Wittels, *An End to Poverty*, trans. Eden and Cedar Paul (London, 1925), pp. 14, 86–87, 125, 136. See also Society for the History of Czechoslovak Jews, *The Jews of Czechoslovakia: Historical Studies and Surveys* (Philadelphia, 1971), vol. 2, pp. 215, 445.

190. Freud observed too that Popper-Lynkeus "had been a friend of Ernst Mach." See Sigmund Freud, "My Contact with Josef Popper-Lynkeus," *International Journal of Psycho-Analysis*, vol. 23 (1942), pp. 85 ff. Also see *Letters of Sigmund Freud*, ed. Ernest L. Freud, trans. Tania and James Stern (New York, 1964), pp. 315, 376.

191. Frank, *Einstein: His Life and Times*, p. 176.

192. Henry I. Wachtel, ed., *Security for All and Free Enterprise: A Summary of the Social Philosophy of Josef Popper-Lynkeus*, introd. Albert Einstein (New York, 1955), p. vii. The book also contains a biography of Popper-Lynkeus.

193. Otto Nathan and Heinz Norden, eds., *Einstein on Peace*, p. 613.

194. Ibid., p. 133.

195. Albert Einstein et al. *The Principle of Relativity*, trans. W. Perrett and G. B. Jeffery (London, 1923), pp. 42–43.

196. Paul Carus, "The Principle of Relativity," *The Monist*, vol. 22 (1912), p. 188.

197. Einstein et al. *The Principle of Relativity*, p. 41.

198. Schilpp, *Albert Einstein: Philosopher-Scientist*, p. 99. Cf. Ernst Cassirer, *Substance and Function and Einstein's Theory of Relativity*, trans. W. C. and M. C. Swabey (Chicago, 1923), p. 380. V. F. Lenzen thought that a "better name" for the theory would have been "the theory of the covariance of the equations of mathematical physics" ("The Philosophical Aspects of the Theory of Relativity," *University of California Publications in Philosophy*, vol. 4, [1923], p. 163).

199. "Le principe de relativité—ou mieux de covariance—à lui seul serait une base beaucoup trop générale pour qu'on puisse établir sur ce seul fondement l'édifice de la physique théorique" (Einstein, "A Propos de 'La Déduction Relativiste' de M. Emile Meyerson," *Revue Philosophique*, vol. 105 [1928], p. 163).

200. The terminology of "invariant," "covariant," and "contravariant," so important in the development of Einstein's ideas, was invented by the British mathematician, James Joseph Sylvester. For many years Sylvester was unable to work within the British academic community because of his Jewish background. At that time, William Whewell, the Master of Trinity College, Cambridge University, directed the "strict inquisition" of a gentleman to see if he was a Jew, "as in that case he must refuse to admit him as a student of the College." Cf. Major MacMahan, "Professor Sylvester," *Johns Hopkins University Circular*, vol. 16, no. 129 (April 1897), pp. 1–2; David S. Blondheim, "A Brilliant and Eccentric Mathematician," *The Johns Hopkins Alumni Magazine* vol. 9 (1920–21), pp. 119–140; Fabian Franklin, "Address Commemorative of Professor Sylvester," *Johns Hopkins University Circular*, vol. 16, no. 130, (June 1897), pp. 53–55; James Joseph Sylvester, *Collected Mathematical Papers*, (Cambridge, 1909), vol. 3, p. 81. Though some Jewish students were admitted to Cambridge before 1856, it was only in that year that Parliament authorized the granting of degrees to them. Cf. Israel Finestein, "The Dominant English—Jewish Aristocrats." *Congress Bi-Weekly*, vol. 39, no. 13 (November 24, 1972), p. 21.

201. Philipp Frank, Einstein's biographer, in a letter to the author (May 21, 1956) described the atmosphere among the Viennese students at the time:

> I have grown up in a very pagan atmosphere. I think that Freud's background was not very different. It was only one generation prior to mine. The views, and still more, the feeling, among Viennese intellectuals, Jews and Catholics alike, were extremely "gentile" in the old sense of the word, and not in the distorted sense in which it is used in this country. One should always interpret Freud against this "pagan" or "gentile" background. I remember that e.g. Otto Neurath denied any meaning to the term "ethics," except as an interpretation of tribal habits or as a part of sociology.

Frank was Einstein's successor as professor of physics at the German University at Prague.

202. Professor Edward Teller writes that it might be of particular interest to note that while the theory of special relativity carries a name which has been "rightly criticized," the name "general relativity" is in his opinion, "totally unjustified." "This most ingenious and clearly great accomplishment of Einstein's should be known by the designation 'theory of gravitation.'" To call it relativity is in his view "even more artificial" than in the case of the special theory. He therefore tend to concur with the judgment that the word "relativity" has "more an emotional than an objective significance" (Letter to the author, September 17, 1968).

203. Albert Einstein, "Foreword," in Philipp Frank, *Relativity: A Richer Truth* (London, 1951), p. 10.

204. C. P. Snow, *Variety of Men* (London, 1967), p. 76.

205. Leopold Infeld, *Albert Einstein*, rev. ed. (New York, 1950), p. 76.

206. Morris R. Cohen, *The Faith of a Liberal* (New York, 1946), p. 56.

207. Frank, *Einstein: His Life and Times*, p. 206.

208. Schilpp, *Albert Einstein: Philosopher-Scientist*, pp. 13, 678.

209. Infeld, *Albert Einstein*, pp. 41, 48.

210. Seelig, *Einstein: A Documentary Biography*, pp. 70–71. Einstein's backwardness as a child led to fears that he was abnormal. Albert as a boy learned to talk with "much difficulty." "His parents thought he was abnormal. The hired governess called the still, backward, slow-speaking Albert 'Pater Langweil' (Father Bore)" (Anton Reiser, *Albert Einstein*, p. 27).

211. Schilpp, *Albert Einstein: Philosopher-Scientist*, p. 13.

212. David Reichinstein, professor of chemistry at the University of Zurich, and a friend of Einstein when he taught there, wrote:

> The genesis of Einstein's sense for reality can be found in the disharmony which he felt in his surroundings during childhood and youth. Later, when he recognized social injustice and when at the same time the feeling of insecurity developed in him, he appealed instinctively to the firm ground of concrete reality and its unbreakable laws (*Albert Einstein: Sein Lebensbild und seine Weltanschauung*, 2d ed. [Prague, 1935], pp. 37–38; trans. M. Juers and D. Sigmund, *Albert Einstein: A Picture of His Life and His Conception of the World* [Prague, 1934], p. 35. I have slightly emended their translation).

213. See, *The Autobiography of Bertrand Russell, 1872–1914* (New York, 1967; reprint ed., 1968), p. 171.

214. Poincaré, *Science and Hypothesis*, p. 50.

215. *Mathematische Annalen*, Jahrg. 58 (1903).

216. *Beilage zum Programm der Thurganischen Kantonschule, 1903–04. Archiv der Mathematik und Physik*, Reihe III 9 (1903).

217. *Schweiz pädagogische Zeitschrift*, 1909.

218. Seelig, *Einstein: A Documentary Biography*, p. 62.

219. Ibid., p. 68.

220. Ibid., p. 132.

221. Frank, *Einstein: His Life and Times*, p. 206.

222. Michelmore, *Einstein: Profile of the Man*, p. 60.

223. Seelig, *Einstein: A Documentary Biography*, p. 108.

224. Albert Einstein, Arnold Sommerfeld, *Briefwechsel*, (Basle, 1968), p. 30.

225. Seelig, *Einstein: A Documentary Biography*, pp. 108, 208.

226. Leonard Woolf, *Sowing: an Autobiography of the Years 1880 to 1904* (New York, 1960), p. 161; *Beginning Again: An Autobiography of the Years 1911–1918* (London, 1964), pp. 24–25.

227. Bertrand Russell, "Is Position in Time and Space Absolute or Relative?", *Mind*, n.s., vol. 10 (1901), p. 307.

228. John Maynard Keynes, *Two Memoirs* (New York, 1949), p. 82.

229. Ibid., p. 81, 83, 86, 97; Woolf, *Sowing*, p. 147. See also *The Autobiography of Bertrand Russell, 1872–1914*, pp. 83–87, 198–199; J. K. Johnstone, *The Bloomsbury Group* (London, 1954), pp. 31–32; Irma Rantavaara, *Virginia Woolf and Bloomsbury* (Helsinki, 1953), pp. 1, 33 ff.

230. Russell, "Is Position in Time and Space Absolute or Relative?", pp. 317, 306, 299.

231. Victor Lowe, "The Development of Whitehead's Philosophy," in *The Philosophy of Alfred North Whitehead*, ed. P. A. Schilpp (Evanston, 1944), pp. 32, 43, 50; *Understanding Whitehead* (Baltimore, 1962), p. 157 f. Cf. E. A. Milne, *Sir James Jeans: A Biography* (Cambridge, 1952, pp. 4, 19; A. Vibert Douglas, *The Life of Arthur Stanley Eddington* (London, 1956), pp. 10, 128.

232. It is remarkable to what an extent the scientific, political, and philosophical establishments were linked in England by ties of family and marriage. The physicist Lord Rayleigh and the philosopher Henry Sidgwick were both, for instance, brothers-in-law of the Conservative statesman Arthur Balfour. See Lord Rayleigh. *Lord Balfour in his Relation to Science*, (Cambridge, 1930), pp. 9–10. See also N. G. Annan, "The Intellectual Aristocracy," in *Studies in Social History: A Tribute to G. M. Trevelyan*, ed. J. H. Plumb (London, 1955), p. 284.

233. Walter B. Houghton, *The Victorian Frame of Mind, 1830–1870* (New Haven, 1957), pp. 98, 337, 85. The mathematician William Kingdon Clifford, wrote: "We have felt with utter loneliness that the Great Companion is dead. Our children, it may be hoped, will know that sorrow only by the reflex light of a wondering compassion" (*Lectures and Essays*, 2d ed. [London, 1886], p. 389).

234. Charles Richet, *Thirty Years of Psychical Research*, trans. Stanley De Brath (New York, 1923), pp. 32–33. C. D. Broad, *Lectures on Psychical Research* (New York, 1962), p. v. Sir J. J. Thomson, *Recollections and Reflections* (New York, 1937), p. 147 ff. H. W. Blood Ryan, "William Crookes—The Man and his Work: 1832–1919," *The Chemical News*, vol. 144 (June 17, 1932), p. 391, 393. Sir Oliver Lodge, *Past Years: An Autobiography* (London, 1931), p. 270 f. Crookes said that the evidence in favor of levitation was "stronger than the evidence in favor of almost any natural phenomenon the British Association could investigate." When Stokes, the secretary of the Royal Society, declined Crookes' invitation to examine the phenomena with his own eyes, the feeling among psychical researchers was that he had "placed himself in exactly the same position as those cardinals who would not look at the moons of Jupiter through Galileo's telescope." There were indeed threats in 1874 to deprive Crookes of his Fellowship in the Royal Society. See also, Arthur Conan Doyle, *The History of Spiritualism* (New York, 1926), vol. 1, pp. 169, 177, 243, vol. 2, p. 298 ff. E. E. Fournier D'Albe, *The Life of Sir William Crookes* (London, 1923), p. 174 f. Geoffrey K. Nelson, *Spiritualism and Society* (London, 1969), pp. 90, 94, 265.

235. Ernst Mach wrote:

> The hobgoblin practices of modern spiritualism are ample evidence that the conceptions of paganism have not been overcome even by the cultured society of to-day. It is natural that these ideas so obstinately assert themselves. Of the many impulses that rule man with demoniacal power, that nourish, preserve, and propagate him, without his knowledge and supervision, of these impulses of which the middle ages present such great pathological excesses, only the smallest part is accessible to scientific analysis . . . (*The Science of Mechanics*, 3d ed. [Chicago, 1907], p. 463).

236. See Sir Edmund Whittaker, *A History of the Theories of Aether and Electricity: The Modern Theories, 1900–1926* (London, 1953), p. 30 f; G. H. Keswani, "Origin and Concept of Relativity (1)," *British Journal for the Philosophy of Science*, vol. 15 (1965), pp. 281–306. See also Henri Poincaré, *The Foundations of Science*, trans. G. B. Halstead (Lancaster, 1913), pp. 208, 305, 316, 320; "The Principles of Mathematical Physics," *The Monist*, vol. 15 (1905), pp. 5, 12, 16.

237. Max Born, *Physics in My Generation* (London, 1956), p. 193.

238. Ibid., p. 195.

239. Ibid., p. 189. For the reaction of the elder generation to Einstein's "extraordinary postulate," see Sir Oliver Lodge, *Past Years: An Autobiography* (London, 1931), pp. 206–207; Robert John Strutt, Fourth Baron Rayleigh, *John William Strutt: Third Baron Rayleigh* (London, 1924), p. 349; Earl of Birkenhead, *The Professor and the Prime Minister*, pp. 25, 165. H. A. Lorentz unequivocally emphasized Einstein's originality:

It is comprehensible that a person could not have arrived at such a far-reaching change of view by continuing to follow the old beaten paths, but only by introducing some sort of new idea. Indeed, Einstein arrived at his theory through a train of thought of great originality (*The Einstein Theory of Relativity* [New York, 1920], p. 29. He referred to "his [Einstein's] principle of relativity." Ibid., p. 14).

240. Professor Herbert Dingle has written: "On examination it becomes clear that what Poincaré means by relativity is not Einstein's postulate that nature knows no absolute standard of rest, but the much older postulate that the state of motion of a system cannot be found from observations *within the system*" (Viscount Samuel and Herbert Dingle, *A Threefold Cord: Philosophy, Science, Religion* [London, 1961], p. 68).

241. Auguste Comte, *A General View of Positivism*, trans. J. H. Bridges (New York, 1957), pp. 51, 63. Comte also insisted on a historical and ethical relativism (ibid., p. 61).

242. Paul Langevin, "L'Oeuvre d'Henri Poincaré: Le Physicien," *Revue de Métaphysique et de Morale*, vol. 21 (1913), p. 703.

243. The ideas of Poincaré entered into the culture of the Olympia circle, probably with their discussion of the "law of relativity." Solovine tells how the Olympia Academy included Poincaré's *Science and Hypothesis* among its readings:

> We read together, after Pearson, the *Analysis of Sensations* and the *Mechanics* of Mach, which Einstein had already gone through before, Mill's *Logic*, Hume's *Treatise of Human Nature*, Spinoza's *Ethics*, some essays and lectures of Helmholtz, several chapters of André-Marie Ampère's *Essay on the Philosophy of the Sciences*, Riemann's *On the Hypotheses Which Lie at the Base of Geometry*, several chapters of Avenarius' *Critique of Pure Experience*, Clifford's *On the Nature of Things in Themselves*, Dedekind's *What Are Numbers and Their Use?*, Poincaré's *Science and Hypothesis*, a book which impressed us profoundly and held us breathless during long weeks . . . " (Einstein, *Lettres à Maurice Solovine*, p. viii).

Poincaré's book had a discussion in his words of "what we shall call, for the sake of abbreviation, the law of relativity" (trans. W. J. Greenstreet, *Science and Hypothesis* [New York, 1905], p. 76). Einstein, in response to a specific query, wrote that he was not familiar in 1905 with "the consecutive investigations by Poincaré. In this sense my work was independent" (Max Born, *Physics in My Generation*, pp. 193–194). Cornelius Lanczos, an assistant of Einstein's, writes that Einstein was evidently ignorant of a lecture in 1900 in which Poincaré had used the expression "the principle of relativity" (*Albert Einstein and the Cosmic World Order* [New York, 1965], p. 40). Lanczos meant perhaps to refer to Poincaré's lecture in 1904 in St. Louis. We can infer that the reading of Poincaré brought the phrase "the law of relativity" to the consciousness of the young Olympians. Einstein gave it a new connotation, and turned it into the distinctive tenet and heuristic principle of the new generational rebels.

244. Langevin, "L'Oeuvre d'Henri Poincaré," p. 702.

245. Einstein's letter to Heinrich Zangger, quoted in B. Kuznetsov, *Einstein*, trans. V. Talmy (Moscow, 1965), p. 186. Kuznetsov's biography of Einstein has a liberality of spirit which is rare in Soviet writing on the history of science; it stands in marked contrast to the crudities which Soviet ideologists have published on the subject. There are still, however, both lacunae and misinterpretations in Kuznetsov's book which evidently derive from its Soviet setting. It understates Einstein's admiration for Adler, and fails altogether to mention that Einstein volunteered to be a character-witness at Adler's murder trial. It similarly understates the influence of Mach on Einstein, and omits recording Einstein's chief tribute to him. Lenin, who, we must remember, was hostile to both Adler and Mach, still stands as a supreme authority in Soviet thought. Kuznetsov also makes no reference to the correspondence between Einstein and Freud on war, and says nothing of Einstein's virtual disillusionment with humanity.

The role of Otto Nathan, whom Einstein chose as a trustee for his estate, is similarly unmentioned. Clarifying these matters would have confronted the biographer with ideological hazards.

246. The Ecole Polytechnique was closely connected with the military and political institutions of the country, whereas the Ecole Normale was more receptive to revolutionary opinions. Although it was a base for Saint Simonian technocratic ideas, the Polytechnique was less involved in political plots and secret societies, and was later at odds with the Paris Commune. Unlike the Normale, its students could detect nonstudent agitators by asking them for the derivative of sin x or log x, as they did in December 1830, when they came into conflict with the Left Republicans. By an overwhelming majority, the Polytechnicians on March 20, 1871 condemned the Paris Communard movement. Their School escaped being burnt "almost by a miracle." As G. Pinet writes: "The Ecole Polytechnique came to regard itself as a fourth power in the state" (*Histoire de l'Ecole Polytechnique* [Paris, 1887], pp. 99, 116, 136, 187, 192, 296, 313).

247. Paul Appell, *Henri Poincaré* (Paris, 1925), pp. 11, 22, 31, 79.

248. Henri Poincaré, *Savants et Ecrivains* (Paris, 1910), p. 266.

249. Gaston Darboux, "Eloge Historique d'Henri Poincaré," in *Oeuvres de Henri Poincaré* (Paris, 1916), vol. 2, p. lx.

250. Appell, *Poincaré*, pp. 99–102.

251. Jacques Kayser, *The Dreyfus Affair*, trans. Nora Bickley (New York, 1931), p. 386.

252. *L'Oeuvre de Léon Blum* (Paris, 1958), vol. 3, p. 439.

253. Appell, *Poincaré*, pp. 106–108.

254. Poincaré, *Savants et Ecrivains*, pp. 23–24. See also Emile Boutroux, "Henri Poincaré," *La Revue de Paris, Année* 20 (1913), p. 702.

255. Poincaré, *Mathematics and Science: Last Essays* (1913), trans. John W. Bolduc (New York, 1963), pp. 112, 115. Appell, *Poincaré*, p. 105.

256. Poincaré, *Savants et Ecrivains*, p. 170.

257. Ibid., p. vii.

258. Arthur Schuster, "H. Poincaré: Science and Hypothesis," *Nature*, vol. 73 (February 1, 1906), p. 313.

259. Poincaré, *Science and Hypothesis*, (London, 1905), p. xxiii.

260. Mach, *The Science of Mechanics*, p. 293. The "relativists" were regarded as a distinctive school. Cf. V. I. Lenin, *Materialism and Empirio-Criticism*, p. 108. Josiah Royce, "Introductory Note," in Federigo Enriques, *Problems of Science* (Chicago 1914), p. xii. Lenin wrongly classified Poincaré as a "French Machian," and also assailed "the principle of relativity" (Lenin, *Materialism and Empirio-Criticism*, pp. 248, 264).

261. Robert Gilpin, *France in the Age of the Scientific State* (Princeton, 1968), pp. 84, 86, 97, 107, 111, 119–120, 364, 370. See also F. A. Hayek, *The Counter-Revolution of Science: Studies on the Abuse of Reason* (Glencoe, Ill., 1955; reprint ed., 1964), pp. 110–118.

262. Einstein et al., *The Principle of Relativity*, p. 75.

263. Ibid., p. 83.

264. Frank, *Einstein: His Life and Times*, p. 20. Schilpp, *Albert Einstein: Philosopher-Scientist*, p. 15.

265. Joseph Samuel, "A Forerunner of Albert Einstein: Hermann Minkowski," *The Jewish Tribune* (September 23, 1927), pp. 48–49. Mark Zborowski and Elizabeth Herzog, *Life Is with People: The Jewish Little-Town of Eastern Europe* (New York), 1952. Sidney Osborne, *Germany and Her Jews* (London, 1939), pp. 214–215.

266. A. V. Vasiliev, *Space, Time, Motion: An Historical Introduction to the General Theory of Relativity*, trans. H. M. Lucas and C. P. Sanger (London, 1924), p. 165. On Vasiliev himself, see, T. Rainoff, "Alexander Vassilievic Vassilieff," *Isis*, vol. 14 (1930), pp. 342–348.

267. Shmarya Levin, *Youth in Revolt*, trans. Maurice Samuel (New York, 1930), pp. 279, 288. Abraham A. Fraenkel, *Lebenskreise: Aus den Erinnerungen eines Jüdischen Mathematikers* (Stuttgart, 1967), p. 87.

268. Harald Bohr, *Collected Mathematical Works* (Copenhagen, 1952), pp. xxi, xxii.

269. David Hilbert, "Hermann Minkowski," *Mathematische Annalen*, vol. 68 (1910), pp. 469–470.

270. Einstein et al., *The Principle of Relativity*, pp. 76, 91. Leibniz's idea of a pre-established harmony was, as historians of philosophy have noted, probably derived from Spinoza's conception as stated in a famous proposition of the *Ethics*: "The order and connection of ideas is the same as the order and connection of things."

271. Infeld, *Albert Einstein*, p. 122. Lanczos, *Albert Einstein and the Cosmic World Order*, p. 119. Schilpp, *Albert Einstein: Philosopher-Scientist*, pp. 236–237.

272. Schilpp, *Albert Einstein: Philosopher-Scientist*, p. 103. Infeld, *Quest*, pp. 270, 271, 279. A. A. Fraenkel, *Lebenskreise*, p. 172.

Miss Helen Dukas, Einstein's secretary, has narrated how his aphorism came to be inscribed on the chimney at Fine Hall, Princeton University. In May 1921 Einstein gave a series of lectures at Princeton on the theory of relativity. At a reception for him he was asked what he thought of the experiments that D. C. Miller was making; the latter were said to support the "ether-drift" hypothesis and to cast doubt on Einstein's theory. Einstein then answered his questioner: "Raffiniert ist der Herr Gott aber boshaft ist er nicht." Oswald Veblen, the eminent Princeton mathematician, happened to hear this interchange. Then in April 1930 Veblen wrote to Einstein reminding him of the incident, and asking his permission to use the phrase for the mantle-piece of the new Faculty Lounge of the Princeton Mathematics and Physics Departments. In his reply Einstein said that what he meant was: "Die Natur verbirgt ihr Geheimnis durch die Erhabenheit ihres Wesens, aber nicht durch List" (Nature hides its secret through the sublimity of its being, but not through cunning [Note sent to the author, February 25, 1972]).

Cornelius Lanczos has seen in the movement of Einstein's thought from positivism to a Spinozist metaphysics the emotive—philosophical source of the General Theory of Relativity (*Judaism and Science* [Leeds, 1970], pp. 13, 17).

273. *The Born-Einstein Letters*, trans. Irene Born (London, 1971), p. 192.

274. Einstein, *Out of My Later Years* (New York, 1950), p. 25. See also, *Einstein on Peace*, ed. Nathan and Norden, pp. 218, 226, 253.

275. Reiser, *Einstein: A Biographical Portrait*, pp. 28–29.

276. Albert Einstein, "Introduction," in Rudolf Kayser, *Spinoza: The Life of a Spiritual Hero* (New York, 1946), pp. ix–xi.

277. Curiously, the excellent recent English translation by Irene Born renders the phrase "der wahre Jakob" as "the real thing" thereby losing the whole import of the Biblical metaphor. (*The Born-Einstein Letters*, p. 91. Albert Einstein, Hedwig und Max Born," *Briefwechsel, 1916–1955* (Munich, 1969), pp. 129–130.

278. *The Holy Scriptures*, Genesis 27:22, trans. Jewish Publication Society (Philadelphia, 1917), p. 32.

279. Einstein, Hedwig and Max Born, *Briefwechsel 1916–1955*, pp. 204–206.

280. Jorgen Schleimann, "The Life and Work of Willi Münzenberg," *Survey*, no. 55 (April 1965), pp. 72, 74. Werner T. Angress, *Stillborn Revolution: The Communist Bid for Power in Germany, 1921–1923* (Princeton, 1963), pp. 346–347. Arthur Koestler, *The Invisible Writing: An Autobiography* (Boston, 1955), p. 206.

281. Nathan and Norden, *Einstein on Peace*, p. 134.

282. Einstein, Hedwig and Max Born, *Brieswechsel*, pp. 204–206.

283. *The Letters of Sigmund Freud and Arnold Zweig*, ed. Ernst L. Freud, trans. Prof. and Mrs. W. D. Robson-Scott (London, 1970), p. 6.

284. Upton Sinclair, *Mental Radio: Does It Work, and How?* (London, 1930). Many years later in 1941 Einstein was prepared to give a careful hearing to the strange claims of Wilhelm Reich on behalf of the "orgone accumulator" (Ilse Ollendorff Reich, *Wilhelm Reich: A Personal Biography* [New York, 1970]), pp. 84–87.

285. Sinclair, *My Lifetime in Letters* (Columbia, Missouri, 1960), pp. 352–362; *The Autobiography of Upton Sinclair* (New York, 1962), pp. 254–259, 279–280, 292, 305, 326.

286. Sinclair, "Albert Einstein: As I Remember Him," *Saturday Review*, vol. 39, no. 15 (April 14, 1956), p. 56.

287. Einstein quarrelled sharply with the administration of the Hebrew University in Jerusalem because he feared that under the influence of the American model and "high finance," it would become an organization for the "mass production of academic degrees." He wanted it to become a center for "only students, with the ability and will to participate in scientific work," in short, another Olympia Academy writ large. The dissension was so strong that Einstein resigned from the Board of Governors. See Herbert Parzen, "The Magnes-Weizmann-Einstein Controversy," *Jewish Social Studies*, vol. 32 (1970), pp. 199, 203.

288. *The Impossible Takes Longer: The Memoirs of Vera Weizmann*, as told to David Titaev (London, 1967), p. 102.

289. This author's translation. Einstein's original German versions follow:

Weisheit des Dialektischen Materialismus

Durch Schweiss und Mühe ohnegleichen
Ein Körnchen Wahrheit zu erreichen?
Ein Narr, wer sich so kläglich schinden muss
Wir *schaffen's einfach durch Parteibeschluss.*
Und denen, die zu zweifeln wagen
Wird flugs der Schädel eingeschlagen.
Ja, so erzieht man, wie noch nie,
Der kühnen Geister Harmonie

Aufschrift für das Marx Engels-Institut

Im Reiche der Wahrheit-Sucher gibt es keine menschliche Autorität. Wer es da versucht, Obrigkeit zu spielen, scheitert am Gelächter der Götter.

Einstein apparently intended these lines for publication in a volume to honor the former Rabbi of Berlin Leo Baeck on his eightieth birthday. Only a revised version of the last sentences was included in his "Neun Aphorismen," in *Essays Presented to Leo Baeck on the Occasion of his Eightieth Birthday* (London, 1954), pp. 26–27. "Wer es unternimmt, auf dem Gebiete der Wahrheit und der Erkenntnis als Autorität, scheitert an dem Gelächter der Götter (He who undertakes the role of authority in the domain of truth and knowledge will run afoul of the laughter of the gods)."

290. Antonina Vallentin, *The Drama of Albert Einstein*, trans. Moura Budberg (New York, 1954), p. 253.

291. Einstein, *Out of My Later Years*, pp. 216–217.

292. Vallentin, *The Drama of Albert Einstein*, p. 253.

293. Max Planck, *Scientific Autobiography*, pp. 33–34.

294. Reiser, *Einstein: A Biographical Portrait*, pp. 65, 89.

295. Nathan and Norden, *Einstein on Peace*, pp. 238–239.

296. Reiser, *Einstein: A Biographical Portrait*, p. 38.

297. The Ethics of the Fathers, 2:2.

298. Julius Braunthal, *The Tragedy of Austria* (London, 1948), p. 137.

2

Social, Generational, and Philosophical Sources of Quantum Theory

Niels Bohr: The Ekliptika *Circle and the Kierkegaardian Spirit*

The culture of a revolutionary student milieu permeated by the ideas of Mach and Marx was, as we have seen, the supporting social environment of the young Albert Einstein. The student revolutionaries, estranged from the dominent bourgeois culture, were generational rebels. Are alienation and sharp, overt generational rebellion always the mainspring for great acts of scientific orginality? The remarkable facts concerning the pioneering work of Einstein's younger contemporary, Niels Bohr, the inspiring guide of the "Copenhagen School," fall into no such simple pattern. Whereas Einstein drew sustenance from the critiques of Mach and Hume, Niels Bohr was stirred by the religious existentialism of Søren Kierkegaard. Whereas Einstein and his circle were political radicals at odds with the establishment, jobless or marginally employed, impecunious, cosmopolitan strangers from many lands, Niels Bohr and his Danish friends remained in their homeland and wholly identified with it. Einstein was a clumsy lad, virtually expelled from high school, and banished from his university's library; Niels Bohr was an all-around boy, an adored, popular athlete. While Einstein brooded gloomily about the determinism of men, finding the deepest insight in Dostoyevsky's *The Brothers Karamazov*, the serene Bohr delighted throughout his life in the little-known novel of

Møller, *Adventures of a Danish Student*, which whimsically explored the paradoxes of human freedom. Einstein sought to subsume all reality within a system; Bohr denied that such a system was possible, and wondered whether "all reality" had a meaning. Perhaps never in the history of scientific ideas have two outstanding contemporaries been so emotionally and philosophically unlike each other. They represented two distinct intellectual strands, or *isoemotional lines*, in European intellectual life, which intersected occasionally, sometimes only tangentially. Let us delineate the chief traits in Bohr's social environment that helped to shape his emotional-intellectual standpoint.

The "Ekliptika" Circle

Copenhagen at the turn of the century had much in common with Zurich. During the nineteenth century, under the outstanding leadership of Bishop Grundtvig, Denmark had become the model for Europe of a "peasant democracy."[2] The innovation of folk high schools which Grundtvig had inspired was probably the only successful national experiment in adult education that the world has ever seen. Though Danish society was primarily agricultural rather than industrial, it achieved the unusual status of having, in proportion to its population, the largest number of Nobel Laureates in science.[3] Its socialist movement, too, associated with cooperatives and self-governing societies for mutual aid, had a philosophy much like that of the Swiss socialist movement. A distinguished American observer wrote in 1921:

> The socialist movement in Denmark has an intellectual quality. I know of no country where the socialists and trade unionists have developed educational, cultural, and press agencies as they have in Denmark. The University of Copenhagen has entered into sympathetic relationship with trade unionists and has provided educational work to meet their demands.[4]

In 1905 when Einstein's "revolutionary" articles were published, a group of students at the University of Copenhagen met regularly for discussions. They called themselves *Ekliptika* because their number was limited to twelve. All were children of the Danish political, social, and intellectual elite, and could be confident of lifetimes of respected and worthwhile achievement. The circle included the two Bohr brothers, Niels and Harald; Peter Skov, who later became Denmark's ambassador to the Soviet Union; Edgar Rubin, who in 1922 was named professor of experimental psychology at the University of Copenhagen and director of the Psychological Laboratory; his researches, much in keeping with the Gestalt mode of thought, were recognized when he was chosen as president of the International Congress of Psychology in 1932; Niels Erik Nørlund, awarded the university's gold medal in mathematics in 1908, in 1922 appointed professor of mathematics, and subsequently, as director of the Geodetical Institute, the author of an authoritative work on Danish topography; and his brother, Poul Nørlund, who made many contributions to the archaeology of Norse antiquity and helped develop the work of Danish museums. Other members were: Einar Cohn, who pursued a career as an economic statistician through a series of important governmental and research appointments, serving for many years as secretary of the National Economic Society and permanent undersecretary in the Danish government;[5] Viggo Brøndal, who in 1928 became professor of Romance languages and literatures at the University of Copenhagen; Vilhelm Slomann, who had broad-ranging interests in the history of art, embracing such recondite topics as Indian influences on European furniture, and who became director of the Museum of Applied Art; Kaj Harriksen, an entomologist, later curator of the Zoological Museum;[6] and two women, Astrid Lund and Lis Jacobsen the latter of whom took her doctorate in philology and became editor of the *Dictionary of the Danish Language* and president of the Society for Danish Language and Literature.[7]

The brothers Bohr were especially warm and affectionate toward each other. Harald, endowed with rare mathematical talent, was the more famous athlete of the two; he was on the soccer team that won second place for Denmark at the Olympic games in London in 1908. He was not yet twenty-three years old when he made his public defense of his doctoral thesis; "it was, however, not mathematicians who filled the hall but for the most part football enthusiasts who had come to see one of their favorites in this role, so unusual for a star athlete." So the newspapers told. To his sports fans he was "little Bohr"; he and Niels all their lives were companions in idea and feeling: "Each was the other's closest friend; and they shared everything between them, thoughts, interests, and worldly goods."[8] Niels too was a "keen footballer, but he did not get any further than becoming reserve goalkeeper for the AB team. . . .[9]

The members of *Ekliptika* used to discuss "mainly questions of philosophy and epistemology" which were provoked by the lectures of their famed professor of philosophy, Harald Høffding. As Vilhelm Slomann recalled: "Their way of thinking seemed to be coordinated; one improved on the others or his own expressions, or defended in a heated yet at the same time good-humored manner his choice of words." With Høffding's teachings they would engage and disengage, absorb and reject, thereby acquiring a sense of their own intellectual character. To Høffding, therefore, we shall turn.

Harald Høffding

Harald Høffding occupied a place in Danish thought analogous to that of his friend William James in America. Born into a family of wealthy merchants which traced its forebears to the seventeenth century, Høffding at first studied theology, but became estranged from the church and was drawn to philosophy, particularly by the influence of Søren Kierkegaard. From Kierkegaard, Denmark's greatest intellectual figure, Høffding learned that existence is too diverse and many-sided to be encased within a system. In 1904 Høffding visited Harvard University where William James invited him to address his students. Høffding said:

> Pluralism makes the world new for us and necessitates a revision of our categories, our principles and our methods. . . . We are inclined to suppose that we can develop—or perhaps already have developed—thoughts in which all existence can be expressed. But, as a Danish thinker, Søren Kierkegaard, has said, we live forward, but we understand backward. . . . Only when life is closed can it be thoroughly understood. This is our tragi-comical situation. Even a divine thinker could only understand the world when the life of the world was finished.[10]

James was so taken with the sentence from Kierkegaard that a few years later in his famous *Pragmatism* (1907), he cited it to explain his intent.[11] The passage indeed was James's favorite philosophical quotation; it appeared as well in three of his other works in 1905 and 1909. James also relished Høffding's humorous sally against the absolutists. He wrote of the story to their philosophical ally, F. C. S. Schiller: An American child asked its mother whether God had made the world in six days. "Yes," answered the mother. "The whole of it?" pursued the child. "Yes."—"Then, it is finished, all done?"—"Yes."—"Then in what business now is God?" asked the child. James felt the answer to be given was that He was sitting for his portrait by absolutist metaphysicians.[12]

Høffding was lecturing in 1904 on the "wonderful contradiction" in the great rationalistic systems of Plato, Spinoza, and Hegel, namely, "that they cannot explain the striving and struggling thought whose work these systems themselves are." "An absolute systematization of our knowledge is not possible," said Høffding, and indeed, this was "no evil." The great systems had only a pragmatic usefulness; they were "projections, electric search-lights, with whose help we try to explore the dark." But when one sought to extend any analogy based on a part of existence to the whole, said Høffding, "no verification is possible." There was an "incommensurability" between thought and reality which would "always be experienced anew." Høffding had been led to this standpoint by Kierkegaard: "In my youth, the influence of the Danish philosopher and religious thinker, Søren Kierkegaard, was decisive for me. He waged a passionate war against speculation, with strong accentuation of the conditions of thought and of the value of the single, real, personal life."[13]

This sense of Høffding's that no analogy, no "type-phenomenon" (as he called it, after Goethe's original expression), no mode of thought would be adequate to encompass the totality of reality led William James to characterize him as "one of the wisest, as well as one of the most learned of living philosophers." As James wrote in his preface to Høffding's *The Problems of Philosophy*:

> No one better than Professor Høffding in these pages has shown how all our attempted definitions of the Whole of things, are made by conceiving it as analogous to some one of its parts which we treat as a type phenomenon. No one has traced better the logical limitations of this sort of speculation. We never can absolutely prove its validity. We can only paint more or less plausible pictures . . .[14]

The alternation of analogies was, in Høffding's view, the source of the advancement of scientific thought. Every analogy (today called *model* or *paradigm*) was based on a type-phenomenon which evoked in us "the feeling that we stand at the limit of our powers." The notions of a principle of conservation and an atomic hypothesis were such basic interdomainal analogies. But no analogy could pretend to provide the model for the totality of Being since there were always competing type-phenomena: "Around this point the battle of the different world-views revolves." Continuity and discontinuity, said Høffding, were standpoints based on competing type-phenomena: "Thus it would appear that neither of the two elements is the only accredited one." Positivists and idealists, wrote Høffding, attached a greater significance to "the unavoidable discontinuity of our cognition," whereas the realists were continuity-minded: "So the schools change places in the great arena on which the battle of truth must be fought out. Whenever a point of view ceases to yield significant results, the inquiry involuntarily seeks out a new one. . . ."[15] Høffding's friend William James was indeed at this time trying to fashion a new hypothesis of discontinuity, an atomistic theory of space and time according to which there were "minimal amounts of time, space, and change". Change, from the standpoint of the "discontinuity-theory," took place by "finite buds or drops."[16]

What, then, decided one's choice of type-phenomenon as the guiding explanatory model in one's research? According to Høffding, the individual personality expressed itself in the choice of method; personality and scientific research were thus co-involved. On this point, Høffding went beyond Kierkegaard, extending the enclave of choice, the "either-or" into scientific method itself. Kierkegaard had regarded personality and scientific research as antagonistic to each other; he who would think "scientifically," said Kierkegaard, had to "doff his personality." Høffding, however declared in 1905 that Kierkegaard had been wrong in this respect, that there was thus a "qualitative dialectic," an "existential choice" (as the phrase goes today) among the competing analogies. He was no longer enthralled by the notion that a neutral, impersonal method governed the work of science. "I myself, in my youth," wrote Høffding, "swore allegiance to such a view, under the overpowering influence of S. Kierkegaard. If I have finally broken its hold on me, I venture to say that this is because I have become better acquainted with both science and the personal life."[17] The conflict in science between the standpoints of individuality and totality was thus the outcome of divergences in personality: "The great question is whether out of this strife the elements or the totalities (the solar system, organisms, souls, human societies) will come off victorious." Such conflicts pervaded every branch of life—the humanities as well as the scientific. The "fierce struggles" between Tolstoyan altruism with its self-surrender and the Nietzschean self-assertion were one such instance: "In our actual human life, there is apparently a constant oscillation going on between these two antagonistic poles."[18] The same strife of standpoints, according to Høffding, divided the rival conceptions of scientific method.

Beneath the level of the conscious personality, furthermore, said Høffding, there were the hidden regions of the unconscious from which the warring philosophies emanated:

> Behind our clearest consciousness there are always unconscious tendencies and dispositions, which can perhaps be recognized later on, but which for long periods of time can function as latent coefficients. Our consciousness works itself out of a dark chaos, and its sporadic elements are combined through an involuntary synthetical process.

There were sociological elements that entered into the unconscious: "The philosopher cannot avoid being a child of his time and his nation. In his most independent moments he is often guided by an endeavor to find expression for peculiarities and tendencies of which he is not conscious. . . ." But Høffding was not a sociological determinist; apart from the social factors affecting the thinker, there were the purely individual, residual components of his choice: "The individuality of the philosopher, his personal equation, must inevitably have a great influence in that undefined region where metaphysical hypotheses live and move. . . ." All understanding is an utterance of the *personal life*, whose manifold forms and stages the psychologist ought to follow, step by step. . . ."[19]

When Høffding wrote that "scientific work is a work of personality," he was affirming that a personal choice was as operative in the ultimate methodological commitments of scientists and in their analogies, as it was among religious philosophers and metaphysicians. However, when the limitations of a favored analogy were forgotten and the analogy imposed on the totality of existence, such a personal choice led to anthropomorphism; for every anthropomorphism is, in essence, an image that claims to be more than an image. That was why every concept of God was the concept of an idol. The basic problems of philosophy were then, according to Høffding, insoluble; one could not hope for answers but at most, in Poul Møller's words, "the knowledge that these questions themselves, since they are based on untrue concepts must vanish away." Against this indeterminate background, one made one's choice among complementary alternatives. Høffding referred to his friend James who held "there are cases where faith creates its own verification. . . ."[20]

Høffding had a personal predilection for the themes of individuality and discontinuity. He lectured in 1904 on the new discontinuities in existence—the mutations which the Dutch botanist, Hugo de Vries, was investigating. "We have no right to suppose," said Høffding, "that the fact that we can not understand phenomena, if we can find no connection or continuity, should be without ground in reality itself." Such discontinuities in existence might be uneliminable by human science, wrote Høffding. "As a critical monist," he said, he was prepared to acknowledge that reality might not be a complete, continuous system.[21] Nevertheless, he thought that science would always be trying to remove the surd discontinuities.

Yet, Høffding's unique contribution to scientific thought was his insistence on the heuristic potentialities of the notion of discontinuities in existence. It was "discontinuity," he said, "which more than anything else brings new content, releases locked powers, and opens up the greatest tasks in the realm of life no less than in the realm of science."[22] He conceded to Kant that a longing for continuity was intrinsic to what we desired in causal explanation, but he felt that such a longing was a psychological fact about ourselves which in no way implied that the desired principle was "objectively valid". A "purely psychological epistemology," therefore, would never satisfy a scientific investigator.[23] Psychological existence itself presented one with radical discontinuities: "between the different individual consciousnesses there is always an abrupt and striking discontinuity." Therefore Høffding forebore from affirming any world-formula for relationships of continuity and discontinuity; he felt that to do so would mislead one into mythology. He defined the "fundamental problem of philosophy" as "concerned with the continuity of existence in relation to the special and individual forms of existence."[24] But every mythology and every metaphysics that tried to resolve this problem would necessarily "leave unanswered questions both as to that which preceded what is described as the beginning in their drama and that which follows after its close."[25]

Høffding likewise felt that ethical absolutes did violence to the variations in human individuals. There was an Ibsenite flavor to his defense of a "law of relativity in ethics." Individuals, he said, differed so much with regard to their sexual instincts that sexual ethics had to be individualized too; otherwise, to decree a strict universal commandment would place an unfair, unequal burden on some unfortunates.[26] Unlike the Hegelians who were drawn to "wholes" and "totalities," Høffding wrote that "the future of the race can only be saved then through individual variations."[27] The greatness of a period's intellectual life depended on "the force and independence which single personalities possess" as well as on the community as a whole. Political philosophy and natural science thus converged around the crucial type-phenomenon—individuality; such was the basis of Høffding's philosophy of personality.

In Kierkegaardian fashion, Høffding rejected the notion held by Marxists and Spencerians that ethics was a branch of sociology. "Sociology looks backward to discover the laws of social evolution. Sociology comes after reality." But "the moral philosopher takes his place in the battle itself." Such was Høffding's message when he spoke on November 11, 1904, at the London School of Economics and Political Science, with the distinguished sociologist L. T. Hobhouse in the chair.[28] He thus admired the exuberant and embattled Rousseau for his "strong emphasis on the personal, subjective aspect of every serious view of life." More profound than either the dogmatic encyclopaedists or the theologians, Rousseau was a "subjective thinker" as Pascal before him and Kierkegaard after, whose "view of life comes as an expression and a result of personal experience of the need and yearning, the anguish and pain, the guilt and exaltation, that the subject has experienced within him."[29]

Always, however, Høffding saw the respective complementary truths in standpoints that others usually took as contradictory. He was stirred by Rousseau's personal, subjective standpoint, but also felt the grandeur in Spinoza's conception of nature as a geometrically determined order.[30]

The student Niels Bohr was, of course, well aware that his professor of philosophy lacked an intimate knowledge of science. Indeed, when Bohr attended Høffding's course on formal logic, he "noticed some serious errors in the professor's exposition, which with his wonted candor he pointed out to him." Høffding took his pupil's criticisms gracefully, and later submitted to Niels the proof of the amended course for his approval.[31]

Niels Bohr had listened to Harald Høffding's ideas long before he was formally Høffding's student. For Høffding was a close friend of Niel's father, Christian Bohr, and during his boyhood Niels heard the discussions between Høffding, his father, and several other friends who gathered at the Bohr home after sessions of the Danish Academy. The men had originally met in cafés, but tiring of that, began regular Friday evening discussions at each others' homes.

These intellectual dialogues profoundly affected Niels Bohr,[32] and he and Høffding grew to be lifelong friends. As Niels progressed in his career as a physicist, Høffding took an ever greater pride in their friendship. In March 1927, shortly after his eighty-fifth birthday, Høffding wrote to Emile Meyerson, the philosopher of science, that of the articles of praise published that day in the newspaper "an article of M. Niels Bohr especially brought me a great joy. M. Bohr declares that he found in my books the ideas which have helped scientists in 'the understanding of their work and have been thereby of real assistance to them.' "[33] Then, when Høffding became seriously ill, he wrote of the comfort Niels Bohr had brought him:

> I have had very interesting conversations with M. Niels Bohr especially on the irrationality which the theory of quanta has brought into physics. M. Niels Bohr is not only a great physicist, he is interested also in philosophy and literature, and he envelops his friends with a warm sympathy. Several months ago I was sick and in bed for several days, and I think spoke of my illness as very dangerous; then M. Bohr came several evenings to spend hours by my bedside. He spoke to me of his work and of other interesting subjects, and he read to me several works from one of his favorite poets, while Madame Bohr talked with my wife. It was a great encouragement in our uneventful, isolated life.[34]

A year later, in October 1929, Høffding mentioned again the "visits of Niels Bohr," and the great pleasure they were giving him; Bohr would explain his scientific research, "and then," said Høffding, "we discuss their philosophical consequences."[35]

To the old philosopher it seemed that Bohr's researches were confirming his own and the Kierkegaardian standpoint—that no all-inclusive system, no all-subsuming analogy was possible. When he wrote a book on the concept of analogy, his illustrative model was Bohr's work. The approbation of his pupil "my friend M. Niels Bohr, the celebrated investigator of the world of electrons" made him feel that his labor would not perish with the next generation.[36] When Bohr explained to Høffding that one could not decide "whether the electron is a wave motion (in this case one could avoid discontinuity) or a particle (with discontinuity between the particles)," and that "certain equations lead us to the first conception, certain others to the second," the old professor felt that his physicist-pupil had put into practice what he himself had expounded in theory: The new science seemed to be confirming his theory of knowledge, Høffding thought, when Bohr said: "No image, no word can answer to all the equations."[37] Høffding's misgivings, in February 1926, about the title of his book, *Philosophical Relativity*, vanished when Bohr congratualted him on it.[38] Behind his own conception of science, Høffding reiterated to Meyerson, was still a Kierkegaardian teaching: the idea of a postulate grounded in personal need—a perspective on the universe, inevitably partial in its character and arising from psychological sources, yet shaping our notion of things.[39]

Høffding became a revered, patriarchal figure in Western philosophy. In 1909, an honorary degree was conferred upon him by Cambridge University. Raising the question of what Høffding's central message had been in *History of Philosophy* which was translated into so many languages, the chancellor said: "I will answer: the personality of each philosophy." What William James affirmed in a provocative, impressionistic way, Høffding made the theme of his life's scholarship.[40]

The centenary of Høffding's birth was celebrated on March 11, 1943 during the Nazi occupation in World War II. The Danish Academy placed a wreath on Høffding's grave. A few days later, on March 19, Niels Bohr spoke in his memory; he submitted for publication a last essay of Høffding's aptly entitled, "Psychology and Autobiography."[41]

That something of the Jamesian pluralistic and personalistic message was imparted to the young Niels Bohr through the teachings of Høffding we can surmise from Høffding's memorable account of his visit with James in 1904. In August 1904, together with his friend Otto Jespersen, the eminent linguist, Høffding sailed for America to attend the International Congress of Arts and Sciences which had been convened at St. Louis to help celebrate the centenary of the Louisiana Purchase. A letter from James had done much to persuade him to go. Their first meeting took place in September on James's farm in New Hampshire. Høffding recounted:

When I came out on the porch, a short man, wearing shorts, and in shirtsleeves, and carrying oars over his shoulders, came across the yard. He had a fine intelligent face, a little nervous perhaps, but with beautiful eyes. And thus I stood in front of the man whose works had given me such a lively and warm impression of himself—one of the finest descriptive psychologists of our age. . . . In the course of our discussions the next day we found points where we agreed but also points where we disagreed. I had made my reservations clear the year before about some of James's ideas in a University course of 1903 (dealing with religious-philosophical experiences in modern times) which was published in the second volume of my *Minor Works*. The description of James was later used in foreign editions of *Modern Philosophers*, and I know James valued its characterization.[42]

Above all, Høffding was moved by this example of a thinker trying to surmount his acute neurosis through a philosophical decision.

James's mildness and peace of mind were, however, no gifts from nature. During an earlier period of his life he had had violent attacks of neurasthenia accompanied by feelings of anguish. Everything in life seemed insecure and dangerous. He had to grasp at the most severe Biblical phrases to avoid becoming mentally destroyed. He has described this himself in his famous book, *The Varieties of Religious Experience.*

Although at that time Høffding did not know the details of James's case history, he was aware that James was one of the "sick souls" he described. He saw firsthand James's struggles with himself: "When I visited him, he was suffering from a violent depression." And it was James's conviction, based on experience, "that such a sick soul has a right to grasp at that which has the power to keep it alive in the moment of crisis." From this conviction, wrote Høffding,

both his intensely religious character, inspired by a longing for peace and power over the inner darkness, and his total freedom from any dogmatism, can be understood. His position with respect to the religious problem was that of a spiritual doctor, employing the means and cures which experience has taught him, and without any concern about systems.[43]

Such were the narratives and ideas of William James that young Niels Bohr probably heard from Professor Høffding on the Friday evenings at home among his father's friends, after Høffding's return from America. In the lecture hall at the university he heard Høffding dwell on the bond between a man's personality and his philosophy. To the young Bohr, brooding about the anxieties of decision and wondering whether any rationalistic, complete system was possible, James's message on the open possibilities, the indeterminacies of existence, and the action that could shape the nature of that existence to an unpredictable direction chosen from among the options always inherent in reality, must have reinforced his own direction.

Most important, however, was Harald Høffding's role as sympathetic expositor and critic of Denmark's philosopher, Kierkegaard.[44] His lectures and writings presented Kierkegaard in a daring and significant light for young scientists. Kierkegaard emerged as an alternative to the materialism and determinism that had enveloped so many young European intellectuals. Young Marxists at that time accepted the doctrine of Ludwig Feuerbach, the philosophic mentor of Marx and Engels, that religious and theological ideas were simply projections upon a supernatural realm of man's subjective, psychological longings, and were presumably devoid of any objective truth. Høffding, however, as he counterposed Kierkegaard to Feuerbach, restored a charter for personal decision in matters of truth. He wrote: "Søren Kierkegaard expounded in his own fashion the same idea that Ludwig Feuerbach had a few years earlier established in *The Essence of Christianity*, namely, that theology is psychology."[45] But Kierkegaard went on to say that this psychological pole was precisely the important thing, this subjectivity, this state of striving for truth was all-important; whereas the objective pole, the objective reality, to which it stood in unknowable relation, was of no consequence. "The whole significance is displaced from the objective (dogmatic) pole to the subjective (psychological) pole."[46] With Lessing, Kierkegaard said that our relationship to truth exists only in an eternal search for it; existence as experienced was imbued with an objective uncertainty that was proportionate to the contradictory character of its content.

From the proposition that theology is psychology, Kierkegaard (as Høffding expounded his views) argued that we must simply drop the idea of truth-in-itself, the objective truth; all that we can have is psychological truth, hence *subjectivity is truth*. Here was a phenomenological conception that went far beyond Ernst Mach's. Although Mach polemicized against absolute space and absolute time, he still left one with the feeling that there was an absolute truth, that certain functional relations, did, as a matter of fact, characterize the world elements. But with Kierkegaard, absolute truth itself was displaced. Truth could be grasped only in the form of subjectivity; only that can be "true" which is grasped with subjective energy and passion. Thus were sown the first seeds of the notion that theoretical perspectives on physical experience which seemed contradictory were but complementary standpoints, depending not, as in the theory of relativity, on the state of uniform motion of an observer's frame of reference but rather on the personal decisions of an experimenter, on the particular, experimental arrangement he devised to measure and report on his physical experiences.

Confronted with a chasm between man's strivings for truth and any ultimate knowledge of God, an "absolute difference" between man and God such that God was not even conceivable by extrapolation as man's ideality raised to highest degree, Kierkegaard fell back on the notion that God was a postulation, "a sheet-anchor," as Høffding wrote, "which we have need of on the wild sea of existence and which is continually being agitated below the swell."[47] Renunciation was a keynote for Kierkegaard; one had to renounce the hope of assurance of certain knowledge of God and live heroically, even masochistically. Kierkegaard made the final yardstick for his personal acceptance of Christian belief the fact simply (as Høffding put it) that "it calls up the strongest contradiction, the greatest suffering, the highest passion in the subjectivity."[48] Niels Bohr was not to go to that extreme. Yet he imbibed from Kierkegaard the centrality of the idea of renunciation which, as we shall see, was a cornerstone of his philosophy of science.

As he studied Kierkegaard, Høffding came to his conclusion that in the history of thought, "two types of thinkers stand opposite to one another"; one type sought unity, connections, continuity, gradual transitions in reality; the other looked for qualitative novelty, abrupt transitions, sudden turns, sharp boundaries. These were universal psychological types in all human intellectual life: "The two types and directions collide in a slough of great problems in the area of natural science, as well as the arts."[49] They contest with each other over the unity of the forces of nature as against their differentation, over the relationship of willful decision to the law of causality; they lead to counter dicta in their approach to existence. Kierkegaard, wrote Høffding, "belongs decisively to the second type"; he taught us, said Høffding, that it is no achievement to proclaim a unity, or synthesis, when opposing elements are simply overlooked or by-passed.[50] In what he called his "qualitative dialectic," as distinct from Hegel's dialectic, Kierkegaard insisted on the irreducible, uneliminable boundaries between different modes of existence and life. Little children in the street used to run after Kierkegaard, calling him "Either-Or." Yet this slogan was indeed Kierkegaard's signature to reality; for all life, all existence, he claimed, was leading, over and over again to crossroads and going forward by repeated jumps and jolts.

But Kierkegaard rigorously restricted the "qualitative dialectic" to the realm of subjective, psychological experience. He firmly refused to extend his dialectic to the natural or physical sciences. He would not follow the way of Schelling, the German metaphysician, into constructing a nature dialectic. As Høffding stated:

According to Kierkegaard, the life of the spirit is subject to determinants completely different from those of the life of nature. In the most definite fashion he denies the existence of any analogy between spiritual and organic development. Qualitative metamorphosis, in which the new stage, the higher state, comes into existence as the result of a decisive jolt or jump, is characteristic of spiritual life, whereas organic development, as for instance, the growth of a plant is a successive unfolding of what is already at hand in the seed.[51]

From Kierkegaard's standpoint, no causal explanation whatsoever could be provided for the basic transitions in human existence. Consequently, "the genetic or evolutionary view is simply missing altogether in Kierkegaard's representation. He therefore keeps separate those forces which might in one stage be an indication of the next. Because of this—in full accord with the theory of the leap—every single stage appears as something totally complete, closed, which points neither forward nor backward. . . .[52]

The young Niels Bohr, imbibing the theory of the leap, evidently felt that there was no warrant for confining its domain to the psychological. The isomorpheme, the canonical form, of the qualitative dialectic might be exemplified not only in stages of human existence but also in stages of atomic existence. Years later Bohr was to hold that his principle of complementarity had a universal application to every branch of science and human activity. Høffding, moreover, expressed misgivings concerning Kierkegaard's theory of abrupt transitions, which in part pointed to the direction Bohr took. Høffding felt that Kierkegaard should not even call his configurations of subjective existence "stages," because "one thinks of a stage as part of a development, but of Kierkegaard's stages one does not hear where they came from, and basically . . . one does not know either where they are going."[53] Was Kierkegaard, moreover, able to exclude dogmatically the possibility of a causal approach to subjective, psychological transitions? Høffding ventured to suggest that perhaps beneath the level of consciousness, in which decisions seemed abrupt and discontinuous, there was a causal line, proceeding "half-consciously or unconsciously.[54] The sense of a discontinuous transition would then be superseded by the more underlying reality of a continuous one: "As a result of a natural psychological deception we are inclined, looking at our decisions later, to give our conscious decision more importance than it has." Indeed, Høffding wrote, there are alternative ways in which transitions can take place: "Water moves forward in a flowing stream as well as in the cascading water-fall. . . ."[55] Kierkegaard, he felt,

> has no eye for the small things, for the small increment which over and over connects quietly and often though unnoticed also achieves that which is praised as great and high. Only he who recognizes that life has many ways and types for its development, only he gives life its full due. Why force it into the straitjacket of a method?[56]

Indeed, according to Høffding, "upon closer inspection" of a jolt, one finds that "even here the continuity is not broken off. . . ." Thus Høffding, while heralding the contribution of Kierkegaard's "qualitative dialectic," persisted in maintaining that it should not be regarded as the controlling, exclusive form for understanding subjective experience. And here too, one can see the beginnings of that standpoint leading to the principle of complementarity, envisaging equally valid, alternative modes of representing events depending on one's mode of experimental arrangement. Kierkegaard, with the bias of a theologian and humanist, had averred that the principle of personality stood in contrast to scientific research, which effaced personality; Høffding, however, disagreed, the entire history of science he wrote, shows that "scientific work never stops being a work of personality."[57]

Kierkegaard himself (in Høffding's view) had verged toward such a complementarist outlook. "In the *Concept of Dread*," wrote Høffding, "Kierkegaard had become unfaithful to his slogan 'Either-Or,' and undertaken the attempt of developing an 'As-well-as.' More precisely, he wanted to have both an 'Either-Or' and an 'As-well-as.' He did not want to choose between jump and continuity, but instead wanted to combine both."[58] Nonetheless, the qualitative dialectic still remained dominant in Kierkegaard; "a transition is a break for him at all times."[59] Kierkegaard could not even talk, as Hegelians and Marxists do, of crises developing immanently; "there can actually be no talk of crises according to his view-point," wrote Høffding. "Kierkegaard has clearly let himself be led too far in his polemic against the smooth transitions of speculative philosophy. . . ."[60]

Such were the issues of the discontinuity of stages of existence, and the plurality of modes of analyses that Høffding raised for his students, Niels Bohr and the *Ekliptika* circle, as he lectured and wrote on Kierkegaard. To be sure, Bohr was not the only member of the *Ekliptika* circle whose lifework was to echo the Kierkegaardian themes of Høffding's teaching. His friend Edgar Rubin developed a conception of psychology highly analogous to Bohr's conceptions in physics. According to Rubin, psychology had been misled by the "monistic tendency"; he advocated instead a critical tendency which would recognize the fragmentary character of psychological knowledge: . . . "in psychology we have scientific knowledge of only a limited number of fragments." It was the power of religious and artistic motives that "seduced" psychologists into thinking they could construct theories "which concern a whole, especially the whole domain of psychology." The Kierkegaardian philosophy survived in Rubin's warning against the acceptance of the "inheritance from that sort of philosophy which has as its end the giving of a well-rounded picture of the world. . . ." Such world-hypotheses were not required by the scientist: "It is a common experience that really productive scientific work can be done without any use of all-embracing hypotheses." Rubin voiced his misgivings concerning the dogmas borrowed from daily life and philosophy that permeated psychology. Their influence was powerful "because the power of custom is great; the case is here as it can be with the religion of childhood." But he was suspicious, as a Kierkegaardian might be, of the orthodoxy that so-called schools of thought in science imposed. They wanted to confer hegemony on their respective modes of thought. "Psychological schools are a nuisance. As a rule the association of the members with each other serves as a help towards the continued belief in some unproved favorite notions which characterize the school." One should not through definitional devices or general assertions "prohibit others' from developing their own methods and interests: "In scientific psychology we have no use for popes."

Also in a Kierkegaardian vein, Rubin affirmed the reality of the "concrete individual" whose character could not to be subsumed under deterministic or dualistic systems: "Like every other concrete thing, the individual cannot be defined." The individual, furthermore could not be separated in his psychological existence from interaction with the physical world. A new kind of physics, Rubin speculated, might replace the old—a novel physics in which psychical entities played a role:

Physics has evolved in accordance with the rule of Spinoza and does not know anything about psychic life. Psychology has never kept to the rule of Spinoza, but has always regarded the psychic world in connection with the physical world. . . . It may be that physics, incorporating the psychic, will give up the rule of Spinoza as no longer fruitful, and at the same time be profoundly altered with regard to basic concepts.[61]

A partial step in this direction was achieved in Bohr's principle of complementarity. The Kierkegaardian isoemotional line, which through Høffding's teaching, helped define the *Ekliptika* circle persisted in the lives of both the physicist and psychologist. Furthermore, it was Edgar Rubin who had led Niels Bohr to read the work of William James:

I was a close friend of Rubin, and therefore I read actually the work of William James. William James is really wonderful in the way he makes it clear . . . that it is quite impossible to analyze things in terms of—I don't know what to call it, not atoms. I mean simply, if you have some things . . . they are so connected that if you try to separate them from each other, it just has nothing to do with the actual situation. . . . Rubin advised me to read something of William James, and I thought he was most wonderful.[62]

Clearly, something more than intellectual curiosity inspirited in Niels Bohr's concern for these issues suggested by Høffding and discussed among the members of the *Ekliptika* circle. His concern was deeply personal and emotional, and is best exemplified in his reaction to a novel that became his favorite.

"The Adventures of a Danish Student"

No scientist has ever felt his lifework to be linked with a particular novel to the extent that Bohr felt his work associated with Poul Martin Møller's *Adventures of a Danish Student*. This novel, which, like its author, is virtually unknown today, is perhaps the only novel that has had a powerful influence on the history of science. It clarified in Bohr's consciousness certain deep, underlying emotional and intellectual conflicts: anxieties of choice, action, and his self-definition.

Bohr's daring adaptation of the quantum notion to the electronic transitions in the atom was not the result of neutral, dispassionate hypothesizing; the discontinuous leaps of the electrons formed the kind of hypothesis for which his emotional character was longing—a resolution of his personal anxieties projected onto the atomic world. Bohr felt furthermore that the psychological need for a principle of complementarity was indicated in the discussion of the characters in Møller's novel. Bohr indeed made it a scriptural reading in his University Institute for Theoretical Physics; it became a document of the Copenhagen spirit. The Belgian physicist, Leon Rosenfeld, Bohr's co-worker, tells:

> Every one of those who came into closer contact with Bohr at the Institute, as soon as he showed himself sufficiently proficient in the Danish language, was acquainted with the little book; it was part of his initiation. Bohr would point to those scenes in which the licentiate describes how he loses the count of his many egos, or disserts on the impossibility of formulating a thought, and from these fanciful antinomies he would lead his interlocutor along paths Poul Martin Møller never dreamt of—to the heart of the problem of unambiguous communication of experience. . . .[53]

Mogens Pihl, Bohr's friend, remarked that Bohr "does not forget to remind us of that Danish literary work which he valued most of all, and he often referred to particularly in conversations, namely Poul Martin Møller's 'The Adventures of a Danish Student.' " Bohr's son, Hans, tells how his father "often returned" to Møller's novel: "In the description of the licentiate's difficulties in coming to a decision, he saw a perfect example of his ideas about complementary features in psychology."[64]

In the dialogue of the novel Bohr found the model for his own thought processes as he coped with his primary scientific problem. When Bohr wanted to explain the principle of complementarity, he referred spontaneously to this novel which had concurred with his innermost feelings. He wrote:

> In order to illustrate this important point, I shall allow myself to quote a Danish poet and philosopher, Poul Martin Møller, who lived about a hundred years ago and left behind an unfinished novel still read with delight by the older as well as the younger generation in this country. In his novel, called *The Adventures of a Danish Student*, the author gives a remarkably vivid and suggestive account of the interplay between the various aspects of our position, illuminated by discussions within a circle of students with different characters and divergent attitudes to life.[65]

Who was Møller, and why did his novel, born of a student culture, so enthrall Niels Bohr? Poul Martin Møller read his manuscript *En dank Students Eventyr* in 1824 to the students' union at Copenhagen. Møller was an intellectual seeker, almost a wanderer; he had voyaged for two years in the Orient in the guise of a ship's clergyman. His Asian travels had "burst for him the attractive bubble of the exotic," awaking in him a greater appreciation and understanding of Danish life. He wrote poems to celebrate the homeland:

If such things be poverty's true measure,
Silk-clad eastern prince, I understand;
Then I break my Danish bread at leisure,
Thanking God, I too exclaim with pleasure,
"Denmark is a little, beggar land."

According to the literary historian P. M. Mitchell, Møller represented the beginning of a "bourgeois realism in Danish poetry." Danish culture as a whole was at that time "moving towards the realization of bourgeois ideals."[66]

A few years after his return from the Orient, Møller began a career as a teacher of philosophy, serving first as lecturer at Christiana University, and then from 1831 until his death in 1838, as professor of philosophy at the University of Copenhagen. As a university student, Kierkegaard, then in his period of "full revolt against the world," found in Møller a man who could understand him because they had experienced the same striving. Møller often warned Kierkegaard not to try to circumscribe the universe, "not to lay out too big a plan of study," to restrain the daimonea that threatened his work. It is said, in fact, that Møller evoked Kierkegaard's "awakening" in 1836 by advising him against the "perfectly frightful" way in which he was "so through and through polemicalized." "Remember ye the far-travelled man," he told Kierkegaard. On Møller's death, Kierkegaard wrote: "Yes, now he is far-travelled but I at least shall still remember him."[67] And in 1844, six years after Møller's death, Kierkegaard dedicated *The Concept of Dread* with words of devotion that he rarely used:

To the Deceased Professor Poul Martin Møller,

The fortunate lover of Greek literature, Homer's admirer, Socrates' intimate, Aristotle's interpreter—Denmark's joy in "Joy over Denmark, far travelled, always remembered in the Danish summer,"—my (youth's enthusiasm) admiration; (the mighty trumpet of my awakening, my sentiment's desired object, the confidant of my beginning, my departed friend), my loss (lost reader), this work is dedicated.[68]

Møller's novel *The Adventures of a Danish Student* tried to transcend the polemical confrontation of life-standpoints. Published posthumously, unfinished and fragmentary, it had a frank, naive freshness and was the first Danish novel to treat contemporary life. The main characters are two cousins, the licentiate and the magister; gaily, wittily, against the setting of student life, they dispute the meaning of things. The licentiate, charming, irresponsible, always falling in love, is counterposed to the magister endowed with erudition but with a purely practical, "soberly efficient" type of mind.[69] When the "philistine" rebukes the licentiate, "addicted to remote philosophical activities," for not finding a job, the licentiate replies (in a conversation that Bohr quoted at length):

My endless inquiries make it impossible for me to achieve anything. Furthermore, I get to think about my own thoughts of the situation in which I find myself. I even think that I think of it, and divide myself into an infinite retrogressive sequence of "I"s who consider each other. I do not know at which "I" to stop as the actual, and in the moment I stop at one, there is indeed an "I" which stops at it. I become confused and feel a dizziness as if I were looking down into a bottomless abyss, and my ponderings result finally in a terrible headache.

The Philistine then answers:

I cannot in any way help you in sorting your many "I"s. It is quite outside my sphere of action, and I should either be or become as mad as you if I let myself in for your superhuman reveries. My line is to stick to palpable things and walk along the broad highway of common sense; therefore my "I"s never get tangled up.[70]

Which was his real "I"? Where could he find himself? Here was the prototype problem that was sublimated in Bohr's physical theory. It would not be easy, Bohr wrote, "to give a more pertinent account of essential aspects of the situation with which we are all faced."

"Poul Møller had little imagined," commented Leon Rosenfeld, "that his light-handed banter would one day start a train of thought leading to the elucidation of the most fundamental aspects of atomic theory and the renovation of the philosophy of science. Yet it is hardly an exaggeration to say that the perplexities of this licentiate, especially his struggle with his many egoes, were the only object lesson in dialectical thinking that Bohr ever received, and the only link between his highly original reflection and philosophical tradition." These "youthful meditations," Rosenfeld adds, foreshadowed

> all the main themes that recur, steadily amplified, in his later epistemological work: the use of language for the objective communication of experience; the ensuing necessity of fixing the unambiguous meaning of words by reference to situations of common experience; and finally the possible occurrence of a duality of aspects requiring special caution in the use of language to secure unambiguous communication.[71]

In truth, Rosenfeld magnifies the undoubtedly important role of Møller's novel in Niels Bohr's work; it was not Bohr's only link to philosophical thought, for Høffding's teachings and Kierkegaard's books were also influential. Yet Bohr's interest in Møller's novel points most clearly to the basic generative anxiety in Bohr's unconscious.

According to Werner Heisenberg, the formulator of the principle of indeterminacy, Bohr "was primarily a philosopher, not a physicist," though he adds that Bohr understood that natural philosophy in our age carries weight only if sustained by the "inexorable test of experiment."[72] And every philosophical standpoint reflects underlying emotional anxieties and conflicts, together with the attempt to resolve them. Bohr found the academic philosophies of his time lifeless and uncongenial. He could muster little interest in axiomatics or Bertrand Russell's logical formulae; he felt that the logical positivists such as Otto Neurath, who was dedicated to tracking down the blunders of metaphysicians, were "trivial."[73] Bohr sensed, however, that the whole tenor of his life's scientific work was toward resolving the psychological and personal philosophical anxieties of which the licentiate was the spokesman. There was the anxiety: How can I find my true, real self? There were the preoccupations: How can I possibly act? How can I break the entrapment of the infinite regress, the unfathomable abysses, that surround every idea and decision I entertain?

The licentiate undertakes to explain to his cousin why it takes him such a long time to write his dissertation:

Certainly [says the licentiate] I have seen before thoughts put on paper; but since I have come distinctly to perceive the contradiction implied in such an action, I feel completely incapable of forming a single written sentence. And although experience has shown innumerable times that it can be done, I torture myself to solve the unaccountable puzzle, how can one think, talk, or write. You see, my friend, a movement presupposes a direction. The mind cannot proceed without moving along a certain line; but before following this line, it must already have thought of it. Therefore one has already thought every thought before one thinks it. This every thought, which seems the work of a minute, presupposes an eternity. This could almost drive me to madness. How could then any thought arise, since it must have existed before it is produced? When you write a sentence, you must have thought it, otherwise how could you know that a sentence can be produced? And before you think of it you must have had an idea of it, otherwise how could it have occurred to you to think of it? And so it goes on to infinity, and this infinity is enclosed in an instant.

"Bless me," said Frits with indifference, "while you are proving that thoughts cannot move, yours are proceeding briskly forth!"

"That is just the knot," replied the licentiate. "This increases the hopeless mix-up, which no mortal can ever sort out. The insight into the impossibility of thinking contains itself an impossibility, the recognition of which again implies an inexplicable contradiction."[74]

To the licentiate in his anxiety-neurosis, no autonomous, independent action seemed possible; no thought could exist in its own independent, segmental reality; the past became an enveloping morass from which a present could never emerge; one's self became divided, and the actor lost his capacity to act when he allowed himself to become a spectator of his action:

"Then on many occasions man divides himself into two persons, one of whom tries to fool the other, while a third one, who in fact is the same as the other two, is filled with wonder at this confusion. In short, thinking becomes dramatic and quietly acts the most complicated plots with itself and for itself; and the spectator again and again becomes actor."[75]

The Kierkegaardian Model

Bohr found liberation from what we might call his anxiety of infinite decision in the qualitative dialectic of Søren Kierkegaard. In 1909, as Bohr was completing his thesis, he sent a copy of one of

Kierkegaard's books to his brother Harald on the occasion of Harald's twenty-second birthday:

I am sending you herewith (besides what Mother is so good as to send you in my name) Kierkegaard's "Stages on Life's Journey." It is the only thing that I have to send, but I do not believe that it would be easy to find anything better. In any case I have had very much pleasure in reading it; I even think that it is one of the most delightful things I have ever read.

When his first paper was published, Bohr wrote again about both his doubts and enthusiasm concerning Kierkegaard's book: "When you have finally read the *Stages* . . . you shall hear a little about it from me. I have written a few notes about it (not in agreement with K.) but I do not intend to be so banal as to attempt, with my poor nonsense, to interfere with your impression of so nice a book."[76] The impact of Kierkegaard's book on Niels Bohr was more profound than he first avowed. Many years later, in 1933, he told Professor J. Rud Neilson:

He (Kierkegaard) made a powerful impression upon me when I wrote my dissertation in a parsonage in Funen, and I read his works night and day. His honesty and willingness to think the problems through to their very limit is what is great. And his language is wonderful, often sublime. There is of course much in Kierkegaard that I cannot accept. I ascribe that to the time in which he lived. But I admire his intensity and perseverance, his analysis to the utmost limit, and the fact that through these qualities, he turned misfortune and suffering into something good. . .

A great resurgence of interest in Kierkegaard indeed took place half a century after his death during Niels Bohr's student years in Copenhagen. Four books about him, written from diverse viewpoints, appeared in 1898; one argued, for instance that Kierkegaard's philosophy because of his excessive individualism, could never become the religion of socialism. In 1903, when Bohr was nearly eighteen years old, a book was published which stressed the significance of Kierkegaard's doctrine of "paradox" and his break with the Hegelian system. In the same year, A. B. Drachmann, editor of Kierkegaard's works, lectured to the students' union on "Paganism and Christianity in Søren Kierkegaard." In 1909, 1910, and 1911, the first three volumes of Kierkegaard's papers were published.

Young Danish intellectuals began to identify with Kierkegaard's outlook. A persistent theme in their discussions was the psychological basis of Kierkegaard's philosophy. Thus it was said that Kierkegaard's relations with his father, the man who once cursed God on a desolate moor, had been central to his philosophical decisions.[78] Kierkegaard's dispute with the Danish church, which would have seemed utterly parochial, trivial, and ridiculous to Einstein and Friedrich Adler was of supreme moment to Kierkegaard's Danish followers. Høffding said that it was a dispute "to which our literature knows no equal." "It is a quarrel of the same kind as that between Plato and the Sophists, and between Pascal and the Jesuits," in the course of which "mighty words" were spoken on "the relation between ideal and self-deception which have a message to every one, whatever his stand."[79]

What characterized the young Kierkegaardians of Niels Bohr's generation was that they took a pre-eminently psychological approach to Kierkegaard's philosophy. As Aage Henriksen writes: "No argument against a purely personal historical and genetic explanation of his [Kierkegaard's] production has been able to deter them from adopting the psychological view."[80] It was a novel mode for understanding and interpreting philosophy. The first literature on the Kierkegaardian philosophy, from 1854 to 1870, had been concerned with religious polemics. Then in 1877 George Brandes wrote an influential book that recognized Kierkegaard's literary genius. Brandes declared that Kierkegaard was the Scandinavian thinker who "has had the greatest share in the intellectual education of the younger generation." But as Brandes told Nietzsche in 1888, his object in writing his "polemical pamphlet" had been to "counteract" Kierkegaard's influence.[81] For Brandes was an empiricist in the Millite mode and therefore, despite Kierkegaard's power, regarded him as a philosophical reactionary. Brandes believed that had death not intervened prematurely, Kierkegaard would have come to his senses and discarded Christianity and the Scriptures in favor of a pagan humanism.[82]

The third generations of Kierkegaardians that flourished during the time of Niels Bohr's youth saw him rather as the exemplar of honesty and utter sincerity, as a man who, probing into pretensions to knowledge and objectivity, made clear that they concealed chasms of ignorance and decisions that were subjective. To the average Dane, Bishop Grundtvig remained the representative national thinker, its most imposing scholar, and inspiring teacher. He was beloved for reviving the Danish church by introducing "animated and popular preaching, hearty singing and frequent communions with a new and excellent hymn-book for general use. . . ."[83] But for the restless Danish intellectuals who were in some way spiritually excommunicate, Kierkegaard become the guiding prophet.

In the minds of the younger generation, moreover, the Kierkegaardian philosophy was enlivened through Ibsen's plays which, over a period of several decades, proclaimed the primacy of the individual in detachment from the solidarity of any social group.[84] In 1881, Ibsen's *Ghosts* had shaken the intellectual life of the Copenhagen community; the liberal and conservative press alike raged against the drama. The playwright replied with his searching play, *An Enemy of the People*, one of the most devastating critiques of democracy ever written. The message once again was Kierkegaardian: "The strongest man in the world is he who stands [most] alone." Brandes commented: "Not since he wrote *Brand* had Ibsen followed so closely in Kierkegaard's footsteps as he does here." Together with Kierkegaard, Ibsen shared the conviction that in every human being there dwelled "a mighty soul, an unconquerable power."[85] The circumpressures of society, organizations, families, waged a relentless war against the individual. Only he was free who withstood these circumpressures, and stood free with his own self.[86] Here was no Parisian positivism exalting social determinism nor Marxist materialism with its credo of the creative instincts of the masses. This was a philosophy of young Danish intellectuals who defied any social determination of their lives and thoughts. In 1905 when Niels Bohr was a university student, *An Enemy of the People* was once more a subject of debates among young social philosophers. In Moscow and Petrograd, on the eve of revolution in 1905, the play, as the famed director Constantin Stanislavsky wrote, was the "favorite play of the revolutionists." In their eyes, what could be more revolutionary than telling the truth? On the day when the Czarist troops fired on the crowd in Kazansky Square, actors and spectators merged at the footlights: "The younger people in the audience jumped to the stage and embraced Dr. Stockmann."[87]

Of the Kierkegaardian themes so widely circulating among young Danish intellectuals, which had special appeal for Niels Bohr?

Qualitative Leaps. In 1841 Søren Kierkegaard went to Berlin to attend the lectures of Schelling, the German founder of *Naturphilosophie.* He was first enthusiastic, "indescribably glad": "When Schelling mentioned the word *Virkelighed* [actual daily life], in connection with the relation of philosophy to *Virkelighed*, thought leaped within me as the babe leaped in Elizabeth." A few months later Kierkegaard had changed his mind, saying that Schelling was driveling on.[88] But Høffding thought that the first experience, when Kierkegaard's thought "leaped," may well have been the beginning of Kierkegaard's theory of qualitative leaps.[89] According to Kierkegaard, the human will and creative spirit act through abrupt, discontinuous transitions, or leaps; continuous transition, he felt, was impossible. He thought that no such leaps occurred in the "quantitative" sciences, that the leap was "qualitative" in character:

> in transition there is a stop and a leap (i.e. a break in continuity). . . . When one thinks abstractly, there is no such break. But neither is there any transition, because abstractly seen, everything *is.* But when existence adds time to movement, and I act accordingly, then a leap is revealed in the only way it can be."[90]

This concept of the leap, of the discontinuous transition from one stage to another, evidently entered into the perspective from which Niels Bohr viewed the problems of atomic theory. The terminology in which Bohr phrased the hypotheses of his 1913 classical papers had a Kierkegaardian ring; indeed, the structure of the hypotheses was isomorphic with the transitions in the *stadia* (stages) of human existence described by Kierkegaard:

> 1. That energy radiation is not emitted (or absorbed) in the continuous way assumed in the ordinary electrodynamics, but only during the passing of the systems between different "stationary" states.
>
> 2. That the dynamical equilibrium of the systems in the stationary states is governed by the ordinary laws of mechanics while these laws do not hold for the passing of the systems between the different stationary states.[91]

Kierkegaard's dialectic of the stadia of life, affirmed that spiritual evolution of the individual involved three stages or spheres of existence: the aesthetic, the ethical, and the religious.[92] Each stage was held to be an independent sphere of life, enclosed, isolated, and exclusive. No rational evolutionary formula, no Hegelian law for the sequence of stages, no necessary pattern for resolving internal contradictions governed the transitions in Kierkegaard's *stadia.* Instead, he posited a leap from one stage to another, a discontinuous transition not to be rationalistically explained.[93] One did not move, for instance, to the religious stage from the ethical stage by reflection on the latter's inadequacies; rather a conversion ensued, an abrupt moment of rupture, a leap into the absurd that brought a radical renewal. The Kierkegaardian disjunction "either-or" emphasized that choice was a matter of discrete alternatives; there were no gradations or gradual transitions in the stadia of human existence. As Høffding described Kierkegaard's conception:

> His leading idea was that the different possible conceptions of life are so sharply opposed to one another that we must make a choice between them, hence his catchword *either-or*; moreover, it must be a choice which each particular person must provide for himself, hence his second catchword *the individual.*[94]

Høffding, as we have seen, had not been able to accept Kierkegaard's conception of the leap. He criticized him for not providing a motivation for the transition from one stage to another, for always describing the transitions as crises, as breaches of continuity.[95] He cavilled at the conception that movement from one stage to another was always contingent, never necessary, and at the notion that there were inexplicable transitions. The phenomena of change, he felt, required a gradualist, quantitative dialectic.[96] When it came to ultimates, Høffding's predilection was still for "continuity," though he avoided dogmatizing on the issue. Given this difference, we can well imagine that the *Ekliptika* circle spent evenings arguing whether Kierkegaard or Høffding was right. The Kierkegaardian model of discontinuous leaps became part of Niels Bohr's deepest emotional-intellectual standpoint. The atom in its "stationary state" was later like one of Kierkegaard's stadia of existence. And the leap of the electrons from one orbit to another was like the abrupt, inexplicable transitions of the self.

Moreover, quite apart from the discontinuous electronic transitions from one orbit to another, Bohr's theory of the atom departed in its total imagery from the old model of the Newtonian, continuous world. According to the classical ideas, the electrons around the nucleus could have traveled in any of an infinity of possible orbits; any circular motion with any given radius would have been satisfactory. But in Bohr's conception, only certain orbits with specific radii were admissible; only those orbits in which an electron's angular momentum would be an integral multiple of a certain constant were physically possible; all others were excluded. The electron in its possible motions was confined to a small specific number of orbits; the atom's physical history was restricted to a set of discontinuous orbits much as the Kierkegaardian self was confined to its fixed set of stadia of existence.

The Kierkegaardian model reached so deeply into Bohr's thinking that he described the atom in its transitions from one stationary state to another as possessing a "free choice" like that of the human subject choosing his qualitative leaps from one stage to another. Bohr summarized the significance of his work as follows:

> the author suggested that every change in the state of an atom should be regarded as an individual process, incapable of more detailed description, by which the atom goes over from one so-called stationary state to another. . . . We are here so far removed from a causal description that an atom in a stationary state may in general even be said to possess a free choice between various possible transitions to other stationary states.[97]

Oskar Klein has written that although Bohr had little interest in formal logic, "his engagement in the old fundamental problem of ethics, concerning the freedom and limitation of will was however much deeper. . . ."[98] Bohr sought to confirm a subatomic world isomorphic with his emotional and philosophical self-definition in the everyday world.

Every great physicist approaches the world of physical phenomena with guiding philosophical analogies that express his innermost emotions and longings. He is fortunate if the objective physical data and problems allow for a fruitful conjuncture with his subjective standpoint. The emotional-intellectual standpoints of creative scientists can be utterly diverse; Newton was enthralled by a vision of a neo-Platonic unity; Einstein was sustained by the spirit of Marxian-Machian rebellion; Bohr felt the dramatic urge of Kierkegaard's qualitative dialectical leaps of the stadia of human existence. When he was in his first creative phase, formulating the quantum theory of the atom, Bohr brought his own emotional-philosophical longings to bear on empirical facts which otherwise seemed lacking in theoretical significance; they had an illuminating power one would never have dreamed possible.

In his "casual reading" Bohr had chanced upon Balmer's rule, a simple arithmetical expression for the wave-lengths of the lines in the hydrogen spectrum,[99] formulated in 1885 by Johann Jakob Balmer, a Swiss teacher adept at combining numbers. Legend has it that Bohr read it in a children's book.[100] To Bohr, Balmer's formula suddenly became "the touchstone he needed": "the theory of atoms which he had been all the time 'dreaming' of (to use his favorite expression) suddenly crystallized in the Balmer Formula."[101] Physicists have marveled at Bohr's amazing empirical intuition. The dream, the unconscious vector, which at this juncture was guiding his intuition, was something like the Kierkegaardian model of stadia of existence, with the self freely, inexplicably choosing its stage, abruptly and discontinuously, in a qualitative leap. Bohr's theory of the hydrogen atom can be seen from a psychological standpoint as a projection of the Kierkegaardian qualitative dialectic.

Thus, pondering the numerical ratios of the frequencies of the spectral lines emitted by the hydrogen atom, Bohr dared in his hypotheses to extend to the structure of the atom itself the quantum condition proposed by Max Planck with respect to the energy radiated by hot bodies. A strange world appeared on which electrons jumped from one orbit into another without any intervening, continuous trajectory. And from these hypotheses, through several steps of algebraic reasoning, there emerged the simple arithmetical rule of Balmer, the Swiss schoolteacher; his empirical formula was a derivable consequence of Bohr's novel hypotheses.

A work of synthesis is always guided by some powerful image of reality arising from deep, personal emotions, often of unconscious origin, but moving the ideas and perceptions of the scientist in strange directions. The guiding emotional image may come from sources quite extraneous to the purely scientific data and theories then obtaining. When Bohr achieved his remarkable hypotheses which unified within one theory the labors of the spectroscopists, the formulae and constants of Balmer and Rydberg, the experiments of J. J. Thomson, which brought forward the existence of electrons, measuring the ratio of their charge and mass, those of Ernest Rutherford on the positive charge localized in atomic nuclei, and Max Planck's theory of energy-quanta, his generative emotion was allied to Kierkegaard's yearning for autonomous selfhood through discontinuous stadia.

The generative emotions of theories are the unconscious causes for the variations in scientific hypotheses; a kind of Darwinian selection by the physical environment then takes place, and those hypotheses which are falsified rather than sustained by the evidence are provisionally eliminated. There was a time, for instance, when Bohr was strongly inclined to lay hands upon the law of conservation of energy and to deny that it held true for the emission of beta (electronic) particles from radioactive atoms; Bohr even conjectured that such a departure from the law of conservation would help explain the ceaseless production of energy by the stars. His emotions were clearly probing for a principle of nonconservation. As George Gamow has written: "This was the era when so many laws of classical physics were rejected under the impact of the newly developed Theory of Relativity and the Quantum Theory that no law of classical physics seemed to be unshakable."[102] Bohr's anticonservative view led to "heated verbal discussions and voluminous but never published correspondence" among him, Paul Ehrenfest, and Wolfgang Pauli who, staunchly adhering to the principle of the conservation of energy, proposed to balance nature's accounts by supposing that unknown particles were being emitted in the beta processes. In later years, experimental evidence accumulated which supported, for instance, the contravening existence of neutrinos. Thus, a generative emotion can find itself contravened by the actual weight of empirical evidence. Indeed in the history of scientific ideas it is probably the case that the overwhelming majority of such generative emotions underlying the variations in ideas, that is, the novel scientific hypotheses are extinguished by the factual, experimental environment. Yet without such generative emotions, the nisus toward scientific creativity would be gone. The great turning points in the history of science have occurred where there has been a coincidence between the generative emotion and the actual stage of the factual evidence in the given theoretical situation.

Renunciation. Apart from the word "complementarity," no conceptual word appears as often in Bohr's writings as "renunciation." When he engaged in his famous philosophical controversy with Albert Einstein, the word Bohr used most often to characterize his methodological decision was "renunciation." Advancing powerful scientific arguments against the notion of strict causality or determinism in physics, Bohr spontaneously used the Kierkegaardian idiom, suffused with Kierkegaardian emotions. Bohr thus wrote of:

> the necessity of a final renunciation of the classical idea of causality and a radical revision of our attitude towards the problem of physical reality.

. . . In quantum mechanics, [he wrote], we are not dealing with an arbitrary renunciation of a more detailed analysis of atomic phenomena, but with a recognition that such an analysis is *in principle* excluded.

. . . We have in fact to deal with new fundamental features of atomicity . . . which have demanded a still more far-reaching renunciation of explanation in terms of a pictorial representation.

. . . A comparison of purely logical aspects of relativistic and complementary argumentation reveals striking similarities as regards the renunciation of the absolute significance of conventional physical attributes of objects.

. . . The question at issue has been whether the renunciation of a causal mode of description of atomic processes involved in the endeavors to cope with the situation should be regarded as a temporary departure from ideals to be ultimately revived or whether we are faced with an irrevocable step. . . .

. . . These ideas . . . on the excitation of spectra by impact of electrons on atoms, involved a further renunciation of the causal mode of description. . . .

. . . With his mastery for co-ordinating apparently contrasting experience, without abandoning continuity and causality, Einstein was perhaps more reluctant to renounce such ideals than someone for whom renunciation in this respect appeared to be the only way open. . . .

. . . Notwithstanding all novelty of approach, causal description is upheld in relatively theory within any given frame of reference, but in quantum theory the uncontrollable interaction between the objects and the measuring instruments force us to a renunciation even in such respect.[103]

This was an essay on both physical theory and the ethic of renunciation; for it conveyed all the emotional valences associated in Bohr's mind with the word "renunciation." It was linked to the affirmation of human freedom. Bohr had written several years earlier that "only a renunciation . . . will enable us to comprehend . . . that harmony which is experienced as free will."[104] Thus, "renunciation as regards the causal space-time coordination of atomic processes" was no purely logical adventure in an alternative mode of analysis; it was a fulfillment of Bohr's emotional cravings.[105] One clearly calls for renunciation only of that which is desired, or which is most instinctive or intuitively perceived; Bohr's essays, in particular, called for renunciation (or repression) of a sense of explanation that had seemed most basic. Einstein acknowledged that Bohr's renunciation was "logically possible," but felt nonetheless that "it is so very contrary to my scientific instinct that I cannot forego the search for a more complete conception." The ethic and outlook of renunciation were indeed foreign to Einstein's nature.

To Bohr, however, the ethic of renunciation was a primary tenet: "It is sheer folly to believe that one can achieve anything in this world without renunciation," he said.[106] He had imbibed the centrality of this notion from Kierkegaard. To move from the ethical stadium of existence to the religious (the highest), an act of renunciation was required. One had to renounce the universal rules of ethics as one chose the absurd; thereby, in losing all, one finally regained everything. This was the calling of the Knight of Faith in Kierkegaard's *Fear and Trembling*; its high exemplar was the patriarch Abraham, prepared to sacrifice his son Isaac even if he thereby renounced all universalist ethics while holding to the faith that his renunciation would bring a higher recompense. Kierkegaard wrote:

> In resignation, I make renunciation of everything; this movement I make by myself, and if I do not make it, it is because I am cowardly and effeminate and without enthusiasm and do not feel the significance of the lofty dignity which is assigned to every man, that of being his own censor. . . . This movement I make by myself, and what I gain in myself is my eternal consciousness. . . . By faith I make renunciation of nothing; on the contrary by faith I acquire everything. . . .

Renunciation brought a sense of self-mastery and entry into a higher realm. Furthermore, Kierkegaard wrote: "I am able by my own strength to renounce everything, and then to find peace and repose in pain. I can stand everything—even though that horrible demon, more dreadful than death, the king of terrors, even though madness were to hold up before my eyes the motley of the fool. . . ." Abraham was the great renunciator because "he brings to naught his joy in the world—he renounces everything." To renounce was the ultimate imperative of religion: "What every religion in which there is any truth aims at, and what Christianity aims at decisively, is a total transformation in a man, to wrest from him through renunciation and self-denial all that, and precisely that, to which he immediately clings. . . ."[107] For Bohr the renunciation of strict causality, despite the hardship it entailed, was the essential step to the higher truth of complementarity.

Subjectivity and Physical Reality. When Niels Bohr formulated his principle of complementarity in 1926, he proposed that physicists renounce the hope of achieving a system of theory based on one model, either particle or wave. An uneliminable dualism pervaded all physical theory; it was inherent in the choices made in experimental operations. Thus, too, if one devised experiments for ascertaining "the position of a particle," one could not speak in the same context of "the velocity of a particle" since experiments designed to measure the first precluded any operations which could measure precisely the latter. An experimental choice thus seemed to partition the world of potential physical phenomena. The principle of complementarity affirmed that at any given time one could choose between dual complementary descriptions, but that doing so would thereby exclude the other alternative. No description could combine or synthesize the complementary alternatives.

The source of the principle of complementarity, as Bohr often reiterated, was in his own "psychical experience," which provided him with the guiding analogy for physical theory. "This conception of complementarity," notes Leon Rosenfeld, "inspired Bohr all the time from the days of his youthful meditations. . . ."[108] This duality of complementaries seemed to Bohr "a fundamental feature" in the nature of all human knowledge. He felt that the "laws of psychology" showed the traits of complementarity so clearly, and were therein so analogous to the principles of quantum theory, that knowing them made it "easier to adjust ourselves to the new situation in physics."[109] Above all, it seemed to Bohr that the decisions of the person, the subject, whether as observer or participant, or indeed as both, partially determined the character of his external reality, the structure of the object. Bohr had early arrived at the view that "as a universal feature" the "contrasted idealizations" of thought had only restricted domains of validity, that they were complementary to each other.[110] But now the Kierkegaardian motif further asserted itself. The dominant scientific realists and materialists held that there was an objective truth, independent of all observers. Kierkegaard, however, maintained that truth is subjectivity:

If only the object to which he is related is the truth, the subject is accounted to be in the truth. When the question of truth is raised subjectively, reflection is directed subjectively to the nature of the individual relationship: if only the mode of this relationship is in the truth, the individual is in the truth, even if he should happen to be thus related to what is not true.[111]

Thus Bohr came to the view that scientific truth and physical reality themselves rested on subjective decisions, that the complementary physical realities corresponded to complementary subjective decisions, and that, therefore, no clear dividing line existed between subject and object. "The impossibility of distinguishing in our customary way between physical phenomena and their observation places us, indeed," Bohr wrote, "in a position quite similar to that which is so familiar in psychology where we are continually reminded of the *difficulty of distinguishing between subject and object*."[112] Our forms of perception had failed," said Bohr, because of "the impossibility of a strict separation of phenomena and means of observation, and the general limits of man's capacity to create concepts, which have their roots in our differentiation between subject and object."[113] It was necessary therefore, he continued, to take recourse "to a complementary, or reciprocal, mode of description," one "perhaps most familiar to us from psychological problems." The physical sciences, "the so-called exact sciences" had tried, in Bohr's view, "to attain to uniqueness by avoiding all reference to the perceiving subject," but now, quantum theory had, in the footsteps of the theory of relativity, finally shown "the subjective character of all the concepts of classical physics."[114]

Somehow, furthermore, the whole experience of human freedom —the sense of choice and of open alternatives—was allied in Bohr's mind to the choice of a physical reality which was made by the observer's decision among complementary experimental arrangements. The feeling of free will itself, Bohr wrote, was founded on "the impossibility in introspection of sharply distinguishing between subject and object as is essential to the ideal of causality." This was a human situation, he emphasized, which presented "in several respects a formal similarity with which, to the great disquietude of many physicists and philosophers, we have met in atomic physics."[115] It was not a disquietude that Bohr shared; for him, there was a Kierkegaardian sense of human liberation, of the breaking of boundaries to the freedom of the human subject. Hitherto, "the feeling of free will, which governs the psychic life," had been menaced by the "apparently uninterrupted causal chain" of physiological processes; now, however, "an unvisualizable relation of complementarity" seemed to dissolve the tormenting anxiety of determinism counterposed to free will; each mode of description was valid within its own limits.[116] Above all, whether to adopt a materialist or idealist philosophy or a wave or particle description of physical reality was a matter of human choice. The "facts" themselves never determined the subjective choice; instead, a subjective ingredient partially determined the class of facts.

The Danish student of Møller's novel—trying vainly to catch his consciousness in its moment of thinking or decision but always finding it deflected by his effort, his real self remaining elusive—hovered in the formative imagery of the principle of complementarity. As Bohr described the psychological basis of complementarity:

> We are thinking here of the well-known characteristics of emotion and volition which are quite incapable of being represented by visualized pictures. In particular, the apparent contrast between the continuous onward flow of associative thinking and the preservation of the unity of personality exhibits a suggestive analogy with the relation between the wave description of the motions of material particles . . . and their indestructible individuality. The unavoidable influence on atomic phenomena caused by observing them here corresponds to the well-known change of the tinge of psychological experiences which accompanies any direction of attention to one of their various elements.[117]

Thus a kinship obtained between Bohr's principle of complementarity and Kierkegaard's concept of dread. To Kierkegaard, dread was the point of intersection of the nature-determined world and the world of the individual free spirit; the person chose between them.[118] The physicist choosing between complementary descriptions was enacting a Kierkegaardian drama in quantum theory. The Kierkegaardian "either-or" was implicit in the notion that the choice of one set of experimental conditions determined a corresponding set of theoretical ideas and excluded another set which was founded on a correspondingly different experimental decision.

With a Kierkegaardian view of existence, it seemed to Niels Bohr that no human model or set of words could ever grasp the meaning of the universe. If we tried to do so, we would be entrapped in self-contradiction. To Kierkegaard's remark that he found himself with 70,000 fathoms of water underneath him, Bohr "used to add with a twinkle in his eyes: 'It is much worse—we are suspended over a bottomless pit—caught in our own words.' "[119] Bohr felt the brooding sense of a question over things, challenging all pretenders to certainty. He said that "there is no meaning to saying that life has no meaning."[120] He often used to say: "Every sentence I say must be understood not as an affirmation, but as a question," thus echoing the words of Henrik Ibsen: "My calling is to question, not to answer." But it was the Kierkegaardian Ibsen whom Bohr found congenial, the Ibsen whose *Brand* when it first appeared, was regarded by many people in Denmark as Kierkegaard in verse form.[121] Of Bohr too, one could say with Brand:

Be what you are with all your heart,
And not by pieces and in part.

This questioning sense of the limitations of human models and language, this feeling that no system fashioned by humans could be complete and adequate to existence, formed the emotional basis of the principle of complementarity. "Conplementarity," writes L. Rosenfeld, "is no system, no doctrine with ready-made precepts. There is no via regia to it; no formal definition of it can be found in Bohr's writings, and this worries many people."[22] There were many indeed who spoke of "complementarity" as " 'just philosophy' ".[123] Indeed, we might say that it was a guiding, heuristic principle born, as all such principles are, of the deepest emotional perspectives. Complementarity expressed the emotional conviction that no man-made words, image, formula, model, canonical form, or concatenation of symbols could ever be adequate to reality. Curiously, complementarity did not open itself to a plurality of perspectives; rather it made central to its standpoint the notion that a duality of perspectives was inherent in the interactions between human perceivers and physical objects. The coat of arms that Niels Bohr hung from his house, only a few feet from the king's residence at Frederiksborg Castle, bore the ancient Chinese symbols of *yang* and *yin*, the ultimate duality in all being, and as such, the basis of the principle of complementarity.[124]

However, Bohr's complementarity always remained linked to a sense of human alternatives; as a Kierkegaardian, he could never accept a deterministic framework for his dualities. For a long time Bohr searched in classical philosophic writings for a guiding analogy for his complementarist leanings. At one time he believed he had found it in Spinoza's notion of psychological parallelism which seemed to express the complementarity of psychological events with physical ones. In time, however, as his ideas matured, Bohr came to reject Spinoza's conception precisely because it was deterministic; it also involved an equivalence between mental and physical phenomena that was foreign to Bohr's spirit. Interestingly, the thoroughgoing determinism that Bohr found so uncongenial in Spinoza was valued by Einstein.[125]

Niels Bohr: Gentle Revolutionist

Niels Bohr wrote his first paper in the mansion called Naerumgaard where his grandparents and father had lived. He had been delaying writing up his work because he was always noticing "new details." "At last my father sent me out here, away from the laboratory, and I had to write up the paper," said Niels, recalling his father's peremptory demand for quick completion so alien to his own habits of work.[126] His creativity was not fully evoked under his father's guidance. From his youth, Bohr's meditations were drawn to the philosophical problem of the surd status of human decision and self-observation with respect to strict determinism, a problem which tends especially to preoccupy those who, like the young John Stuart Mill, have experienced some traumatic encounter with paternal authority. The influence of Harald Høffding was, in Niels Bohr's case, as we have seen, partially that of a teacher who in the eyes of the rebel replaces the father with a novel outlook and doctrine.[127] It was not Høffding's formal teaching of logic which impressed Niels, but rather an influence "more marked in the general attitude to human problems."[128] Høffding had taught him that discontinuities were real and possibly uneliminable, and that a distinctive personality expresses itself in every philosopy. The first tenet helped guide Bohr in his first years of fashioning the quantum model of the atom, the second was most important when he was fashioning the principle of complementarity.

No one, however, would have said that a generational rebel dwelled in the soul of Niels Bohr. Unlike Einstein, Niels Bohr was born into the scientific establishment and, moreover, regarded himself as its continuator. Niels' father Christian Bohr was a noted scientist and professor of physiology at the University of Copenhagen. He had done experimental research yet he always argued against the mechanistic standpoint in biology. One would fail to comprehend living processes, he held, unless one were guided by teleological conceptions of the functions of organs. Evidently, Niels Bohr felt in his latter years that his principle of complementarity made its peace with his father's standpoint. Thus in 1932, when he expounded the consequences of complementarity for biology, "this had a special appeal to him; he had been deeply impressed by his father's views on the subject, and he was visibly happy at being now able to take them up and give them a more adequate formulation."[129] The father's reflections had "come as near as one could expect at the time to establishing a relation of complementarity between the physico-chemical side of the vital processes . . . and the properly functional aspect of these processes, dominated by teleological or finalistic causality."[130] The principle of complementarity with its conjunction of two opposed standpoints was "quite in the spirit of his father's ideas."[131] Clearly Niels Bohr in his middle age did indeed find a mode of thought, an isomorpheme that seemed to promise the overcoming of generational tension between himself and his father. Thus a measure of pluralistic conciliation entered Bohr's principle of complementarity. Whereas for Kierkegaard the choice of a stage of life was entire, and one's personal decision gave one's self entirely to either an aesthetic, ethical, or religious stadium, the physical experimentalist could live with alternative world-pictures and could move from one to the other as practical requirements indicated. Something of the Jamesian "and" complemented the Kierkegaardian "either-or." James had written that "radical empiricism insists that conjunctions between them [sensations] are just as immediately given as disjunctions. . . ."[132] It is strange how the most profound philosophical convictions revolved around the emotions educed by such words as "and," "either-or," and "not." Yet in such symbolic nuclei were locked the emotional energies that give birth to novel theoretical products. The principle of complementarity authorized coexistent world-pictures, competitive isomorphemes, as long as the limitations of each resulting from their defining experimental conditions were always borne in mind.

Niels Bohr, a man alienated from neither his land nor his people, grew up loving to play on teams and work with teams. Where as Einstein wrote: "I am a horse for single harness not cut out for tandem or team work. I have never belonged wholeheartedly to country or state, to my circle of friends, or even to my own family,"[133] Niels Bohr could say the opposite. He delighted in working with his colleagues; the Copenhagen spirit nurtured in Bohr's institute made Copenhagen a center of scientific theorizing such as the world had not seen since Padua. Bohr, moreover, never wavered in his identification with his family, country, and friends; he was in no conventional sense an alienated intellectual. His achievement in science was compatible with his sense of solidarity with his existing society; he neither saw its downfall nor called for its reconstruction.

"In Denmark I was born, and there my home is," Bohr took pleasure in quoting from a poem by Hans Christian Andersen. He would continue, "From there my world begins," conveying to his friends that in Denmark was found "the right atmosphere" for combining tradition with receptivity to new ideas.[134] Denmark was a country devoid of revolutionary tradition or radical political ideas. There had been no Danish counterparts to the revolutionary movements in America and France during the nineteenth century. The ferment aroused by the Declaration of Independence and the Declaration of the Rights of Man wracked other lands, but not Denmark, where these documents exerted little influence. "Copenhagen, of course," writes K. B. Andersen, "had its clubs, where the members met round the punch bowl and sang songs condemning the 'Bigwigs,' but the spirit of revolt expressed in such speeches and songs was generally very mild." The famed Grundtvig, who affected Danish life more deeply than any other thinker, was "rather a supporter of absolutism."[135] Nonetheless, Grundtvig inspired the "Danish Community," an unparalleled movement for the people's self-education in history and the humanities. The Danish Constitution of 1849, with its democratic reforms, came without an era of ordeal. The Danish cooperative movement as well as the labor movement later in the century worked within the framework of democratic traditions. Copenhagen, peaceful and tolerant like Zurich, hosted the assemblies of the Socialist International.[136] The serenity of Danish life, however, could provide the soil in which germinated Bohr's path-opening vision. Niels Bohr quoted the words of Grundtvig to his young, foreign colleagues at his institute: "The right of Danes only by birth, but the right of all by the laws of hospitality."[137] He imparted to others something of the Danish heritage. Generational tensions in Danish society could express themselves in ways which did not involve their exacerbation into hatred.

For somehow even the rebels in Denmark felt a security in a rebellion located in its gentle setting. Søren Kierkegaard, for all his stormy declamations, could write of Copenhagen as affectionately as Socrates and Pericles had spoken of Athens:

Some of my countrymen are maybe of the opinion that Copenhagen is a tiresome town and a small town. It seems to me, on the contrary that, refreshed as it is by the sea . . . and without being able even in winter to dismiss the recollections of its beech forests, it is as favorable a place as I could desire to dwell in. Big enough to be a great city, small enough to have no market price set upon by men, where the tabulated comfort one has in Paris that there are so and so many suicides, where the tabulated joy one has in Paris that there are so and so many persons of distinction, cannot penetrate disturbingly and whirl the individual away like foam, so that life acquires no significance . . . because everything dashes off into space without content or too full of it. Some of my countrymen find the people who dwell in this town not vivacious enough. It does not seem so to me. The speed with which in Paris thousands form a mob around one person may indeed be flattering to the one about whom they collect, but I wonder if that makes up for the loss of the quiet mind which permits the individual to feel that after all he too has some importance. Precisely because the individuals have not totally fallen in price . . . precisely for such reasons is life in this capital so entertaining for him who knows how to find in men a delight which is more enduring. . . .[138]

Likewise for Bohr, the Danish culture had a harmony all its own. It was cosmopolitan both by necessity and by choice; it was too small to indulge in fantasies of imperialist cultural superiority, yet its spiritual resources were rich and rational enough so that it had an active, discriminating receptivity with respect to rival foreign influences from Britain, France, and Germany. As Bohr wrote: "What most decidedly characterizes Danish culture might be just that immediate willingness to absorb learning brought to us from abroad or which we ourselves bring home, and the retention of that outlook on life which is conditioned by our heritage. . . ." The Danish language, the medium of expression of a small people, was withal adequate to scientific needs, and kept a tender devotion on the part of its speakers which was utterly dissociated from any notion of linguistic imperialism. Niels Bohr could thus warmly endorse Kierkegaard's linguistic patriotism:

As a starting-point in defending the Danish language against the allegation that the better-known cultural languages have greater merit, he (Kierkegaard) expresses with such warmth and intensity his feelings for the richness and beauty of our mother tongue and employs such artistry and profundity that his observations actually hold good not only for Danish but for all other languages, thus instructing us in the infinitely refined and effective features of language.[139]

The Copenhagen spirit in physics thus was a remarkable emotional corollary of Bohr's sense of the Danish cultural heritage; it was a style in physics created by Bohr, "a style," writes Victor Weisskopf,

> of a very special character which he imposed onto physics. We see him, the greatest among his colleagues, acting, talking, living as an equal in a group of young, optimistic, jocular, enthusiastic people, approaching the deepest riddles of nature with a spirit of attack, a spirit of freedom from conventional bonds, and a spirit of joy which can hardly be described.[140]

Was the creativity of Niels Bohr wholly exempt from the patterns of generational revolt that characterized the Zurich-Berne circle? Were his creative energies, encouraged as they were by his family and community, immune to any sense of alienation? Were they in no way a transmuted form of the energies borne by vectors of generational rebellion?

The energies of generational tension are as universal, variegated, and fluctuating as those of sexuality. Like atomic energies, under conditions of high excitation they can lead to violent dissociations; in more ordinary interactions, they can provoke leaps from orbit to orbit or group to group. Such energies may bind the comrades of generational circles or cause them to bombard the nuclei of parental and social authority. The analogy reaches into our linguistic usage, for we have all heard the nuclear family spoken of as the basic unit of society and generational communes referred to as the nuclei of a new society. From these energies of generational tension arises discontent; the received ideas are then searched for their vulnerable sections. Were such energies lacking, the moving discontent would be gone.

To be sure, pure curiosity, which is the heart of science, is hardly reducible to generational tension. Pure curiosity may have its source in sublimated sexual energies, or perhaps, in the child's wish to overcome its powerlessness by understanding the world; the spirit of the scientific enterprise is founded on such underlying motivations. They are reflected too in corresponding criteria for the adequacy of theories. The first suggests the use of "beauty" as a criterion for a theory's adequacy (as Einstein and Dirac have done). The second regards the pragmatic control of events as the measure of a theory's success; it sees theories not as replications of a

Divine Intelligence but as man-made products or man-made tools. But whichever psychological motivation is primary, the fact remains that the succession of scientific ideas as a social phenomenon is marked by the conflict of scientific generations. For each generation, moved by curiosity, still seeks to satisfy that curiosity in a novel fashion. Dress, for example, is founded on generic human needs, yet we are still governed, at least partially, by changing fashions and each period seeks to differentiate itself by fresh departures in literature and the arts as well, where the idioms of the old inevitably fail to speak for the young. This same phenomenon occurs in the sciences. As the physicist George P. Thomson has written: "The really deeply penetrating theories of physics, those that deal with fundamentals, are curiously a matter of fashion."[141]

Generational tensions can issue in conflicts, violent, disrespectful, self-destructive, and full of the impulse to reject. Or they can propel a quest for new forms, transfiguring the urge to parricide, blending the old and the new with a wholeness of judgment, and the awareness that though each must try to realize its favored forms, each has its mission for achievement. It bespeaks Einstein's greatness that at the height of his enthusiasm for Ernst Mach, he also acknowledged the huge segment of realities which were grasped by the methods of his opponents, Boltzmann and Planck. In Niels Bohr, too, one perceives generational tensions as a mainspring in the search for novel modes of thought—though without the manifestation of the self-declared generational revolutionist.

Niels Bohr's great burst of "scientific" originality came, interestingly enough, shortly after his father's death in 1911. More importantly, Bohr had gone through an emotional-intellectual journey which led him from Goethe, of whom his father was a "keen worshipper," to Kierkegaard. What significance did this transition have in terms of the development of Bohr's scientific conceptions? With regard to science, Goethe was above all famous for his antagonism to Newton's experiments with the refraction of light; it seemed to Goethe a rape of Nature to apply a prism to a ray of light. Nature was feminine; Nature was a goddess who indulged in friendly play with men: *Mit allen treibt sie ein freudliches Spiel.* Light must remain inscrutable, a feminine ingredient in God who was identical with Nature, for Goethe held to the "pure, deep, innate . . . view that God is inseparably within Nature, and Nature within God."[144] This was a curious conjunction of a feminine pantheism with an antiexperimental bias; it aroused considerable discussion among scientists in the nineteenth century. Von Helmholtz, the revered chief of science in Germany, wrote of "this view of Nature" which

> accounts for the war which Goethe continued to wage against complicated experimental researches. . . . He would have us believe Nature resists the interference of the experimenter who tortures her and disturbs her; and, in revenge, misleads the impertinent killjoy by a distorted image of herself. Accordingly, in his attack upon Newton he often sneers at spectra, tortured through a number of narrow slits and glasses, and commends the experiments that can be made in the open air under a bright sun. . . .[145]

Yet Niels Bohr chose as his lifework to deal with spectra and refracted light, and, moreover, to take them as clues to otherwise inaccessible nuclear realities. Thus in his choice of problem in physical investigation, Niels Bohr broke with his father's Goethean metaphysics. That Christian Bohr, an experimental physiologist, endorsed all of Goethe's antiexperimental views is hardly likely; nonetheless, he was an enthusiast for Goethe's philosophy. Furthermore, as an enthusiast for the *Naturphilosophie* of Schelling and Hegel, Goethe had an antimathematical bias. Explanations of natural phenomena, according to Goethe, should be plainly visualizable; they should possess the property of *Anschaulichkeit* (visual intuitiveness).[146] The Romantic writers frequently affirmed that only the inferior understanding thinks in terms of quantities and numbers, and that such categories are inapplicable to the higher realities which are accessible only to the superior faculty of reason. A disciple of Goethe would thus find himself committed both to antiexperimental and antimathematical biases.[147] From such dogmas too, Niels Bohr departed. His model of electronic transitions was unvisualizable; his derivations of the possible electronic orbits were mathematical. Above all, Bohr did not share Goethe's pantheistic optimism; instead, he experienced the Kierkegaardian, unfathomable abyss.

Let us now turn to the question of the psychological motive that propelled Niels Bohr to reject his father's Goethean enthusiasm, and to venture forth on his own journey toward Kierkegaard. It is the licentiate in Møller's *Adventures of a Danish Student* who casts most light on the innermost generative anxieties of Niels Bohr. The licentiate fears a paralysis of his powers of thinking and acting; his fears have all the symptoms of "obsessive thinking." The obsessional person, like the licentiate, is prone to a kind of regression in which "preparatory acts become substituted for the final decision, thinking replaces acting, and, instead of the substitutive act, some thought preliminary to it asserts itself with all the forces of compulsion."[148] Thus the licentiate finds himself driven to conjuring up an indefinite series of thoughts of thoughts of thoughts . . . ad infinitum, which seems to preclude his ever enunciating a conclusion, and analogously, ever making a decision. The licentiate, endlessly interposing a series of reflections on reflections on his proposed action, compulsively engenders a Zenonian chasm between his intention and any outcome in action for his aim or goal. Furthermore, the obsessional person becomes a compulsive skeptical epistemologist in the everyday activities of life; he broods on the insecurity of his premises and the weak spots in his reasoning. He wonders, for instance, whether memory is at all trustworthy, and his doubts spread infectiously, pervasively, so that they apply to actions completed in the past as well as to the time domain of present and future. These compulsive, obsessive thought processes arise because of an underlying conflict of opposing impulses; such a conflict inhibits energies from following their normal paths of action. Instead, such energies are expended in endless stages of preparatory thought; the needed act is repressed and lost as a shadowy limit which the infinite series can never approach. Behind the obsessional pattern, underlying the inhibition from concluding a thought or action, is the chronic coexistence of a love and hatred toward the same person. These polar opposites never find their synthesis; rather the emotional conflict multiplies the member thoughts in the indefinite sequence because no member resolves it.[149] Unable to suppress or eradicate one of the disjunctive pair of contending emotions, the person finds himself inhibited from any conclusion which affirms one or the other. Hence arises the compulsive need for uncertainty, the hesitance to confront realities.

To Freud it seemed that the histories of obsessional patients almost invariably reveal an early development and premature repression of the sexual instinct of looking and knowing. When the desire for knowing was dominant in the constitution of an obsessional person, "brooding becomes the principal symptom of the neurosis. The thought-process itself becomes sexualized, for the sexual pleasure which is normally attached to the content of thought becomes shifted on to the act of thinking itself. . . .[150] With the licentiate, indeed, a pleasure seemed to be found in the indefinitely prolonged, intellectual preparatory processes themselves, and presumably, in his early childhood, he would have had experienced intense repressions and not known early satisfactions of sexual curiosity.

Above all, the licentiate's obsessional thinking might well have had its origin in the opposition within his psyche of a powerful hatred equal in energy to an intense love. The neurotic equilibrium of indecision was finally founded on such a balance of emotional vectors. And if we ask what was the source of the tremendous attraction that Kierkegaard's writings had for Niels Bohr, it was that they strengthened his will, emphasized the arbitrary, discontinuous decision, the leap from one *stadium* to another without intervening Zenonian continuities. The qualitative dialectic was the liberating answer to infinite regressions; the human will could assert its primacy.

The strange protomodel of his emotional life that Bohr found in Møller's licentiate provides us with our primary insight into Bohr's unconscious—a kind of X-ray emanating from the emotional-intellectual nucleus of Bohr's psyche. That Bohr was aware of his departure from his father's notions in his early years as a scientist is shown by his conception of himself in later years as returning to his father's ideas and making his peace with them.[151]

In an unusual literary allusion, Niels Bohr indicated some feeling of generational rebellion to which he never referred otherwise. On meeting the young student Werner Heisenberg, Bohr was very curious to know what had led Heisenberg to participate in the German Youth Movement. Bohr remarked: "We in Denmark prefer the heroes of the Icelandic sagas, the poet Egill, son of Skallagrim, who at the tender age of three defied his father. . . ."[1] Among the many Icelandic sagas Bohr chose one that began with a young son's defiance of his father; evidently, this free association was a primary element in Bohr's unconscious. The saga tells how Skallagrim, a grim and taciturn man, flees to Iceland from Norway after some bloody slayings. His precocious son Egill has the gifts of both skald and warrior; at the age of three he composes verses, and at the age of seven he kills an eleven-year-old playmate. Egill is a "problem child." On a visit to the Norwegian homeland, he swallows inordinate quantities of beer at a drinking orgy and then stabs his host, the king.[2] Further episodes continue the tale of Egill's career as a murderer, arsonist, and chieftain. He is heathen but heroic. Like Niels Bohr, he was a second son, described as "exceedingly ugly and like his father, black of hair. And when he was three winters old, then was he great and strong, even as those other boys that were six winters old or seven. He was soon a chatterbox, and skilled in words." When the father, mother, and older son were summoned to a "bidding" with the father-in-law, Egill was supposed to remain behind, too young presumably to comport himself well amidst the "great drinkings." "But Egill was ill content with his lot. He went out of the garth and found a certain draught horse . . . fared a-back of him, and rode after Skallagrim and his." The drinking party welcomed the youngster kindly, whereupon he recited a poem ending with lines not weighted with modesty:

Find a three-year ode-smith
Better than me.

His mother was exceedingly proud of Egill when at the age of seven he accomplished his first murder; it proved, she said, that he had "Viking stuff":

> But when they fell to playing together, then was Egill overmatched for strength . . . [His opponent Grim was four years older] Egill became wroth, and heaved up the bat and smote Grim; but Grim laid hande on him . . . [Thereupon Egill secured an axe] Then leapt Egill at Grim and drove the axe into the head of him, so that straightway it stood in his brain.

Egill duly reported his accomplishments to his parents. "Skallagrim made as if he found little to be pleased with in this. But Bera said that Egill was of Viking stuff, and said that that would be his lot, as soon as he had age thereto, that that should find him warships." Again the adorable seven-year-old murderer broke out into poetry:

My mother told me
For me they should buy
Fleet keel and fair oars
To fare abroad with Vikings;
To stand up in the stem there,
Steer the good ship,
Hold her so to harbour,
Hew a man or twain.[154]

Egill, son of Skallagrim, proponent in the Icelandic sagas of the child's self-affirmation against the father, was Niels Bohr's spontaneous favorite among an innumerable cast of characters.

Lastly, one is left wondering about the little-mentioned cultural duality in Niels Bohr's background: the Jewish background of his mother Ellen, and his father's Danish Lutheran heritage. Niels was deeply attached to his mother, whom all describe as a gracious, kind woman. She had evidently acquiesced in renouncing her own Jewish culture completely, for no effort was made to transmit any part of it to her sons. Yet Ellen Bohr's father, the banker David Baruch Adler, had been a leading personage in the Jewish community. Born in 1826, David Adler had served as a member of the board of representatives of the Jewish congregation in Copenhagen. To his achievements in politics and commerce he had added a variety of philanthropic ventures; he established a fund for the widows and daughters of impoverished merchants. In politics, David Baruch Adler adhered to the left wing of the National Liberal party, was a founder of the Free Trade Society, and served in both the lower and upper houses of parliament. After Adler's death, Ellen married Christian Bohr.[155] To what extent some sense of emotional strain, renunciation of part of herself, or underlying melancholy was transmitted from Ellen Bohr to her son Niels, we can only speculate. Certainly in Ellen's childhood the novels and writings of the Danish Jewish author Meir Aaron Goldschmidt were still significant; he had depicted with resentment the many unfairnesses which beset Danish Jews in the nineteenth century.[156]

The Jewish community in Denmark was small; there were only 3,476 Jews in the country in 1901 (1.4 per thousand of the population), and during the years from 1894 to 1903, mixed marriages among the Jews were as high as forty-five percent.[157] Niels Bohr was neither involved nor identified with any activity that related to the Jews until 1933, when together with his brother Harald, he founded the Committee for the Support of Intellectual Refugees, designed primarily for Jewish scholars.[158] During the Nazi occupation, when a Danish Anti-Jewish League was organized, its journal attacked Bohr in May 1941 as a Jew who had risen to prominence and influence in Danish life.[159] In the summer of 1943, Bohr refused an invitation to escape to England because he feared his flight might entail serious consequences for the whole Jewish community.[160] In September, however, he learned that he was to be arrested as part of the "deportation" of Jews. Thereupon he escaped by boat to Sweden, where his first thought on arrival was how to help the Danish Jews. Until that time he had never engaged in politics. Largely through his influence and moral suasion upon the king and foreign minister of Sweden, that government declared its readiness to receive all Danish Jews who were encouraged to organize their mass flight.[161]

Among those, however, who were captured while trying to escape the Nazi pogrom was Niels Bohr's aunt, the famed Danish educator, Hanna Adler, who was then eighty-four years old.[162] Niels had always admired and loved her and spent much time with her. Insofar as the sense of a Jewish culture was imparted to him, it probably would have been through his aunt's thoughts, feelings, and example. As the headmistress of a school, she practiced her progressive ideas in education. To save her from the Germans, the ministry of education, the mayor of Copenhagen, the rector of the university, and four hundred of her former pupils signed a petition on her behalf, saying that "Mrs. Adler's life work was one of such a character that the whole of Danish society is exceptionally indebted to her . . . The deportation of Mrs. Adler would result in a shock for a large number of Danes."[163]

Werner Heisenberg wrote discerningly that "Niels grew up at a time in which great strength of character was needed to slough off nineteenth-century middle-class ideas and the teachings of Christian philosophy in particular."[164] In that placid Copenhagen setting, the energies for this transition of stages had to derive primarily from one's own striving. Within the harmonious Bohr family the tensions, whatever their sources, were nonetheless sufficient to cause Niels to take a philosophic-scientific direction very different from his father's.

Anxiety is linked to some situation of danger, real or imaginary, associated with the person's relative helplessness. Repressed in the unconscious, it may affect the person with a kind of paralysis. He may be unable to move his arm in space or, like the licentiate, fear the movement, the direction, of his thoughts. As we have seen, Bohr's identification with the licentiate suggested a fear on Bohr's part as to what direction his own ideas might take; hence his anxiety. But there is a tremendous distinction between an anxiety expressed and an anxiety resolved. The latter involves an act of freedom which converts dissipated energies into creative ones. Freud once remarked: "An abnormal person can become an accurate physicist."[165] It would be truer to say that anxiety resolved can give birth to a creative physicist. For in resolving one's anxiety, by "objectifying" it, or finding traits in the external world isomorphic with one's own, some features of reality, hitherto not apprehended, may be perceived, defined, and clarified.

Werner Heisenberg and the Revolt against Determinism

The German Youth Movement

"The engaging of one's interest in a certain subject and in certain directions," wrote the physicist Erwin Schrödinger in 1932, "must necessarily be influenced by the environment, or what may be called the cultural milieu or the spirit of the age in which one lives." In nearly every branch of human activity, he observed, a desire for change had arisen, "a profound skepticism in regard to traditionally accepted principles"; the spirit of radicalism was no longer confined to "a cranky and noisy minority."[166] The cataclysm of World War I had ensued between 1913, the year in which Niels Bohr first propounded his quantum theory of the atom, and 1932, when Schrödinger made his observation.

The economist John Maynard Keynes wrote a celebrated and influential tract called *The Economic Consequences of the Peace.* Yet the more elusive, intellectual consequences of the world wars have scarcely been explored. The bloodshed, crises, and social dislocations which were the consequence of World War I produced or intensified novel emotions and unconscious motivations among scientists and artists alike. A revolt began against the doctrine of determinism; logicians, physicists, painters, and poets took part in the uprising. Among novelists and philosophers during the nineteenth century, the movement of ideas had oscillated between determinism and freedom. One generation would adhere to determinism, the next to free will, and its successor would return to determinism. But until 1914, this cycle does not seem to have included it its domain the modes of physical theory. Since the time of Isaac Newton, working physicists held to a strict determinism; whatever the prevailing fashion in literary and philosophical ideas, scientists persisted in holding to the heuristic principle of determinism. French poets and political philosophers of the late 1800s, for instance, joined in the Bergsonian fashion of antideterminism and free will, but their greatest man of science, Henri Poincaré, spoke for his co-workers when he enunciated his unswerving adherence to determinism.

War, violence, and deceit in Europe, the collapse of societies, the older generation's loss of moral authority and the disintegration of moral codes, the randomness which individuals' movements seemed to acquire, the uncertainties of one's social position from one day to the next—all this significantly affected the emotional unconscious of a new generation of scientists. Their most eminent figures later sought understanding and inspiration from Niels Bohr at his institute. The Copenhagen spirit was, indeed, a mediating one, seeking to preserve sanity and humor while absorbing what was the valid contribution of the new emotional-intellectual temper. The principle of complementarity urged a reconciliation between the determinist and indeterminist modes of thought, a new, stable equilibrium after the disorienting intellectual consequences of the world war.

The formative years of Werner Heisenberg, the youthful pioneer of the principle of indeterminacy, were lived under the impact of the vicissitudes of the world war and its aftermath. It is well to inquire how these events shaped his scientific standpoint, for the immediate postwar generation emerged disenchanted and disillusioned with the inherited order of existence. Many years later Heisenberg recalled the emotional mood of 1920:

> When I left school in 1920 in order to attend the University of Munich, the position of our youth as citizens was very similar to what it is today. Our defeat in the first world war had produced a deep mistrust of all the ideals which had been used during the war and which had lost us that war. They seemed hollow now and we wanted to find out for ourselves what was of value in this world and what was not: we did not want to rely on our parents or our teachers. Apart from many other values we re-discovered science in this process.[167]

The details of the emotional formation of Heisenberg's scientific standpoint are a rich psychological-historical romance. He came from a tradition-laden family. His father, August Heisenberg, held the chair in Byzantine studies at the University of Munich. It was the only position of its kind in all of Germany. He worked indefatigably at his subject, edited the *Byzantinische Zeitschrift*, and established the existence of obscure but enlightened commentators in the twelfth century, namely, the brothers Mesaritai, and Blemmydes, whose textbooks of logic and physics, August Heisenberg claimed, "have really created an immortal name for the author." The elder Heisenberg lived vicariously in that world of order which had lasted until the mid-fifteenth century. He blamed "the avarice of western Europe" for the fact that "the strength of the Byzantine world was forever broken down," and that it had been deprived of a renaissance. He found solace in recondite studies of its sepulchres and churches and called for a broad, historical approach that would unify the materials of art, social life, and language.[168]

August Heisenberg had been a captain in the reserves; when World War I broke out he entered the German army and fought on the western front. He was wounded in 1916 and sent home. "During the last year of the war," tells his son, "I worked as a farm laborer in the Lower Bavarian Alps, to keep body and soul together."[169] His experience as a farm laborer had a certain importance for young Werner. He worked in the fields with a pair of oxen, doing his best "to cut the meadow so straight that no strip of grass, or, as the farmer called it, no 'pig' was left behind."[170] He would look back on these days; even in the joy of being a student of physics a few years later, on glimpsing the blossoming meadows and the mowing machines, he would feel "sorry" that he was no longer a farm worker on the Grossthalerhof in Miesbach. He felt himself part and parcel of the German soil, a creature of its lakes and mountains; thus the nationalistic cult of the youth movement came as naturally to him as his vocation for theoretical physics. He saw the collapse of his father's world of middle and modern Greek. Other scientific sons have been moved by rebellious feelings against the world-outlooks their fathers represented. But in Werner's case, the world itself had defeated and wounded his father. And there is psychological evidence that some sons who see their fathers humiliated will superimpose on their own rebellious feelings the powerful emotion of loyalty to a defeated father. Einstein and Freud were bound by Jewish loyalties that originated in their perceptions of their fathers as having been persecuted. Werner Heisenberg was stirred to German loyalties by the sight of his wounded father-soldier. Thus he was drawn to the German Youth Movement by feelings far more compelling than the counterlogic in the crude, materialistic formulae and urban proletarian arguments of the Marxists.

Then came the turbulent events of 1919, which culminated in Munich in the brief episode of a soviet republic. For a few weeks, the

Kantian socialist, Kurt Eisner, the anarchist Gustav Landauer, and the poet Ernst Toller hoped to lead humanity to a higher stage. But Eisner was assassinated and the powers of irrationality proved so much greater than the socialists had calculated that they failed. In retrospect, their attempt itself seems to have been irrational. Young Werner Heisenberg did more than witness these events; he served with an anticommunist military detachment of schoolboys to help suppress the Munich Soviet. Heisenberg has written more than once of these circumstances which led to the generational event, the turning point in his life:

> At that time Munich was in the throes of a revolution. The inner town was still occupied by the Communists, and I, then seventeen years old, had been assigned with some school comrades as auxiliaries to a military unit which had its headquarters opposite the University, in the Theological Seminary. Why all this happened is no longer quite clear to me. . . . In the Ludwig Strasse we fetched our meals from a field-kitchen in the University courtyard.[171]

In an essay written later, however, Heisenberg recalled more clearly the circumstances of his 1919 involvement in the anticommunist force; there were emotions surrounding them which might well have accounted for a momentary strain of memory:

> In order to make it understandable that the memory of my study of the "Timaeus" meant a great deal to me at this moment, one would probably also have to be aware of the remarkable circumstances under which I did the reading. In the spring of 1919 conditions in Munich were rather chaotic. There was shooting in the streets, although one never knew exactly who the combatants were. The governmental power changed hands among persons and institutions whose names one hardly knew. Looting and robbery, of which I had once been a victim myself, made the expression "Soviet Republic" seem to be a synonym for a state of lawlessness. When finally a new Bavarian government was organized outside Munich, and committed its forces to conquering the city, we hoped that orderly conditions would be restored. The father of a friend whom I had once helped with his schoolwork took command of a company of volunteers who wanted to take part in the conquest of the city. He bade us, that is, the adolescent friends of his son, help the troops with their familiarity with the city as orderlies. Thus it came about that we were assigned to a staff called the Cavalry Rifle Command 11, which had set up its quarters from the university. Here than I did my service, or

more accurately, here we led together a most unfettered life of adventure; we were liberated from school . . . and we wanted our freedom to get to know new sides of the world. The circle of friends, with which a year later I wandered over the hills along the Starnberg Lake, was organized in nucleus in this very place.[172]

As Einstein grew to intellectual maturity in the friendships of the Olympia Academy and Niels Bohr in the *Ekliptika* circle, Werner Heisenberg grew to manhood in the Cavalry Rifle Command 11. The social environment of Munich, with governments changing hands, sides fighting confusedly, and bullets aimed uncertainly and striking by chance was a sociological protomodel of physical indeterminacy—the disintegrating society like a disintegrating atom, with bullets fired, like electrons, in uncertain trajectories.

This was July 1919; it was "a warm summer." Werner Heisenberg, on sentry duty, would frequently rise at dawn, "withdraw," as he tells, "to the roof of the Theological Seminary and lie down there to warm myself in the sun, any old book in my hands. . . ." There, like Descartes on military service three centuries earlier, Heisenberg experienced a sudden insight. It came to him during his reading of Plato's dialogue *Timaeus*:

On one such occasion, it occurred to me to take a volume of Plato on the roof, for I wanted to read something different from the books we were supposed to study in school. With my somewhat modest Greek knowledge, I came upon the dialogue called *Timaeus*, where for the first time and from the original source I read something about Greek atomic philosophy. This reading made the basic thoughts of atomic theory much clearer to me than they had been. . . .[173]

Heisenberg felt he now had "an inkling of the reasons that had in the first place caused Greek philosophy to conceive of these smallest indivisible building-stones of matter." True, he did not find Plato's thesis that atoms were regular solids "fully convincing, "but at least I was happy to learn that they did not have hooks and eyes."[174] Here we meet the "hooks and eyes" which were for young Heisenberg the symbol of all that he found repugnant in the atomic theory he read in physics textbooks. He had his first encounter with the "hooks and eyes" when he was a student in the upper two classes at the gymnasium; it was only then, he writes, "that I acquired a special liking for physics—oddly enough because of a fortuitous encounter with a fragment of modern physics."[175] What was this "fragment" which ignited Heisenberg's vocation as a physicist? Heisenberg's narrative continues:

> At that time we were using a rather good textbook of physics in which, quite understandably, modern physics was treated in a somewhat off-hand manner. However, the last few pages of the book dealt briefly with atoms and I distinctly remember an illustration depicting a large number of them. The picture was obviously meant to represent the state of a gas on a large scale. Some of the atoms were clustered in groups and were connected by means of hooks and eyes supposedly representing their chemical bonds. On the other hand, the text itself stated that according to the concepts of the Greek philosophers atoms were the smallest indivisible building-stones of matter. I was greatly put off by this illustration, and I was enraged by the fact that such idiotic things should be presented in a textbook of physics. . . .[176]

There was an inordinate emotional vehemence in the reaction of the adolescent Werner Heisenberg to the diagrammatic use of hooks and eyes for chemical bonds. He was much upset by the notion that perhaps there were such mechanisms within atoms. Later, when Burkhard Drude, a student friend at Göttingen, defending perceptual models, argued that modern technology might yet construct a microscope with such great resolving power that the structure of the atom could be perceived, "so my objections to perceptual models would finally be dispelled," "this argument," writes Heisenberg, "disquieted me deeply. I was afraid that this imaginary microscope might well reveal the hooks and eyes of my physics textbook. . . ."[177] Thus, a strong emotional aversion, quite apart from the experimental facts, an emotional rather than logical a priori deriving from unconscious sources, filled Heisenberg with an anxiety for which the hooks and eyes were the symbol. He feared that experiment might confirm their existence. A strong unconscious motive to banish such imagery from physics was evidently at work within him.

In the spring of 1920 while he was hiking with his comrades of the youth movement, Werner found support for his antagonism to hooks and eyes from a friend who was a more advanced student of philosophy. His friend, even more appalled by visual models of atoms, had arrived at the "firm conviction that the whole of modern atomic physics was false. . . . At that time our judgments were obviously very much rasher and more dogmatic than they are today."[178]

The hooks and eyes seemed moreover to have become for Heisenberg a symbol for the materialism of the defeated elder generation. He links his attitude toward them with the restless, defiant, new-community-seeking, and self-defining youth movement. The traditional romantic philosophy of Germany with its dislike for atoms, which were seen as Newtonian, bourgeois mechanical entities, blended with the juvenocratic ideology. Heisenberg recalls:

It may have been around the spring of 1920. The end of the First World War had brought the youth of our country into restless movement. The reins had fallen from the hands of the deeply disappointed elder generation, and the young people were gathering in groups, in communities, smaller and larger, looking for a new way of their own . . . because the old seemed to have been broken.[179]

The old guidelines were gone: "we must perhaps be reminded that the protection of parents and schools which surrounds youth in peaceful times, had largely been lost in the time's vicissitudes, and that, partially as a substitute, a spirit of independent opinion had sprung up among the youth. . . ."

On this particular spring hike with ten to twenty companions of his movement "remarkably enough," writes Heisenberg, "we had that first talk about the world of atoms which proved meaningful in my later development as a natural scientist."[180] Werner told his friend Kurt, a good sportsman from a Protestant officer's family who wished to become an engineer, of his misgivings about the hooks and eyes in the physics textbook. Werner could confide in his friend because there was complete trust between them; the preceding year when Munich had been under siege and their families had consumed their last piece of bread, Werner, his brother, and Kurt had made a trip through the lines of the combatants and come back with a pack full of bread, butter, and bacon. Now the two friends discussed the natural sciences. "I told Kurt," writes Heisenberg, "that in my physics textbook I had come upon a diagram which seemed to me completely senseless."[181] Kurt explained how it depicted the formation of a carbonic acid molecule. Heisenberg recounts:

In order to explain further just why one atom of carbon and two atoms of oxygen form one carbonic acid molecule, the artist had given the atoms hooks and eyes (*Haken und Osen*). . . . To me this seemed wholly senseless, because hooks and eyes are—as I saw it—quite arbitrary forms, which can be chosen in different ways according to their technical usefulness. But atoms are supposed to be the result of natural laws, and guided by them in forming molecules. In this, I believed, there could be no sort of arbitrariness. . . .[182]

Kurt replied that the hooks and eyes did indeed seem suspect, but nonetheless they scarcely troubled him, for he felt, they did exhibit certain relations which were known experimentally to hold: "He [the artist] drew hooks and eyes in order to show as straightforwardly as possible that there are forms which can lead to two oxygen atoms, but not three, combining with a carbon atom." Nonetheless, the important point remained for Werner that "the hooks and eyes are nonsense," and that neither he, his friend, nor the artist really knew the atomic forms which accounted for the observed chemical bonds, while the concept of "chemical valence" itself indeed might be no more than words.[183] Then another friend Robert, a student of literature and philosophy, (probably the "advanced student" previously mentioned) joined the conversation. He had been reading Malebranche, and he felt that in the human soul itself there were imprinted certain basic structures, and that when we apprehended "the form" of the atom, the experimental facts were perceived in conformity with these superimposed structures.[184] The hooks and eyes in the textbook diagram were not to be taken seriously. Again Werner found his memory returning to his reading of Plato's *Timaeus* the year before. At that time, he had not been able to understand it. But now he felt it illumined by Robert's words: "I was for the first time able to understand—even if for the meanwhile still unclearly—that one could arrive at such unusual thought constructions about the smallest particles. . . ."

Again and again throughout his life, Heisenberg recalled that experience on the roof of the Theological Seminary, "lying in the trough of the eaves, warmed by the first rays of the sun," and tenaciously grappling with Plato's account of the mathematical forms of "the smallest particles of matter." Plato had conjoined the Pythagorean discovery of the five regular solids with Empedocles' doctrine of the four elements: earth, fire, air, and water. The "corpuscles" of the respective "simples" were to be composed of corresponding regular solids: the earth corpuscles of cubes, the particles of fire of tetrahedrons, those of air of octahedrons, and water of icosahedrons. These in turn were to be fashioned in various ways from arrangements and sections of isosceles right-angled triangles and equilateral ones.

Heisenberg felt these notions were "wild speculations," yet he remained fascinated by them. When he returned to the subject in later years, it became clear that what he had largely valued and seen in Plato was an alternative to the materialist philosophy which was then threatening to engulf Germany. For this young volunteer against Marxists and communists, Plato was the answer to Democritus' materialism; Werner's reading of the Platonic text offered hope for his spiritual salvation. In Plato "the smallest parts of matter are not the fundamental forms."[185] As Heisenberg recalled, Plato, "disliked Democritus so much that he wished all his books to be burned."[186] By founding Empedocles' elements on the Pythagorean regular solids, Plato had surmounted the materialism of Empedocles: "The elementary particles in Plato's *Timaeus* are finally not substance but mathematical forms."[187] Similarly, Heisenberg wrote: "In modern quantum theory there can be no doubt that the elementary particles will finally also be mathematical forms. . . ."[188]

Thus, the young Werner Heisenberg sought an answer to his inner restlessness and the restlessness of the time:

> The restlessness remained, and became for me a part of the general restlessness which had taken hold of the youth of Germany. When a philosopher of Plato's stature believed that he recognized structures in the occurrences of nature, which have now been lost to us . . . , what then does the word "structure" or "order" even mean?

Was the notion of order itself always relative to a particular historical period? "We had grown up in a world which had seemed well ordered. Our parents had imbued us with the bourgeois virtues, which constitute the preconditions for the maintenance of that order. . . . Now, however, the old structure of Europe had been broken by defeat." Which way should he go? Were the communists right? Were those "who had given up their lives in the streets of Munich to prevent a return of the old-style order"? Were those right who wanted to proclaim a new order, above nation, embracing all humanity? "In the minds of the young people these questions surged back and forth, but the older ones could no longer give us the answers," Heisenberg writes.

Thus, the problem of atomic structure was much more than an academic issue for the young Werner Heisenberg; the definition of its character and order seemed to pose in microcosm all the anxious alternatives of the disorderly social world. Writing about the collapse of conceptions of order, Heisenberg has said, "I almost suffered physically because of this." He listened to student debates, and thought about the concept of order. One day he heeded a call to go to a youth meeting at Prunn Castle. The convener spoke ominously, in a tone "which until then I had not heard." The youth were going to decide for themselves how things should be. More speeches. Was the fate of our people or of humanity more important? Had defeat rendered meaningless the sacrifice of death? Was inner truthfulness more important than adherence to traditional forms? Most of those in the courtyard were schoolboys, but others were veterans of the trenches. In this world of sociological indeterminacy, amid the collapse of social systems and values, the emotional perspective underlying Heisenberg's principle of indeterminacy was conceived.

The social world of the young Werner Heisenberg was indeed one in which outcomes were unpredictable; sovereignties changed hands, ruling classes were replaced, Jewish prime ministers ousted Christian kings, and the young dictated to the old. Munich in 1919 was far removed from placid Copenhagen or serene Zurich; it was the signal-post of unleashed aggression.

The section of the youth movement to which Werner Heisenberg belonged was the White Knight circle (*der Weisse Ritter*).[189] Its leaders, all of them either students or recent university graduates, were staunch antidemocrats who were contemptuous of humanist and cosmopolitan ideals and the idea of progress. They were authoritarian and anti-Semitic. Their chieftain Martin Voelkel, a young minister from Berlin, "thought it 'intolerable' that a Jew should lead a group in the *Bund*, and another spokesman predicted a terrible fate for German Jews unless they left the country of their own volition." Walter Laqueur, a historian of the German Youth Movement, writes:

> The era of the 'White Knight' is likely to be seen as a negative phase in the history of the German Youth Movement. Youthful idealism, personal integrity, and *elan vital* were in abundance, but these were matched by an excess of hazy and high-flown word-painting, of extravagant and vaporous romanticism.[190]

A close friend of Heisenberg's recalled him in this atmosphere while studying at Göttingen under the eminent physicist Arnold Sommerfeld:

He looked even greener in those days than he really was, for, being a member of the Youth Movement, the moral idealism of which greatly attracted him, he often wore, even after reaching man's estate, an open shirt and walking shorts. He always considered himself constitutionally lucky and this was quite true.[191]

The youth movement indeed gave him a buoyancy and a sense of being among the regenerating elite; moreover, it counterposed the adored myths of medieval romance to the cold, colorless materialism of the Marxists. These youths talked of knights and castles, swords and armor, chivalry and the Holy Grail, instead of class struggle. Their leader, Voelkel, did homage to the Trinity, "God, the I, and the weapon," with which one served the Holy Grail. An apocalytic emotion prevailed, a sense of a proper twilight for younger gods: "A great people ought to have a beautiful and wise end."[192] The White Knight circle, moreover, was exclusively for males; these were no Olympians wandering into lectures with Slavic girls; after all, a female knight was rather hard to conceive. Lastly, here was a circle in which the individual felt himself called to the service of the totality; in a world of lost identities and anomies, he acquired a sense of his role in the historical drama. But the high-flown lines and "stage props" concealed an underlying ugliness; the White Knight circle blended with the Nazis, who in the early 1920s had only a small group in the Reichstag, but led the student votes in many universities.

Munich was from 1918 to 1920 a city of sociological indeterminacy. In November 1918, Kurt Eisner, the socialist journalist, with a detachment of sixty men put an end to the centuries-old hereditary monarchy of Bavaria.[193] Small groups of determined men seemed to have the power to deflect social processes in unpredictable directions. What was expecially unique in the Bavarian revolution, furthermore, was that "unlike its parallels in Berlin and other centers, it was a revolt of the intellectuals," who felt themselves able to make history as a socialist elite notwithstanding the obstacle that most of the people wished neither to end the monarchy nor to establish socialism. And, moreover, it was in Werner Heisenberg's own beloved neighborhood of Schwabing that the revolution was conjured. Schwabing, the northeastern quarter of Munich, was "the gathering place of artists and litterateurs, the Latin Quarter of Munich bohemians," who lived in opposition to the "ties of convention and conformity.' " Schwabing handled past history with an aplomb and disregard that had never before been seen. The avant-garde in art turned out to have its avant-politics.

> Writers, painters, and draftsmen vied with one another in caricaturing and lampooning the state, the church, king and God, fatherland and family, officers and officials—everything, in short, that stood for authority and morality. . . . The most radical were the Jewish writers, most of whom adhered to communism or anarchism. . . .[194]

These artists and writers, turned revolutionaries, were men in cafés. Politically, the most important was the Café Stefanie on the Amalien-strasse "where the 'general staff' of the revolution used to assemble": Eisner, Mühsam, Toller, and Landauer. Their denigrators used to call it the café "Megalomania" and mocked its habitués as dreamers and fools; they little expected that the dreamers would engineer the downfall of the ruling elite. The guardians of the establishment could scarcely credit the Schwabing intellectuals with such pretensions to exchange their café chairs for ministerial ones. The novelist Mühsam, whose spiritual ancestors were the anarchists Bakunin and Max Stirner; the journalist Eisner, who was on bad terms with even his own party; the soldier-poet Toller, who declaimed with wild images of ideals and new men; of such might be the kingdom of heaven, but not of earth.

Brief was the moment of power that was vouchsafed the Marxists, anarchists, and communists in Munich. Kurt Eisner's government was overwhelmingly repudiated in the election of January 1919. Eisner himself was assassinated on February 21, by a nationalist student activist.[195] Then, emulating Lenin's example, the revolutionaries "seized power", in April 1919, proclaiming a Bavarian Soviet Republic. Their "power", however, dissolved. Within the soviet leadership the proponents of terror replaced the gentler voices of Mühsam and Toller. Frightened, the Bavarian middle class responded with a violent, anti-Semitic counterterror. The so-called Thule Society organized groups of volunteers, arming them and coordinating their actions. The *Einwohnerwehren*, the residents' armed forces, was chiefly made up of students—young warriors against Marxism and materialism. Some historians have tried to explain the birth of Hitler's Nazi movement as the consequence of the events of 1918–1920 in Munich. "Without Eisner, no Hitler," they say; "without the soviet republic, no anti-Semitic Nazi dictatorship."[196] Such a sociological law, in which the reaction so far exceeds the action, may afford some comfort to self-extenuating

Nazis who find it convenient to abandon their voluntarism for determinism in attempting to assuage their guilt feelings. It would be more factual, however, to say that without the Munich of 1919, Werner Heisenberg would not have conceived the principle of indeterminacy.

During those hectic days in Munich, it seemed almost that chance decided the side of the barricades on which one would be found. The poet-soldier Ernst Toller was only a few years older than Werner Heisenberg. He too joined student groups, one of which was the so-called Young Germans' Cultural and Political Union. The members of this group had also experienced disillusionment with their elders, but they were led to a rupture with the old system which was far more radical than that of the White Knights. As Toller told:

> Our League was still banned and so too were all the other little groups which had been formed at other Universities. But our society was a portent. We had started to revolt against war. We believed that our voices would be heard the other side of no-man's-land and that the youth of every country would join us in our fight against those whom we accused as the originators of the War: our fathers![197]

Toller ridiculed those who tried to return to the "medieval Mysteries which were to reawaken community feeling," those who felt themselves "like medieval knights, missionaries of the Holy Ghost." Of course, as a child, Ernst Toller had known the ostracism and epithets directed toward Jews; he had also served at the front. He became commander of the army of the Bavarian Soviet Republic. Werner's revolt took a different form—some might call it a counter-revolt.

Werner Heisenberg could not accept his father's orderly world, his longings for the well-regulated Byzantine cosmos. That had been shattered by contemporary events. Yet, he was obviously repelled by the Marxists and communists who jeered at the old Bavarian society. Their materialism was crude and abhorrent; the newspapers reported that they performed unspeakable atrocities on murdered prisoners in a Munich high school. Werner loved the old ways and the classics. His revolt was against the materialistic physics to which the Bolsheviks reduced everything. He would replace the hooks and eyes of the textbook, crude symbols of male sexuality, with the spirituality of Platonic forms.[198] And at the same time, he would strike at the harsh determinism of the vulgar, scientific materialists. The subjective observer would intervene in the physical order to affect its causal chains in unpredictable ways.

Friends of the young Werner Heisenberg wondered why he chose to study physics instead of music. The mother of one of his comrades in the youth movement said to him: "From the way you play and speak about music, I get the impression that you are much more at home with art than with science and technology, that you prefer the muses to scientific instruments, formulae and machinery. If I am right, why ever have you chosen natural science?" Werner replied in effect that he chose physics rather than music because he felt that it was entering a revolutionary stage: "In short, I firmly believe that in atomic physics we are on the track of far more important relations, far more important structures, than in music. But I freely admit that 150 years ago things were the other way round."[199] The subjective, psychological ingredient of participating in the opening of a "vast unexplored field," in the basic "reformulation" of natural laws, was the controlling factor in his choice of vocation, in the channeling of his revolutionary impulses. Young Werner had indeed "never been really at home in the world of [scientific] instruments"; the painstaking labor involved in making even relatively unimportant measurements struck him as "sheer drudgery." He had heard from "my own father" (as he said), often enough the insistence on care in small details.[200] But his own spirit was moved by a metaphysical urge. "I am much more interested in the underlying philosophical ideas than in the rest," he told his future professor, Arnold Sommerfeld, at their first interview. Such an interest in philosophy has indeed always been characteristic of the would-be revolutionary activist in science. For the desire to actualize a philosophical idea in scientific research involves the desire to reconstruct its intellectual basis. When a novel standpoint or isomorpheme is realized, then the system of scientific knowledge itself has been made the vehicle for the realization of new, underlying generational emotions.

At one time, the emotions of the youth movement activist nearly disrupted Werner Heisenberg's scientific vocation. In the summer of 1922, at the suggestion of Sommerfeld, young Werner journeyed to Leipzig to attend the Congress of German Scientists and Physicians. Einstein was scheduled to be one of the speakers. As Werner entered the lecture hall, a young graduate student "pressed a red handbill into my hand, warning me against Einstein and relativity."[201] "The whole theory was said to be nothing but wild speculation, blown up by the Jewish press and entirely alien to the German spirit." At first Werner thought the leaflet was the writing of a lunatic, but when he heard that its author was a renowned experimental physicist, he experienced an intense mental depression: "I felt as if part of my world were collapsing." He had thought, he writes, that "science at least was above the kind of political strife that had led to the civil war in Munich, and of which I wished to have no further part." He sensed that even a celebrated German physicist, could be a man of "weak or pathological character," who enlisted political passions to refute Einstein's theory because he could muster no sound scientific evidence. The incident upset Werner far more than the facts warranted, for his own dual loyalties to the youth movement and theoretical physics were, for once, in violent opposition: "At first, the Leipzig experience left me with a deep sense of disappointment and with doubts about the validity of science in general."[202]

But why should a proto-Nazi leaflet have so shaken Werner's commitment to science? Underlying his emotional response was Werner's own deep identification with the German people, in all their moods, exaltations, even irrationalities. He recalled how he had felt in August 1914, when, at the age of twelve he accompanied his father who was going to report for military duty. "All the railway stations were filled with shouting crowds of excited people; freight cars, decorated with flowers and branches, were packed with soldiers and guns." Never had young Werner experienced such a sense of community: "You could address any stranger you wanted to as if you were old friends; everyone helped everyone else—we had all become brothers in fate. I should not like to eradicate this day from my memory." He held fast to the feelings of "this incredible, this unimaginable day, a day no one who witnessed it could forget. . . ."[203] With the advent of war, "all petty, everyday cares suddenly disappeared" and were superseded by a "broader solidarity." Werner's best friend, a cousin only a few years older than himself, became a soldier and died in France. Now, Marxists and Jews ridiculed the patriotic idealism that had stirred Werner and his cousin. "Do you think he ought to have told himself that the whole war was nonsense, a fever, mass suggestion, and have refused this call on his life?" young Werner asked Niels Bohr. In the youth movement he recaptured that almost mystical experience of 1914. He has written of a youth festival in the Swabian Jura: "I was quite overcome by the forces generated at this spontaneous gathering, much as I was on the first of August, 1914." It was a movement which looked to its peoples' past; in anguish, they recalled their peoples' ordeal in the Thirty Years' War: "A feeling of kinship with that age seems to have seized young people all over Germany."[204] In the spirit of student movements, Werner had engaged in back-to-the-people activities, teaching astronomy in workingmen's classes.[205] And now at a physicists' meeting, he heard that Einstein was one of those who scoffed at patriotism and the war's sacrifices. Werner was torn between his patriotic identification and his scientific integrity; the two pulled him in incompatible directions. He left the conference in despair, giving up the chance offered by Sommerfeld to meet Einstein. Yet helped by Niels Bohr and the Copernhagen spirit, he retained his scientific vocation. Heisenberg's emotions achieved their fulfillment in a world that he shaped in the image of his principle of indeterminacy. The determinist dogma that the Marxists propagated and their harsh materialism were shown to be unverifiable and unnecessary doctrines of ideology. With mathematics and experiment, Heisenberg projected a physical universe that the youth movement could claim as its own.

Not only did Werner Heisenberg reject the Marxist *Weltanschauung*; With his philosophical physics, he also supplanted his father's clerical, medieval, Byzantine rigidity. In his father's household Werner had imbibed a Kantian outlook with its sharp separation of "two realms with the objective and subjective aspects of the world respectively"; the objective was coextensive with the world of natural science whereas the subjective was the domain wherein the groups to which we belonged exercised their influence, "be it our family, nation or culture."[206] Werner confessed to Wolfgang Pauli that he did "not feel altogether happy about this separation." He was moved toward fashioning a world more congenial to the youth movement's impulses, a world in which romantic idealism held true, in which the observer perforce was a participant whose decision molded physical reality itself. If the ego alone could not "posit" external reality, the physicist could still intervene and thwart in part the materialist, determinist hegemony. The world he created was one "no more material than the triangles of Plato's *Timaeus*."[207] It was a world, moreover, in which "alternatives" were even "far more fundamental structures of our thought than triangles."[208] Nature was disenthralled from the determinist demon.

The pioneering generation of physicists at the turn of the century the older Poincaré, the younger Einstein—had tended to look upon the human consciousness as an ineffectual entity. It was at most a recorder of physical events without an autonomous existence. Werner Heisenberg, the romantic activist, could never subscribe to this denigration of the human status. He argued with his physicist friends: "There is no doubt that 'consciousness' does not occur in physics and chemistry, and I cannot see how it could possibly result from quantum mechanics."[209] He listened to his ebullient fellow student, Wolfgang Pauli, meditating that as the "parables and images of the old religions" lost their force, the old ethics would collapse, and "unimaginable horrors will be perpetrated." The horrors came with Adolf Hitler; the youth movement matured into romantic sadism.

In his later years, Heisenberg tried to probe the latent errors in his own youthful philosophy of freedom. "We are without a compass and hence in danger of losing our freedom," he said. And that compass, he came to believe, lay somewhere in the notion of a "central order," a moral order, in the universe. Somehow the youth movement had erred in linking freedom to a kind of human omnipotence; it had found freedom in the defiant assertion of fantasy and illusion. In rejecting materialism it also spurned a sense of human limitation and finitude: "We Germans tend to look upon logic and the facts of nature . . . as a sort of straitjacket. . . . We think that freedom lies only where we can tear this jacket off—in fantasy and dreams, in the intoxication of surrender to some sort of utopia."[210] It was a simile that went to the heart of the romantic activist philosophy—this perception of the facts of nature as a harsh repression of our deepest impulses, this perception of man as inherently insane, and his freedom as the surrender to instinct without the control of reality or reason. The youth movement, with its mythology, had helped to form the mentality that conceived the principle of indeterminacy. But like all revolutionary mythologies, it finally revealed a self-destructive ingredient; with it came the Nazi "myth of the twilight of the gods," in which a mass freedom merged with humanity's collective suicide. Werner Heisenberg, like Einstein, in his latter years, had many second thoughts on the revolutionist's vocation. The revolutionary impulse brought its insight and also its corresponding blindness.

Every revolutionary standpoint is haunted by a penumbral feeling that its rhetoric is hollow, that somehow it has lost its grasp of reality. Heisenberg, whose scientific life had begun with the conviction of the unreality of material objects, finally sensed his own existence threatened. It was a consequence hardly foreseen that the reality of his self should be undermined as well. It was the kind of neurotic state that William James loved to depict. It came to Heisenberg in January 1937, as he stood in the center of Leipzig, engaged in the Nazi campaign to sell "Winter Aid" flags. He writes that this was

> part of the many humiliations and compromises we had to put up with at the time. . . . I was in a state of complete despair as I rattled my box . . . simply because of the utter senselessness and hopelessness of what I was doing and what was happening all around me. Suddenly I was in the throes of a strange and disturbing mental state. The houses in those narrow streets seemed very far away and almost unreal, as if they had already been destroyed and only their pictures remained behind; people seemed transparent, their bodies having, so to speak abandoned the material world so that only their spirits remained.

It had been one thing for the adolescent romantic activist in the fullness of his energies to feel that he could unseat the materiality of the world and subdue it with the ideality of mathematical form. But it was something quite different when he realized that his energies had failed to shape the world. Then it seemed unreal, not because ideal forms had conquered, but because he no longer loved that world. The emotional, libidinal basis of his sense of reality had been disrupted. The cordiality of a few people mitigated Heisenberg's "crisis of reality" to some extent. "But then I was far away again, and began to fear that so much loneliness might well prove more than I could bear."[211]

Meanwhile, however, Heisenberg's principle of indeterminacy had become part of the heritage of theoretical physics. Its discovery had indeed been the outcome of an effort by Werner Heisenberg to alleviate a youthful anxiety. The memory of the conversation he had had with Burkhard Drude, his fellow student at Göttingen, which so "disquieted" him, persisted in his deepest thinking. As we have seen, Drude had said that "it ought to be possible, in principle, to construct a microscope of extraordinarily high resolving power in which one could see or photograph the electron paths inside the atom." Werner felt compelled to show that this could not be possible, that the electron orbits were a myth. His anxiety would be alleviated only when he could show that "not even the best microscope could cross the limits set by the uncertainty principle."[212]

That conversation and Heisenberg's anxiety were echoed in the language of his famous paper of 1927. The microscope of traumatic accuracy was excluded as a metaphysical figment, and the world was opened with possibilities for human choice in the very delineaments of its physical structure:

> When one wishes to be clear concerning what is to be understood by the expression "the location of an object," as for instance of an electron (relative to a given system of reference), then one must specify definite experiments with whose help one can envisage measuring it; otherwise this word has no sense. There is no lack of such experiments which in principle can allow one to determine the "location of an electron" as accurately as one wishes, for instance, one illuminates the electron and observes it under a microscope. The highest attainable accuracy in the determination of location is essentially set by the wave-length of the light employed. One will however in principle perhaps construct a gamma-ray microscope and with this be able to achieve a determination of location as accurate as one wishes. There is however for this determination an additional circumstance: the Compton effect. Every observation of the light scattered by an electron presupposes a photo-electric effect (on the eye, on the photographic plate, on the photocell). . . .[213]

Not only was the comrade Drude rebutted henceforth, but the romantic, voluntarist antideterminist creed of the youth movement was affirmed in the pages of the *Zeitschrift für Physik*!

> But in the strict formulation of the law of causality: "when we know the present accurately, then we can calculate the future," it is not the consequent but the antecedent which is false. We cannot in principle get to know the present in all determinate circumstances. Therefore all perception is a choice from an abundance of possibilities and a limitation of the future possible.[214]

In later years, Werner Heisenberg could see his work as having been linked with the restless, youthful, rebellious, and exploratory spirit of Schwabing, that Greenwich Village-like part of Munich where he continued to live. As a man of Schwabing, home of revolutions, political, artistic, and by analogy, scientific, he wrote:

> Just as the spiritual and intellectual flexibility of the Schwabing artists and their propensity towards everything new rendered fruitful and enriched life at the universities and in the whole, so this mutual acceptance of each other of the people of Schwabing has also determined the concil-

iatory character of the city as a whole, and has created the conditions for a harmonious interplay of all forces. Certainly, Schwabing was also more than just lively and tolerant. Those who have there seen the beginning of the twenties will remember it as a place of exuberant, youthful enthusiasm and joy, filled to the brim with music and poetry, sustained by the force of a few unusual people, who precisely here, were able to enchant the youth. But such years are festive times which cannot endure.[215]

The spirit of Schwabing was carefree, kindly, and liberal. But nearby too was the beer hall in which Adolf Hitler transmuted men into lower states.

The Logical Revolt Against Determinism

The impact of World War I, its dislocation of human lives and social systems, led scientific thinkers to search for modes of thought that would express their disenchantment with determinism. According to Erwin Schrödinger:

It was the experimental physicist, Franz Exner, who for the first time, in 1919, launched a very acute philosophical criticism against the taken-for-granted manner in which the absolute determinism of molecular processes was accepted by everybody. He came to the conclusion that the assertion of determinism was certainly *possible*, yet by no means *necessary*, and when more closely examined *not at all very probable*.[216]

Exner (1849–1926), a professor of physics at Vienna since 1891, had made notable contributions to the study of spectra and atmospheric electricity.[217] Exner argued in the last part of a textbook that all laws of nature were statistical in character, that there were no exact, absolute laws. His argument was much the same as that of the American Charles S. Peirce a quarter of a century earlier. He was essentially, however, a solitary voice among the elder generation; apart from Schrödinger, "other physicists seem to have completely ignored Exner's worth," writes William T. Scott. The quest for indeterminism was to be primarily that of the young. His assistant from 1910 to 1914 was indeed Schrödinger himself.[218]

In war-ravaged Poland, the revolt against determinism led toward the conception of the notion of a "three-valued" logic in which the law of excluded middle would be abrogated. Since ancient times, philosophers had held that a proposition was either true or false, that there was no third alternative (*tertium non datur*). But as the war drew to a close, the Polish logician, Jan Lukasiewicz, felt urgently that in order to promote the notion of indeterminism, it was essential to wrest from the law of excluded middle its privileged status. Lukasiewicz was also a historical scholar who had studied the writings of the Stoic philosophers. He knew that the Stoics, advocates of an ordered, determinate world-empire and natural law, had also held firmly to the logical principle that a proposition is either true or false. But he was also aware that Aristotle, pondering the example that "either there will be a sea battle tomorrow or there will not be a sea battle tomorrow," had felt the indeterminacy of things; even in antiquity it was an episode of war which affected thinkers with a sense of the indeterminate in existence. Aristotle argued that neither alternative concerning tomorrow's sea battle was true today, that propositions dealing with contingent future events were neither true nor false today. Lukasiewicz felt even more strongly a constraining, oppressive incubus in the doctrine of determinism which had to be removed at what he thought was its logical basis.[219]

In 1922–1923 Lukasiewicz explained his animus against the notion of determinism; the latter made people into puppets:

> The determinist looks at the events taking place in the world as if they were a film drama produced in some cinematographic studio in the universe. We are in the middle of the performance and we do not know its ending. . . . In it all our parts, all our adventures and vicissitudes of life, all our decisions and deeds, both good and bad are fixed in advance. We are only puppets in the universal drama.[220]

When he first enunciated these views, says Lukasiewicz, "those facts and theories in the field of atomic physics which subsequently led to the undermining of determinism were still unknown."[221] Under the influence of his antideterminist emotion, it seemed to Lukasiewicz imperative to undermine logical determinacy as the counterpart to philosophical determinism. Therefore he fashioned a three-valued logic whereby a proposition henceforth could be true, false, or indeterminate; the new logic had its own consistent rules and procedures.[222]

Not only did Lukasiewicz want to emancipate the future from the narrow, deterministic alternatives of true or false, he also rebelled against the irrevocability of the past. The past, which in reverie one longed to undo, he wished to render indeterminate, so that that which has happened, which otherwise we can only repress, would retroactively be repealed from the historical record. Lukasiewicz wrote:

> We should not treat the past differently from the future. . . . Facts whose effects have disappeared altogether, and which even an omniscient mind could not infer from those now occurring, belong to the realm of possibility. One cannot say about them that they took place, but only that they were *possible*. It is well that it should be so. There are hard moments of suffering and still harder ones of guilt in everyone's life. We should be glad to be able to erase them not only from our memory but also from existence.[223]

Thus, the antideterminist emotion saw in the three-valued logic an instrument for exorcising past events from existence; the logician was to perform the last collective rites for the erasure of the past. Militants among the younger generation of antideterminist physicists were drawn to the idea of a three-valued logic, and indeed, to the erasure of the irrevocable past. Werner Heisenberg saw an analogous alteration of the past's status issuing as one of the consequences of the principle of indeterminacy. He wrote:

> This knowledge of the past is of a purely speculative character, since it can never (because of the unknown change in momentum caused by the position measurement) be used as an initial condition in any calculation of the future progress of the electron and thus cannot be subjected to experimental verification. It is a matter of personal belief whether such a calculation concerning the past history of the electron can be ascribed any physical reality or not.[224]

It was therefore natural that Heisenberg endorsed the replacement of classical logic with one in which the law of excluded middle was basically changed: "In quantum theory this law 'tertium non datur' is to be modified."[225] Classical logic, he felt, had the same relation to quantum logic as classical physics to quantum physics; it would hold on the level of everyday objects, with which natural language deals, but not on the level of atoms and electrons. In quantum theory, for instance, if one is considering an atom moving in a closed box which is divided by a wall into two equal parts, there are other possibilities besides those two which the classical logic would allow, of the atom being in the left or right compartment. Weizsäcker, Heisenberg's younger friend, also advocated a drastic overhaul of the classical logic. Weizsäcker had at first wanted to be a philosopher and later wrote appreciatively of the existentialist Heidegger. He shifted to the study of physics when Heisenberg told him: "One can't get anywhere in philosophy now without knowing something about modern physics. But you'll have to start on physics pretty soon if you don't want to be too late."[226] The indeterminacy of the existentialist spirit merged with that of subatomic entities.

The masters of the older generation of quantum physicists, however, felt that the scrapping of the classical, two-valued logic was going too far; they felt that their own thought processes had always worked satisfactorily with "truth" and "falsehood," and that a revolution in logic was not called for. Niels Bohr, remarking that "the question has even been raised whether recourse to multivalued logics is needed for a more appropriate representation of the situation," said that "all departures from common language and ordinary logic are entirely avoided by reserving the word 'phenomenon' solely for reference to unambiguously communicable information. . . ."[227] Where an accurate terminology was employed, and where words were used to correspond to measured quantities, the two-valued logic was quite adequate for all situations. According to Bohr, the case for a logical indeterminacy rested on ambiguous language rather than on the physical facts. Max Born likewise expressed his doubts concerning the proposal for a three-valued logic: "I have the feeling that this goes too far. The problem is not one of logic or logistic but of common sense. For the mathematical theory, which is perfectly capable of accounting for the actual observations makes use only of ordinary two-valued logics."[228] And Wolfgang Pauli, who had once staunchly resisted an assault on the law of conservation of energy, also wrote that the physicist was "entitled to go on with the use of the ordinary logic," while rejecting statements about the simultaneous values of incompatible entities as meaningless. Einstein too evidently felt that the problems of quantum mechanics entailed no need for the adoption of a new formal logic.[229]

While Werner Heisenberg was rebelling against the hooks and eyes in his textbook's diagrams of atoms and Jan Lukasiewicz was seeking a logic which would go beyond "yea and nay," the philosopher Ludwig Wittgenstein, serving in the Austrian army, was composing a world-outlook that asserted: "Belief in the causal nexus is superstition."[230] The statement was a kind of manifesto against the received common sense of the sciences; for to any nineteenth-century social anthropologist, it would have been obvious that superstition was precisely the opposite, that it was founded on a denial of the causal nexus in favor of such notions as sympathetic magic. To those, however, whose experience of a world in disorder was of a traumatic character, it was an accurate psychological report to say that "belief in the causal nexus is superstition."

Wittgenstein's close friend during his formative years, Paul Engelmann, noted that the *Tractatus Logico-Philosophicus* cannot be understood apart from "the psychological conditions from which alone such thinking can spring and which must exist, though to a lesser degree, in the reader's mind as well." Wittgenstein was the scion of a wealthy Viennese Jewish family, which had been converted and assimilated into the Austrian Christian culture; most of its brilliant children were later affected by tragic emotional instabilities. Wittgenstein and Engelmann belonged to that part of the "younger generation" that on the eve of war "suffered acutely under the discrepancy between the world as it is and ought to be according to his lights, but who tended also to seek the source of that discrepancy within, rather than outside, himself." These favored children of the upper classes, highly cultivated in the arts, literature, and science, endowed with leisure, suffered guilt over their status, accused themselves, and pronounced themselves doomed; yet they continued to search for a redeeming way of life. They clung to each other, feeling that only personal loyalties were meaningful in a distintegrating world. The stances of social and political rebellion were, however, alien to them; they respected the traditional insignia of authority. Their aggressions, we might say, were less directed toward others than against themselves: "This attitude towards all *genuine* authority was so much second nature with [Wittgenstein] that revolutionary convictions of whatever kind appeared to him throughout his life simply as 'immoral.'" Wittgenstein recognized his own duty to serve in the war "as an overriding obligation"; he reckoned the pacifist opposition of his teacher Bertrand Russell to the world war as "heroism in the wrong place."[231] These young Viennese upper-class, Jewish-descended intellectuals saw themselves as impotent actors in a world that was foundering; there was a "finimundial" spirit about them, a hovering consciousness of the world's end. Karl Kraus, a member of this circle who exerted a decisive and lasting influence on Wittgenstein, wrote a book entitled *Die letzten Tage der Menschheit* (The Last Days of the Human Race).

Linked to the revolt against the causal nexus, which was seen as part of the emotional perspective of a disintegrating world, was the feeling that the basis for moral values had collapsed. Wittgenstein's aphorisms on ethics articulate this emotion.

> If there is any value that does have value, it must lie outside the whole sphere of what happens and is the case. For all that happens and is the case is accidental. . . .
>
> And so it is impossible for there to be propositions of ethics.
>
> It is clear that ethics cannot be put into words.
>
> Ethics is transcendental.[232]

In this disintegrating world, one withdrew into one's self and found refuge there. One cherished one's self in a narcissistic spirit and channeled resentment or aggression toward chaotic existence through a posture of mutism. In a wartime notebook Wittgenstein wrote: "What has history to do with me? Mine is the first and only world!"[233] In silence one awaited the summons of God; such indeed was the meaning of Wittgenstein's assertion that ethics is transcendental:

> When my conscience upsets my equilibrium, then I am not in agreement with Something. But what is this? Is it *the world*?
>
> Certainly it is correct to say: Conscience is the voice of God.[234]

Wittgenstein had his private catechism:

> What do I know about God and the purpose of life? . . .
> That my will penetrates the world.
> That my will is good or evil.
> Therefore that good and evil are somehow connected with
> the meaning of this world.
>
> The meaning of life, i.e. the meaning of the world we can call God,
> And connect with this the comparison of God as a father.[235]

This, for all its eloquence, was in the realm of the unspoken. "*It is clear* that *cannot* be expressed!"[236] The values of life were "unutterable."[237] As Engelmann observes, Wittgenstein felt that "all that really matters in human life is precisely what, in his view, we must be silent about."[238]

When Wittgenstein averred that ethics was in the realm of the inexpressible, he joined those who held that rational human action was out of the question. All doctrines of the unutterable find themselves embarrassed by a plethora of utterances: the conscience, the voice of God (the unuttering one believes) utters so much, and what is more, so many contrary things. At one moment, the unutterable told Wittgenstein to go to Palestine and join the Zionists; at another, it bade him go to Russia and work for Stalin ("the idea of a possible flight to Russia . . . keeps on haunting me"); at another time, it called on Wittgenstein to become a schoolmaster.[239] Was his ethics really "unutterable" or was it rather that Wittgenstein wished to repress the causal sources of his judgments? Wittgenstein's procedure, his friend Engelmann observed, was "anti-psychologistic."[240] There were ingredients in his own psyche that he wanted to repress, that is, render "unutterable". Quite apart from any "transcendental" source, these were evidently determining or conditioning his ethical standpoint. In 1920 Wittgenstein wrote: "I have had a most miserable time lately. Of course only as a result of my own baseness and rottenness. I have continually thought of taking my own life, and the idea still haunts me sometimes."[241] The theme of suicide was recurrent in Wittgenstein's correspondence.[242]

With the transcendental conscience at such odds with itself, imparting so many contrary instructions, one might have expected that some sort of discussion of conflicting claims, causes, and movements would have been in order. Otherwise one simply acted blindly or impulsively, allowing chance to decide whether one became a worker in Stalin's totalitarian system or a dweller in an Israeli kibbutz. To ascertain whether ethics must finally involve a transcendental source, at least a preliminary psychological analysis was required. For the most convincing evidence that ethics is transcendental in origin could be provided only by showing that the psychogenetic account of conscience is inadequate, that it leaves unexplained (possibly inexplicable) lacunae. Then, within the undetermined interstices, a meaning, not of this world, could assert itself.

But the emotions begotten during World War I, with its social climate of disorder, made anticausalism attractive to many young intellectuals. Their emotional a priori rejected a world made up of event-sequences conforming to causal laws. They joined in an isoemotive revolt against determinism.

The Dadaist Revolt against Determinism

The arts were probably the most readable barometer of the revolt against determinism during the war and postwar period. This was especially evident in the turbulent emotional assertions of the Dadaist movement. Here, almost as if in a pure laboratory culture, was the emotional nucleus of the rebellion against causality, order, and logic. "The dada movement," writes Robert Motherwell, "was an organized insulting of European civilization by its middle-class young."[243] The sons of the Zurich bourgeoisie, mostly university students, used to go in 1916 to the Cabaret Voltaire at Number 1 Spiegelgasse, where "nightly orgies of singing, poetry, and dancing" took place, and where Hugo Ball, a poet and conscientious objector who had fled from Munich, improvised the entertainment.[244] Diagonally across the street lived an obscure, undistinguished-looking Russian revolutionist named Lenin; but the Swiss authorities were "much more suspicious of the Dadaists, who were after all capable of perpetrating some new enormity at any moment" than the studious Russian who haunted the library, reading books on economics and philosophy.[245]

The Dadaists exalted chance. Only later did they discover that philosophers and scientists, as Hans Richter says, "were facing the same intractable problem at the same time." "The official belief in the infallibility of reason, logic and causality seemed to us senseless —as senseless as the destruction of the world. . . . We had adopted chance, the voice of the unconscious . . . as a protest against the rigidity of straight-line thinking."[246] Dada, furthermore, had an emotional revulsion against the ordinary, two-valued logic: "Dada took hold of something that can neither be grasped nor explained within the conventional framework of 'either-or.' It was just this conventional 'yes/no' thinking that Dada was trying to blow sky-high."[247] Without any program, indeed at war with all programs and fixed goals, Dada was directionless and ready to move spontaneously in any direction. "This absolute freedom from preconceptions was something quite new in the history of art." They were traditionless, devoid of any sense of generational continuity, "unburdened by gratitude (a debt seldom paid by one generation to another). . . ."[248] "Dada was the freedom-virus, rebellious, anarchic and highly contagious. Taking its orgin from a fever in the mind, it kept that fever alive in new generations of artists."[249]

The young Dadaists felt themselves in rebellion not only against the bourgeois world of their fathers; they also rejected, as we might say, the world of their elder brothers. They wanted to smash the forms of art of the slightly older revolutionaries on which they had been nourished: cubism, futurism, and simultaneism. "The real question was the destruction of values."[250] They felt, as young physicists would soon begin to feel, that the very concept of reality had to be redefined. The rules of perspective used to define the world and the tubed colors to set it down "ran at the heels of things," they said, "and have given up the actual struggle with life; they are shareholders in the cowardly and smug philosophy that belongs to the bourgeoisie." When in this spirit, Picasso gave up perspective, "he felt that it was a set of rules that had been arbitrarily thrown over 'nature,' the parallels which cross on the horizon are a deplorable deception. . . ."[251] In a similar mood, young Werner Heisenberg rebelled not only against the hooks and eyes of the high-school science textbook but also against the notion of the electronic orbits in Bohr's model of the atom. What evidence, asked Heisenberg, was there that electrons traversed orbits around the nucleus like Kepler's planets around the sun? Weren't these orbits a self-deception, a set of images and rules arbitrarily imposed on the observed phenomena? Likewise, the Dadaists craved "a new, direct reality."

Dadaism was the direct expression of a generational insurrection, unrefracted through any political movement or ideology. Its founders aptly named it "Dada" after a children's word they found in a German-French dictionary meaning "wooden hobbyhorse." Hugo Ball had set up a variety show with a new singer at the Cabaret Voltaire; he needed a name, and together with Richard Huelsenbeck, found one. Corresponding to the emotional vector of the time, the name spread rapidly.[252] In Germany especially, in such cities as Berlin, where one never knew where his next meal was coming from, where "fear was in everybody's bone," Dada had its greatest success; it became the terror of the population, which perceived, not inaccurately, that Dada was intent on "smashing the cultural ideology of the Germans," that its votaries were skeptics who were discarding meaning and truth as obsolescent. "Revolution, Dadaism, Nihilism, Action, Gramophone," all of these blended in an amorphous stew of disintegrating civilization. Germany "was seized with the mood that always precedes a so-called idealistic resurrection. . . ."[253] But the resurrection proved to be a regression to the tribalism of the Nazis. When a few years later, Hugo Ball entitled his autobiography *Flight from Time*, he was expressing a motif of regression from ordinary reality—the rejection of the bourgeois reality.[254] It was an age of manifestoes, as when Tristan Tzara, its most articulate spirit, proclaimed in 1918:

> Science throws me off as soon as it pretends to be a philosophical system; for it then loses its useful character. . . . I detest that fat objectivity and harmony with which science finds all in order.[255]

Manifestoes are documents that proclaim the novel emotions of the new generation, their longings, their sense that a rebirth is taking place, that literature, art, science, and politics are given new life. "We were all in our twenties and ready to defy all the fathers in the world in a way that would rejoice the heart of Freud's Oedipus," writes Hans Richter.[256]

Clearly, an isoemotional line linked Heisenberg's indeterminacy to Dadaist striving. "The central experience of Dada, that which marks it off from all preceding artistic movements," writes Hans Richter, was the conviction that chance was the basic stimulus to artistic experience, that it took one closest to the voice of the Unknown within ourselves:

> The new experience gave us new energy and an exhilaration which led, in our private lives, to all sorts of excesses; to insolence, insulting behavior, pointless acts of defiance, fictitious duels, riots—all the things that later came to be regarded as the distinctive signs of Dada. . . . Chance became our trademark. We followed it like a compass. We were entering a realm of which we knew little or nothing. . . . Chance, in the form of more or less free associations, began to play a part in our conversations. . . ."[257]

The Dadaist Hans Arp declared: "The law of chance, which embraced all other laws and is as unfathomable to us as the depths from which all life arises, can only be comprehended by complete surrender to the Unconscious."[258] "All this grew out of the true sense of fellowship that existed among us, the climate of the age, and our professional experimentation."[259] Such was the isomorphic bond between Werner Heisenberg and his youth movement and the Dadaists at the Cabaret Voltaire—the indeterminacy that the young physicist sought in freedom from hooks and eyes, and the Dadaist pursuit of chance.

Within a few years of its inception, the Dadaist movement was absorbed into the surrealist movement. The underlying motivations remained the same—protest against the older generation and rejection of their bourgeois world, its values of order, and its outlook of determinism. The surrealist Artaud declared that surrealism "meant for him pure moral revolt, especially against the coercions of the father." When André Breton, their ideological chieftain, later defected to the communists, Artaud said "the sons of the Surrealist revolution were evidently bent on devouring their 'father.' "[260] In 1922, however, Breton was urging the revaluation of classical conceptions of art and forcefully demanding the repudiation of the merely representational. The efforts of representational art to conform to the familiar patterns of the common-sense, visual world were to the surrealist what models of the classical quantum physicists were to Werner Heisenberg and the new generation; the latter repudiated the efforts to picture the electrons as moving in planetary orbits around the atomic nucleus. André Breton called similarly for the liberation of artists from the limitations of the laws of perspective, gravity, and extension. It was part and parcel of their project to "overturn the foundations of our stupid bourgeois culture," and to "revolutionize our moral and social values." The quest was for a "super-reality."[261]

As Heisenberg was disaffected from the everyday realities of material objects and longed to see them superseded by spiritual mathematical forms, that is, Platonic ideas, so the surrealist strove to "burst the bonds of reason of narrow rationalism," to permeate the world of reality with that of dreams to engender a "super-reality." The classical forms of expression were exhausted; "poetry was dead," "the novel was dead."[262] This younger generation wrote manifestoes, founded new magazines, excommunicated the deviating and recusant, indulged in extravagances and fantasies of violence: "The simplest Surrealist act," wrote Breton, "would consist of going out into the street, revolver in hand, and firing at random into the crowd as long as one could."[263] For in generational movements among artists, unlike those among scientists, the underlying emotions of rebellion state themselves in the rawest, crudest, most manifest, and least sublimated way. Rebellious artists, trying to be most expressive of their emotions, are the maximalists of the unconscious. They are utterly indifferent to the notion of a common-denominatorial world that will test their programs as older ones have been tested. The arts, insofar as they are expressive, are not bound by a criterion of correspondence. Yet the underlying animus of the artists in revolution was similar to that of the avant-garde of scientists. Tristan Tzara, "the great panjandrum of international Dadaism" (as Matthew Josephson called him) said that his technique for writing dadaist poetry was to take a newspaper article, tear it into tiny bits, throw them into his hat, mix them—"then take them out in any order, and presto, there is my poem!"[264] It was all outlandish, this exaggeration of a principle of aesthetic indeterminacy.

The unique convergence between the standpoints of physics and painting was noted by Marie-Antoinette Tonnelat, a collaborator of Louis de Broglie:

> One thus arrives at this quite paradoxical conclusion: a positivism strictly observed which, by a bad play on words, we could characterize as surrealist, can not reach a physically permanent and autonomous reality. In physics as in painting, surrealism denies the possibility of a description which doesn't carry explicitly the stamp of the observer.[265]

Thus in their emotional compulsion to lay bare a world of chance and indeterminacy that would be more real than the everyday order, the proponents of Dadaism and surrealism were brethren in spirit to the young, rebellious seekers in atomic physics.

The functionalist movement in architecture, too as inaugurated by the young Swiss, Charles Edouard Jeanneret, Le Corbusier, was also remarkably isoemotional with the contemporaneous mode of physical theorizing of Heisenberg. "A great epoch has begun! There exists a new spirit!" proclaimed Le Corbusier in his manifesto of 1923, *Vers une Architecture.* Throughout this work, "there is this same revolutionary fervor—the emotional appeal, the audible rolling of drums."[266]

Le Corbusier shared the same antipathy that his young physicist contemporaries had for "mythologic" decorative detail. To Heisenberg and Pauli the electronic orbits were "mythologic"; they longed to dispense with the inessential and to develop a physical theory in which only "observables" appeared, with a bare mathematical formalism, devoid of concessions to such visual imagery as orbits, jumps, and localized entities. Le Corbusier, likewise, said with a forerunner, Perret: "Decoration always hides an error in construction."[267] "A house is a machine for living in," he said; its beauty would be solely that derived from such a machine, not of excrescences.[268] His contempt for the "Decorative Arts," which aroused the resentment of the older school, was analogous to Heisenberg's antipathy to visualizable models.[269] Le Corbusier, like Heisenberg who waxed enthusiastic over Plato's basic mathematical forms, believed that modern industry revealed certain basic forms: "Our eyes are made to see forms in light: cubes, cones, spheres, cylinders or pyramids are the great primary forms."[270] Like Heisenberg, he became "something of an anti-naturalist." The notion of integrating architecture organically with nature, in the manner of *art nouveau* and the later Frank Lloyd Wright was something he repudiated: "The city is man's grip upon nature. It is a human operation, directed against nature."[271] The cubist Braque and Picasso (during his cubist period), had already experimented before World War I with portraying people and objects "in such a way that several ideas of the model, animate or inanimate, were visible simultaneously."[272] Marcel Duchamp, in his famous "Nude Descending the Staircase" had depicted a figure in motion in terms of several successive positions perceived simultaneously. From them too Le Corbusier absorbed a space, not contained and demarcated within fixed cubicles, but "experienced simultaneously from without and within, seen by the observer in passing through, rather than frozen to a single spot in the total composition."[273] "An implacable mathematics and physics reigns over the forms presented to the eye," Le Corbusier said.[274] It is probably not coincidental that his central emotional emphases were common themes shared with the physicist Heisenberg. If so, an isoemotional line connected the construction of physical theories with that of buildings. The physicist, as an architect of theories, was guided by emotions akin to those of the architect of houses. Was it not Charles Peirce who (following Kant) wrote of the "architecture of theories"?[275]

The Philosophical Revolt against Determinism

Bertrand Russell lived longer than any other twentieth-century philosopher and responded most dramatically to the changing emotional directions of the times. The world war, he believed, severed his previous attachment to Platonic theories of timeless essences that existed in aristocratic isolation from the hurly-burly of the ordinary world. The Cantabrigian-Bloomsbury circle, to which he had belonged, adored these eternal entities. As the war drew to an end, Russell was arriving at the view that all the entities of science were "logical fictions."[276] Russell's new emotional perspective was very much involved in this terminology. To call the objects of science "logical fictions" was to suggest that they partook of fantasy, of unreality, though they were differentiated in degree at least from those of insane persons by their logical order. The atoms, ions, and corpuscles of the physicist, Russell argued, "are not the ultimate constituents of matter in any metaphysical sense. Those things are, all all of them, as I think a very little reflection shows, logical fictions. . . ." The "logical atoms," more ultimate than anything physics could discover, were mainly "little patches of colors or sounds, momentary things . . ."[277] As the old, orderly social world collapsed, the philosopher fell back in narcissistic fashion on his own sensations; the latter were cathected as the ultimate, undeniable reals. These momentary, transient, sensory occurrences, this regressive cosmos of the infant, were what remained when the external world was destroyed, partially by its hatreds, and partially too by one's own hatred for it. The distinctive epistemological change for Russell at the war's end was, he said, that he no longer made an inference to an external world, but constructed it out his sensations. Thus epistemological, regressive narcissism was translated into logical atomism.

The revolt against determinism and order was therefore a congenial tenet for Russell in the immediate postwar years. Quantum theory, in his view, had brought such indeterminacies into its account of the world that one might well believe atoms to possess free will. In a later decade Russell ridiculed such opinions, but in 1927 when he wrote *The Analysis of Matter*, he said:

> Nature seems to be full of revolutionary occurrences as to which we can say that, *if* they take place, they will be of several kinds, but we cannot say that they will take place at all, or, if they will, at what time. So far as quantum theory can say at present, atoms might as well be possessed of free will, limited, however, to one of several possible choices.[278]

Russell at this time regarded the world with anarchist emotions. Author in 1919 of *Proposed Roads to Freedom*, admirer of the anarchist Kropotkin, visionary of a guild socialism, imprisoned rebel of 1918, Russell projected his anarchical rebellion into the nature of physical reality. The relativist cosmology, he claimed, was isomorphic with anarchist principles. "In the old theory," he wrote, "the sun was like a despotic government, emitting decrees from the metropolis; in the new, the solar system is like the society of Kropotkin's dreams, in which everybody does what he prefers at each moment, and the result is perfect order." "If nature as protrayed by Einstein is to be our model, it would seem that the anarchists will have the best of the argument," he wrote again in another book.[279] The name of Peter Kropotkin, the anarchist ideologist, had never appeared before nor has it appeared since in a treatise on physics. Thus, logical atomism was isoemotional with logical anarchism.

Philosophies can usually be located in intellectual space by their respective coordinates, that is, their favorite adjectives, nouns, verbs, and connectives. Bertrand Russell's favorite adjective in the postwar years became "higgledy-piggledy." The world, he affirmed in an essay on science, is a "higgledy-piggledy and haphazard affair . . . the universe is all spots and jumps, without unity, without continuity, without coherence or orderliness or any of the other properties that governesses love."[280] In another work, an essay on human happiness, he wrote: "The world is a higgledy-piggledy place, containing things pleasant and things unpleasant in haphazard sequence. And the desire to make an intelligible system or pattern out of it is at bottom an outcome of fear, in fact a kind of agoraphobia or dread of open spaces."[281] In a social world which had lost its moorings, where the classical Victorian structure of values, government, and order was irreparably shaken, the universe seemed pervaded by indeterminacy, by the higgledy-piggledy. If Dada returned to the language of the nursery to express its emotional standpoint, so did the logician Russell.

There was, of course, a deep generational difference between the indeterminacies of the middle-aged Russell and those of the young Dadaists and Werner Heisenberg. The latter found in their indeterminacy a sense of liberation and personal freedom; they felt a sense of open alternatives before them and rejoiced in their community of youth replacing the old. Russell, on the other hand, was filled with an overriding bitterness against his own generation; they had failed, and he hated them and wished they were dead. Yet it was at the same time his own world that had collapsed, and he could never look forward, as the young could, to another one. "The War of 1914–1918," he wrote, "changed everything for me." As an intense anger against humanity and its leaders was ignited in him, he wished the world smashed into smithereens; unable to accomplish this physically, he did so intellectually in his philosophy of logical atomism. "For several weeks I felt that if I should happen to meet Asquith or Grey I should be unable to refrain from murder." When this personal animosity subsided, he wrote in 1916 to a close friend: "I hate the world and almost all the people in it. . . . I hate the planet and the human race,—I am ashamed to belong to such a species. . . ." He felt in 1918 that the time was like that of the barbarian invasions, and that he was like a Roman philosopher bewailing the sack of Rome—the onset of the Dark Ages. He found no solace in the newly organized Soviet Union, which he visited in 1920, writing back to England that he "loathed the Bolsheviks," "an aristocracy . . . insolent and unfeeling, composed of Americanised Jews." His feelings bespoke the aristocrat of the old order who witnesses upstarts and parvenus, "a mixture of Sidney Webb and Rufus Isaacs," replacing his peers. He had left England in despair, feeling like a ghost, he said; soon, curiously, he was finding that the new world which the physicists were describing was equally ghostly: "Now, owing chiefly to two German physicists, Heisenberg and Schrödinger, the last vestiges of the old solid atom have melted away, and matter has become as ghostly as anything in a spiritualist seance." Matter, in the sense of "things," he wrote, "has been replaced by emanations from a locality—the sort of influences that characterize haunted rooms in ghost stories."[282] He found no worthwhile ideals in the decade after the world war to replace the old ones he had had. Science, which he had once regarded as the great redeemer, housed a "corrosive solvent" that promoted "destructive criticism," but undermined at the same time the scientific spirit itself.[283] The world of science, like the social world, was becoming shattered, disorderly, and disunited.

Russell could not, however, always remain a believer in the "higgledy-piggledy" universe. When a philosophical tenet projects a personal emotion on the nature of existence, it will be modified or abandoned as the personal emotion changes. For Russell, the "higgledy-piggledy" universe dramatized his despair; in a world of chance and accident, the human will was impotent. If man was no longer the epiphenomenal, impotent spectator of a determinist collocation of atoms (the image of *A Free Man's Worship*), he was now an equally impotent puppet of chance movements. To Russell's annoyance, however, such distinguished scientists as Arthur Eddington made popular an interpretation of quantum physics which revived the alternative of optimism. Eddington based his notion of a human free will precisely on the physicists' finding that there were atomic indeterminacies. Eddington's argument was logically much the same as Russell's remark in 1927 that atoms behaved as if they had free will. But Russell disliked finding himself in the company of optimistic, religious-minded scientists. Also, he felt scientists were adopting such standpoints chiefly because they were afraid of the communists' materialism. Not only therefore was Russell's will to tragedy disturbed by the notion of a world of open possibilities but, in addition, his own recurrent sympathies for communism were offended.

In the 1930s Russell veered again toward determinism:

> I do not believe that there is any alchemy by which it [the theory of probability] can produce regularity in large numbers out of pure caprice in each single case. . . . We cannot accept the view that ultimate regularities in the world have to do with large numbers of cases, and we shall have to suppose that the statistical laws of atomic behavior are derivative from hitherto undiscovered laws of individual behavior.

Russell was prepared to concede that Eddington's view was "abstractly possible"; nevertheless, he felt that it was much more likely "that new laws will be discovered that will abolish the supposed freedom of the atom." Russell charged that Eddington was trying "to arrive at emotionally agreeable conclusions from the freedom of the atom."[284] But Eddington could clearly respond that Russell was trying to arrive at conclusions that were emotionally agreeable to himself, probably because they were disagreeable to most others and conformed to Russell's awesome attraction to a world of impotent but defiant Byronic men forever defeated.

Thus, the philosophical ideas of indeterminism and determinism waxed and waned in accordance with the emotional climate; such intellectual attractions were far more personal, extralogical responses than they were objective judgments based on evidence. And in the postwar period there was a heightened emotional propensity to see the world as inherently unpredictable.

Indeterminacy as the Premise of a Generation

A generational cleavage arose in the discussion that was provoked by Heisenberg's principle of indeterminacy. The older generation of physicists maintained that its import was purely epistemological; it showed that scientists using light-beams to locate the positions or measure the momenta of electrons could never attain an accuracy that would undo the minimal, unpredictable effects of their measuring instrumentality on the object measured. But that seemed to the elders an extension of older theorems concerning the limits of the resolving power of microscopes. It appeared to them that Heisenberg was simply tracing in his thought-experiments the limitations of his own ideal "electronic microscope." Determinism still seemed to them the working postulate for the theoretical description of nature. A limitation inherent in observational methods, by virtue of their own physical characteristics, should not, they felt, be projected upon physical reality itself. The younger generation, however, drew no such distinction between the respective domains of observable and objective reality. They argued that they had learned Einstein's lesson better than he had; they were the pupils who had grown up with their master's doctrine. Einstein had discarded absolute space because velocities within it could never be ascertained by experimental procedures; they were doing the same with the dogma of physical determinism. Every deterministic explanation presupposed that an exact specification of the initial conditions of objects was possible—their positions, their velocities; Heisenberg had shown that there were limits to such knowledge in the present. With an inherent indeterminacy in the statement of initial conditions, no deterministic explanation was possible. Members of the older generation were realists; those of the younger generation were phenomenalists. As Louis de Broglie described this conflict of generational opinion: "Faced with this complete overthrow of our traditional conceptions which appeared firmly established, a number of physicists, especially amongst the most eminent of the older generations, have declared that it seemed to them impossible to abandon the principle of physical determinism." But this adherence to the principle of determinism, even for elementary entities, "does not appear at all to appeal to many other physicists, particularly amongst those of the younger generations."[285]

This difference of opinion was evidently then as much as an affair of generational style, emotion, and philosophy as it was of a scientific question. Heisenberg's principle of indeterminacy, however, finally achieved an intergenerational acceptance because, like Einstein's principle of relativity, it showed itself to have an immense heuristic power. The elimination of absolute space was accepted by scientists of the most diverse epistemological creeds because the new ideas were so fecund with empirical suggestion; for many, acceptance of the epistemological outlook followed on the predictive successes. In a similar fashion Heisenberg's principle of indeterminacy proved itself a remarkable key for unlocking truths in atomic physics. With its use one could derive such consequences, for instance, as that a particle confined to a small space cannot have zero kinetic energy; one could calculate with its help the order of magnitude of the hydrogen atom and the spread in the wave-length of the light emitted from decaying atoms. From an ultimate philosophical standpoint Heisenberg's principle was compatible with either philosophical determinism or indeterminism. That it was welcomed from the outset by the younger generation was, to begin with, as much an indication of the direction of its emotions as the then weight of experimental considerations. An unarticulated generational premise in this sense was at the basis of the reception of the philosophy of physical indeterminacy. It expressed the isoemotional line of a generational experience, the new emotions of rebellion. Yet it is also true that the intergenerational acceptance of the principle of indeterminacy was finally founded on its empirical predictive and heuristic power. The latter provided the common ground on which all the generations of the scientific community could function as a constitutional body.

Nonetheless, without this emotional, generational, sociological a priori, Werner Heisenberg would not have been led to formulate the principle of indeterminacy. The physicist Leon Rosenfeld declared: *L'idéalisme d'Heisenberg relève de la sociologie, non de la théorie de la connaissance* (The idealism of Heisenberg arises from sociology, not theory of knowledge). He saw no epistemological significance in Heisenberg's reiterated idealism; *c'est le shibboleth d'une classe*, he said.[286] But it was not "the shibboleth of a class"; it was the philosophy of a generation.

That a sociological factor entered into the derivation of a philosophy of indeterminism from Heisenberg's principle of indeterminacy was evident a very simple consideration. Ever since Auguste Comte's time, psychologists and sociologists have been aware that a strict determinist account of human experience was, from an observational standpoint, forever precluded. According to Comte, that was why "interior observation" had never led to a science of psychology: "Interior observation gives rise to almost as many divergent opinions as there are individuals believing themselves to be having it."[287] Comte's argument stated in its generality affirms: Whenever we try to observe our psychological processes as contemporaneously with them as we can, our act of observation alters the magnitude and direction of the observed process; when we try, on the other hand, to observe that process after a time-interval has elapsed, that is, in memory, then we find that the dimensions of the immediate qualitative character of the experience can no longer be precisely recaptured. A haze of refractive experiences has intervened. The logic of Comte's argument, as Niels Bohr acknowledged, was exactly that of Heisenberg's principle of indeterminacy. Bohr's teacher Høffding had also expounded Comte's principle in his *History of Modern Philosophy*, saying that conscious phenomena, from this standpoint, "interrupt the sequence" of their associated material phenomena.[288] The Comtist indeterminacy, to be sure, has never been given a numerical precision since the dimensions of the psychic energy (if it exists) of a minimal act of attention have not yet been specified. But apart from this lack of a mathematical rendition, Comte, like Heisenberg, held that a psychological indeterminacy arose because the act of observation affected the observed process to some indeterminate extent; the process in itself became a metaphysical unknown.

Though the Comtist principle of indeterminacy was generally known and accepted at least in part, it played no role in shaping the decisions of psychologists either to accept or reject the principle of determinism. Freud for instance, was a determinist though he knew all the pitfalls of introspective observation. William James, on the other hand, believed in chance and free will. Himself a most remarkable introspective observer, James did not find congenial Comte's antagonism to introspection. He prefaced a long quotation from Comte in his work, *The Principles of Psychology*, with the remark that "a deliverance of Auguste Comte . . . has been so often quoted as to be almost classical," and that "some reference to it" seemed therefore "indispensable."[289] The passage had some memorable lines:

> Every strong state of passion . . . is necessarily incompatible with the state of observation. . . . The thinker cannot divide himself in two, of whom one reasons while the other observes him reason. . . . For all the two thousand years during which metaphysicians have thus cultivated psychology, they are not agreed about one intelligible and established proposition.

James acknowledged that introspection was indeed "difficult and fallible," but he insisted "the difficulty is simply that of all observation of whatever kind." One tried to do one's best in observation, and hoped that a consensus would emerge as later views corrected earlier ones. For James based his own belief in free will on faith in the veracity of his immediate personal experience. A wholesale attack on the reliability of introspective observation would have imperiled the basis of his philosophy. If at a moment of decision an introspective observer felt a quality of free choice, he would scarcely welcome the suggestion that that quality might be the outcome of the act of attention itself, deflecting what otherwise would have been an unconsciously determined act, and superadding an ingredient of consciousness which the observer confused with an experience of freedom. In short, Comte's principle of psychological indeterminacy was universally disregarded as irrelevant to the truth or falsehood of psychological determinism.

Why then were such momentous philosophical consequences founded on the physical principle of indeterminacy when the corresponding psychological principle had had no such impact? One is led to infer that in part an unarticulated sociological premise governed one's choice in this respect—an extralogical, extrascientific, extrarational motivation. Those who were part of the socioemotional current, the isoemotional line, in the cultural environment after World War I were moved to interpret Heisenberg's principle of indeterminacy as signifying the end of philosophical determinism. Those, often of the older generation, Einstein, Rutherford, Planck, who did not share such emotions wondered that a cosmic consequence could be drawn from a fact of atomic methodology. The emotions of different generations, with their respective historical experiences and locations in the cycles of philosophical opinions, divided them in the debate concerning the philosophy of indeterminacy that arose in the 1920s.[290]

Sociological Premises in Philosophies of Science

Premises having a sociological source thus have pervaded the discussions of the philosophy of quantum theory. For instance, the daring corollary was drawn that the classical law of identity was abrogated by the new physics. As Louis de Broglie described it: Suppose two identical twins resemble each other so much that they cannot be distinguished; then, so long as they walk separately in the town, we can follow them, and call one A, the other B; but if they enter a building from which we are barred, and then come out again, we shall not be able to identify either one as A or B. What then of their individuality? one asks. De Broglie answers:

Contemporary physics has a clear tendency to adopt a phenomenal attitude, and to consider as pseudo-problems those which cannot be in any way determined by experience. If one adopts this point of view, the question of knowing if the individuality of the particulars persists when they cannot be followed must be considered as a pseudo-problem.[291]

The physicist P. W. Bridgman similarly argued that the concept of identifiability applied only to classes of objects that are demarcated by sharp, physical discontinuities; but where we could not through experimental operations follow the histories in space-time of such objects, or where such discontinuities did not exist in the first place, then the concept of identity lost its physical significance. As Bridgman phrased it:

What physical assurance have we that an electron in jumping about in an atom preserves its identifiability in anything like the way that we suppose, or that the identity concept applies here at all? In fact, the identity concept seems to lose all meaning in terms of actual operations on this level of experience.[292]

Curiously, the arguments that physicists directed against the logical principle of identity, A is A, were very much like those the English philosopher F. C. S. Schiller had used before World War I.

That there shall be identity we have good grounds for insisting, but our claim that any A is A may often be frustrated. That therefore every attempted "identification" should come true would be the experience only of an omnipotent being. . . . Only to such a being (if such can be conceived) would it be self-evidently, invariably, and "necessarily" true that "A is A"; in our human thinking, the identities we select may prove to be mistaken.[293]

Very few persons, least of all scientific philosophers, took Schiller's argument seriously, but after World War I it was enunciated as a philosophical corollary of quantum theory by the most eminent physicists. The Aristotelian law of identity was thus circumscribed in its significance with respect to physical entities.

Conservative-minded physicists replied to the revolutionaries de Broglie and Bridgman that though we might be unable to follow the path of an electron among its neighbors, that was no reason for denying the law of identity. If we see a man in a moving mob, but then later fail to identify him, we do not say that the principle of identity has been repealed. We say that we have failed to make a successful identification, but we take it for granted that a man's identity persists, unaffected by our failure. De Broglie's response to this was (as we have seen) that modern physics takes a phenomenalistic standpoint, that is, if we cannot experimentally make certain identifications, we deny that such identities have scientific significance. To this one can reply that such phenomenalism simply begs the question, that myriads of beings—from birds in flocks, to men in mass societies, to ions in chemical mass actions—do not lose their identities because they may not, in certain situations, be experimentally discriminable for consecutive intervals.

Thus, such philosophical arguments can continue, on the purely "logical" plane with no possibility of resolution. What intervenes, however, and gives the strategic advantage to one standpoint is a sociological decision. The law of identity in physical theory became questionable, one observes, at precisely the same time that sociological writers began to dwell on the theme of mass society. There was an obvious preoccupation with the predicament of man in modern industrial society, his anomie in heavily populated centers, the interchangeability of his tastes with those of others in a mass-producing, mass-consuming society, and the amorphousness of public opinion. With the growth of mass parties, marches and parades, the individual voter felt anonymous and apathetic. Plays such as Ernst Toller's tragedy, *Masse Mensch* tried to capture this new spirit. It is likely that this isoemotional line regarding the threat to personal identity, which has become so pervasive a literary and political theme in Western civilization, provided an unarticulated sociological premise for reasonings in the philosophy of science.[294] Where previously a polemic against the principle of identity would have been rejected as utterly implausible, such a critique now seemed in keeping with the cultural climate. For it was in these years that social philosophers and critics, from left to right, liberals and conservatives alike, wrote anxiously of the "mutilated many," the "submergence of the individual," "the lost individual," and the "disintegration of individuality."[295] The mass-man, the statistical fact, had triumphed.

Louis Victor, Prince de Broglie, Aristocratic Revolutionist

In 1925 a young scion of one of the noblest families of France presented his doctoral thesis to the Sorbonne. It propounded a radical new idea, that the electrons moving in the quantum orbits were accompanied by waves whose lengths were equal to Planck's constant divided by the mechanical momentum of the electrons. Such was the hypothesis of Louis Victor, prince de Broglie, concerning the "pilot waves" associated with electrons. Hitherto, quantum theorists had sought to find a discontinuous corpuscular structure in such phenomena as the classically continuous light waves. The young Frenchman reversed the process; he perceived an underlying wave structure in the electrons in their quantum orbits. From this seminal idea grew the imposing theoretical development of wave mechanics.

To the social psychologist of ideas, the drama which was involved in de Broglie's discovery is among the most poignant in the history of science. Hitherto we have described the genesis of ideas among sons of the middle class. To such young men as Einstein and Heisenberg, the vocation for rebellion against the received system of scientific ideas came easily; it was part of the revolutionary student cultures in which they moved. But for the young Louis de Broglie, the path toward autonomous scientific creativity involved a rupture with his background, a crisis of youth, in which he had no support from any circle of comrades in the French radical, bourgeois, or positivist scientific culture.

The Lineage of de Broglie

The family de Broglie traced its ancestry to nobility of the tenth century. In more modern times, they remembered Victor-François, duc de Broglie (1718–1804), a marshal in the Seven Years' War, who, upon the outbreak of the French Revolution, fled to Germany and in 1792 became the commander of the first corps of émigrés. His son Victor-Claude, prince de Broglie (1757–1794), was for a time sympathetic to the Revolution; but when he opposed its more radical decrees, he was condemned to the guillotine by the revolutionary tribunal. He charged his own son to be faithful to the republic, unjust though it was. The orphaned son, Achille-Charles-Victor (1785–1870), became minister of public instruction in the first cabinet of King Louis Philippe, and later minister of foreign affairs and prime minister. He was a moderate liberal who worked, even during the Restoration, for a reconciliation of the old with the new France. His wife was a daughter of the brilliant intellectual, Madame de Staël. When the coup d'état of Louis Napoleon took place in 1851, the duc de Broglie retired with dignity to devote himself to historical studies. The past had always beckoned to the descendants of the de Broglie family. His son, Jacques-Victor-Albert, prince de Broglie (1821–1901), was both a notable historian and diplomat. Albert served as an ambassador to Britain. Most important in the history of the Third Republic, he was the leader of the monarchist opposition and as such had a chief part in ousting Adolphe Thiers from power in 1873. He served the republic twice as prime minister, and also as minister of foreign affairs and minister of justice. Albert tried to effectuate a synthesis between the new liberalism and the old monarchism. He feared the growing power of the centralized state; to safeguard individual liberties, he proposed to restore the local powers of communities and provinces. To temper the inexperience and impressionability of popular assemblies, he proposed the formation of a senate, founded not on the hereditary principle, but on achieved distinction. The constitutional monarch in his view, would add the moderating influence of one thoroughly identified with the long-term interests of the nation as a whole. Otherwise, a chamber of deputies, with its unlimited power, would founder in selfish factionalism. But Albert de Broglie would accept a restoration only on the conditions that it were constitutional and the right man were at hand: "If there are several pretenders," he wrote, "the part of wisdom is to prefer a Republic to a civil war."[296]

Albert de Broglie, the royalist statesman was the grandfather of Louis de Broglie. In his youth, Louis, the future physicist, felt a calling to become a historian of the Middle Ages and eighteenth-century diplomacy. He was drawn by the powerful example of his grandfather. Albert de Broglie, devoted to the values of traditionalism, to Catholic values and ultramontanism, had contested the eighteenth-century skeptical standpoint. He published a work in six volumes, *L'Eglise et l'Empire Romain au* IVe *Siécle* (The Church and the Roman Empire in the Fourth Century), directed against the rationalist view of Edward Gibbon that the triumph of Christianity over paganism was the outcome of political circumstances. His brother, August-Théodore-Paul de Broglie, had dedicated himself to the church; feeling a religious vocation, he resigned his lieutenancy in the French navy to become a priest, and wrote a long series of articles in defense of Catholic dogma against positivist science and other ideological fashions. At the time he died, he was undertaking a book to show the concordance of reason and faith.

Albert, duc de Broglie, the elder and more political brother, wrote volumes on the minutiae of the diplomatic history of the departed monarchy. To write his book, *The King's Secret: Being the Secret Correspondence of Louis XV, with his Diplomatic Agents from 1752 to 1774*, he had, with the help of the keeper of records at the ministry of foreign affairs, exhumed musty documents to supplement dusty familial archives. It was a work in which the family was a continuous wave joined with the national destiny:

I had in my childhood often heard my grand-uncle and his secret relations with Louis XV spoken of, and I felt tolerably sure, from certain indications contained in my father's papers, that the Count's correspondence with the King, as well as that of the other secret agents . . . were in existence.[297]

The monarchist had a sense of the past; his research became an overwhelming reality: "Those with whom historical research is both an instinct and a taste . . . know with what eagerness a discovery of this kind is pursued. . . ." To be sure, his honesty compelled him to acknowledge the "many weaknesses" which emerged in "the character of the chief actor in this little drama," his great-grand-uncle, the Count de Broglie; nevertheless, he felt, there remained "the always pleasing spectacle of a mind to which lofty political views were familiar, and of a soul filled with the passionate desire for the public welfare." And, of course, "miserable truths" would be narrated which to some might seem to discredit the institution of monarchy. To such, the duc de Broglie replied:

I am under no apprehension lest a true picture should harm the great memories of the French monarchy. An institution such as that monarchy, which counts ten centuries of duration and of glory, is strong enough to bear the full light of history; and its discriminating admirers (I hope I am of the number) have no interest in disguising either the faults of the sorry monarch who precipitated its fall, or the evils of that arbitrary power which too often violated its principle and impaired its beneficence.[298]

Albert de Broglie hoped that in the "garrets of more than one ancient château" eager inquirers would pursue researches which "must redound to the honor of their ancestors."

The Vocation of Scientific Revolution

Young Louis de Broglie grew up in an intensely political atmosphere, redolent with loyalties to a bygone era. His father, the Catholic and conservative parliamentary deputy, was together with his mother anti-Dreyfusard "with all their heart." "At their table, at Paris as in the country, they talked only of the 'affair'. . . ." A few relatives were indeed suspected of harboring dangerous doubts concerning the guilt of Dreyfus. In 1898 when Louis was six years old, Dreyfus was brought back from his prison cell in Devil's Island to be retried at Rennes. As Louis' elder sister recalled:

Then came the great day. I still see my mother coming into my room, her dress disordered. . . . She waved in her hand a telegram sent from Rennes . . . crying to all and sundry: 'Thank God! he has been found guilty!' . . . One might have thought that the destiny of France and the future of Christianity hung on the outcome of this trial, on the condemnation of an obscure officer, simply because he was a Jew. . . .[299]

Louis grew up a gay, happy, endlessly chattering, and adored child, dressed in navy blue. Even the severities of the dining table were relaxed in his presence. Brought up, however, in relative solitude, "he read much and lived in unreality. He talked to himself during the hours when he walked far about, putting questions and answers, inventing personages and giving them their replies." From his prodigious memory he declaimed whole scenes of French classical theater.

He seemed to have a taste for history, above all political history. Hearing our family discussing politics, he improvised speeches based on the accounts in the newspapers, and recited the complete list of the ministers, so often changing, of the Third Republic. He had a marvellous gift for imitation, and could mimic the voices and gestures of distinguished people in the funniest way.

His first teacher at home was naturally a priest, with a long beard, a former missionary, who blended with the high sense of mission of the household. Louis' father, very ill with diabetes, took his twelve-year-old son in 1904 to the Chamber of Deputies to watch a session. The famed Jean Jaurès, chief of the Socialists, was giving a speech advocating that compulsory military service be reduced from three years to two. The right wing interrupted furiously and repeatedly, and Jaurès banged on the tribune. One deputy shouted: "Long live the king!" Through it all Louis' father kept his seat serenely, never taking part in the tumult. When the family got home, "Louis mimicked the stormy sessions with astonishing verve, and proposed for discussion some grotesque laws. The Cabinet was overthrown, ministers were changed, I composed proclamations and appeals to the people that I pinned up in the passageways and distributed to the bewildered servants . . . Everybody predicted for Louis a great future as a statesman!"[300]

When Louis' father died in 1906 his mother became head of the family. Then fourteen years old and having had only a succession of priests as his private tutors, Louis was enrolled in a lycée. But he seemed to have no clear vocation though he liked to work, especially at history. Still it was hoped that with his gift for speech and his affable personality, he would yet become a statesman.[301] Certainly he did not seem to be possessed of the demon which had made his older brother Maurice defy their dead grandfather by proposing to resign his commission in the navy and become a physicist. "Science," said the old duke, "an old lady content with the attractions of old men," was not fit for a de Broglie. However, three years after the grandfather died in 1901, Maurice did take an indefinite furlough which evolved into a formal resignation, and he achieved his doctorate in 1908 at the Collège de France.[302]

In a France which had just been riven by the fierce animosities engendered by the Dreyfus affair, the honor of his ancestors must have weighed heavily on young Louis de Broglie. Though at the lycée he first studied mathematics and philosophy, he chose to devote himself to medieval history, and gave himself to paleography and the deciphering of old manuscripts. He took his licentiate, and then at the faculty of letters planned to write a thesis on the effort made by the regent circa 1717 to replace ministers with councils. But he became restless; he found himself uninterested in political history. Moreover, the teaching of history seemed to him preoccupied with trivial erudition, with an archival critique of texts and documents, dissociated from great conceptions and general ideas. The young Louis de Broglie was moved to revolt against the preoccupation with the past that was the heritage of his family. He was thrilled by the vibrant and exploring books that Henri Poincaré had just published—*The Value of Science* and *Science and Hypothesis*. When their father died, the elder brother Maurice, already a physicist of thirty-one, became responsible for supervising his education. Maurice became anxious about Louis' unstable intellectual wandering. ("I was somewhat uneasy about his fluctuation.")[303] Cast in the role of a father, Maurice probably had to endure too the emotional and philosophical rebellions of his Bergsonian brother, who knew more of the philosophy of physics than physics itself. But Louis found his way back to physical reality, and indeed to reconciliation with his brother. With his "crisis of youth" surmounted, with his rebellious impulses finding their satisfaction in a novel creative leap in the ongoing "revolution of physics," wave mechanics could also project a reconciliation of brothers. His brother described the transformation that came over Louis:

> Always it is philosophy, but this time turned toward the sciences; he approaches them then through their profound principles before having assimilated the details. History is therefore abandoned for special mathematics, directed toward the study for the certificates of licentiate in the sciences. . . . All this explains the crisis of youth through which my brother passed toward his twentieth year. Deceived by history toward which he had turned, in part drawn by the example of his nephew, in part to avoid an orientation toward the military or diplomatic careers, which in the preceding period, had seemed in his case the only acceptable ones, plunged into the uncertainties of philosophy, attracted and disconcerted perhaps by the critical examination which Henri Poincaré was making of the bases of mathematical reasonings, he hesitates and is not far from losing confidence in himself; a failure in the certificate for general physics, [curiously, on a question about alternating currents] . . . accentuates his state.[304]

The thought of Henri Poincaré deeply impressed the young Louis de Broglie. He recalls how in 1912, at the age of nineteen, he was tirelessly rereading the courses in mathematical physics and the works on the philosophy of the sciences by Henri Poincaré. Then one day as he was on the train bound for his country vacation, he read of Poincaré's sudden death: "I had the impression of a catastrophe which brutally decapitated French science at the moment when the great revolution which I felt was preparing itself in physics made his presence so necessary."[305] Drawn as he was emotionally to participate in the making of this "revolution," Louis felt reservations about the role of Poincaré as a revolutionary leader. "Poincaré's scepticism," he wrote, "could be discouraging and sterilizing," and he conjectured in later years, that perhaps that was why Poincaré had failed to discover the general character of the principle of relativity, which was the lot instead of "the younger and less sceptical mind of Albert Einstein."[306] To Poincaré it had seemed that there were always a great number of theories that could explain the given phenomena equally well, and that were thus of equal logical cogency. But Louis de Broglie, on the other hand, found this standpoint of skeptical convenience uncongenial; he wrote:

> A few examples don't suffice to prove that there are always an infinite number of possible theories for explaining the same experimental facts, and it seems certain to us that even, when there are a great number of logically equivalent theories, the physicist has the good right to believe that one of them conforms more to underlying physical reality, and is more capable of generalization, more apt to reveal the hidden harmonies.[307]

The Young Bergsonians

Most significant in shaping the mentality of the young would-be scientific revolutionist was the philosophy of Henri Bergson. "Personally, from our early youth, we have been struck by Bergson's very original ideas concerning time, duration, and movement," wrote Louis de Broglie.[308] He was fascinated from his youth by the paradoxes of Zeno, that favorite theme of Bergson. Whereas Einstein in his adolescence had been haunted by the question of what a man's experience would be like if he traveled with the velocity of light, Louis de Broglie pondered on the character and possibility of movement itself. He asked himself, as a student of Bergson, how a movement could be represented by a spatialized vector, and it was "in pressing obstinately against this obstacle," writes Maurice de Broglie, "that my brother, confronting the laws of optics with those of mechanics, arrived at formulating for the first time a relation which underlined the deep interrelatedness of waves and corpuscles in movement."[309]

In the year 1912, Henri Bergson was at the height of his fame and influence.[310] As Einstein's emotions had been stirred by Mach, and Niels Bohr had felt the attractions of Kierkegaard's rupture with continuous transitions and universal systems, and Heisenberg was to feel akin to the spirit of Plato's *Timaeus* which enunciated the primacy of geometrical forms over material entities, so Louis de Broglie was intrigued by Henri Bergson's vision of the nature of ultimate reality.

From 1900 to 1914, the lectures of Henri Bergson at the Collège de France were the center of Parisian intellectual life. The extent to which Henri Bergson led the revolt of a generation against the positivist philosophy that dominated the French universities is perhaps forgotten today. At the Sorbonne, at the Ecole Normale, versions of Comtist postivism were regnant; Emile Durkheim's sociological positivism was well-nigh an official philosophy. A new generation arose in the wake of the Dreyfus case that had had its fill of the bourgeois positivism and "scientism." The phrases of Comte and Taine, duly to be memorized for examinations, seemed to them tired, sterile, and uninspiring. Among those who came to listen to Bergson were Charles Péguy, seeking a new idealism in which to ground his impulses for a freer, nobler society; the young Catholic philosopher Jacques Maritain and Etienne Gilson; even the middle-aged Marxist, Georges Sorel, weary of the "scientific" version of Marxism, and wishing to revitalize the proletarian philosophy with intuition, instinct, the unpredictable, and the captivating imagery of a general strike as the wave of historical movement.[311] Bergson was their leader in the generational revolution of French philosophy; he was the Moses who was to lead them from a drab, pharaonic, bureaucratic scientism to a liberated intuition of freedom and creative evolution. He stood outside the official university system of France, for the Collège de France, at which he lectured, was less a degree-granting institution than a center for public lectures. There Bergson lectured as Socrates would have, not preparing students for examinations, but speaking to all and sundry, to fashionable ladies and unfashionable syndicalists, to those who frequented the market place and those who lived in lonely garrets, to those of both left and right, freethinkers and Catholics, who aspired toward a new vision. Careerist students at the Ecole Normale, their eyes and ears focused only on examination questions, ignored Bergson's lectures at the Collège de France.[312] But there were many like Etienne Gilson, the erudite historian of philosophy, who recalled in later years that philosophy to him had not signified Descartes or Kant:

It was Bergson the genius whose lectures still remain in my memory as so many hours of intellectual transfiguration. Henri Bergson is the only living master in philosophy I ever had, and I consider it as one of my greatest blessings bestowed by God on my philosophical life that, owing to Bergson, I have met philosophical genius both somewhere else and otherwise than in books.[313]

To this whole generation of young idealists, rebelling againt the cold, self-macerations of orthodox materialistic positivism, Bergson was the spiritual guide. They were republicans, but they felt the republic would perish for lack of spiritual sustenance if it remained moored to positivistic ideology.[314] The militant positivism that prevailed at the Sorbonne, not only in the natural sciences but in sociology and philosophy, seemed to them a doctrine that impoverished the spirit and smothered the idealistic resources of youth. It was Jacques Maritain who invented the word "scientism" to denote this ideology of materialism which asserted that "those things alone are intelligible which are materially verifiable," and that all things are governed by mechanical laws.[315] But one "had only to cross the Rue Saint Jacques and to take several steps down the Rue des Ecoles" to the Collège de France, and one was in a different spiritual universe. As Maritain wrote:

A mountain of prejudice and of distrust existed between these two institutions—particularly so on the part of the Sorbonne philosophers with respect to Bergson's teaching. These feelings were so strong that it was almost as difficult for the young students to think of going from the Sorbonne to the Collège de France as from the Sorbonne to the Church of Saint Geneviève, its near neighbor.[316]

The young poet-editor, Charles Péguy, was the leading spirit in helping students who were rebelling against the "historicism" of the Sorbonne to listen to Bergson's teaching. In his magnificent biography of his friend Péguy, the novelist Romain Rolland has written some moving passages on the attraction that Bergson, the man and his philosophy, exerted on Péguy and the younger generation. Bergson was for them the "most illustrious intellectual of our time," "the chief of an avant-garde of the mind." He had challenged, as no one else had, the supremacy of mechanical, determinist categories. The industrial revolution had brought in its wake immense progress, but its philosophical expression had declined into the jejune positivism of Taine and the facile optimism of Spencer. At the Sorbonne its representative was "the militant dogmatism of the Sociological School in France, whose pope was the great official rival of Bergson, Durkheim." But the "free and lucid spirits" wished "to confront the abyss and their anguish. . . . This revolt, among the young, burned with resentment, sarcasm and pain." They heard the liberating phrases of intuition and mystical insight from "the magician of thought around whom all the revolts had grouped: Bergson." Every Friday at the Collège de France, among the three hundred auditors, they heard a truth that was not irrational but rather "transrational." An adversary of Bergson, Julien Benda, said scornfully that "Bergsonism came to tell contemporary society exactly what it wished to hear. . . ."[317] "But," responded Rolland,

> this is precisely what interests us; and far from being in our eyes the sign of weakness in a thought, that it corresponds to the imperious need of an epoch, to what it is waiting for, and to which it appeals;—it is rather the index that it answers to the profound requirements of a necessary hour in the human spirit.[318]

Péguy's journal, *Cahiers de la Quinzaine*, became the organ for the young Bergsonians. In 1913 Péguy wrote sardonically of the positivists: "What they don't forgive Bergson is that he broke our chains."[319]

As Charles Péguy approached his late thirties, he took stock of the generational conflict that was raging in French thought. The crisis in the Sorbonne, he believed, was founded on this rift between the generations; it brought misunderstanding and hostility, yet perhaps also provided the motivation for advances in philosophy, art, and science. In 1911 Péguy devoted a whole number of his *Cahiers de la Quinzaine* to an article that set forth eloquently the complaint of his generation:

> For it is extremely difficult if not impossible [according to the *Cahiers*] for one generation, for one grade which is getting older to believe that others are growing older too. More precisely they are ready to believe that their elders are getting old, and they measure this aging almost geometrically . . . by their advancement in place, power, and temporal authority, but they do not wish to acknowledge that the others, their younger ones, that the youth, alas, that the next generations are

advancing perceptibly with the same speed. All the family crises, *fathers and sons*, come from that. One man doesn't wish to understand that another also, his son, has become a man. And the mothers are generally worse than the fathers. All this crisis of the Sorbonne, which is so profound, (not the Sorbonne, but the crisis)—arises from the fact that a whole generation, which is getting on toward sixty, doesn't want to understand that a whole generation, another, ours, is approaching its forties.

If this perpetual misunderstanding . . . perpetually renewed, springs forth and evolves regularly from generation to generation like a cascade of misunderstanding and non-misunderstanding to the point that it is the law even of aging in the family, in the race, in the people,—in philosophy, metaphysics, in art, in science,—what must it not have produced between the generation which preceded us and ourselves, when one considers the degree of scientific sufficiency which the generation which preceded us attained.

No generation, affirmed the *Cahiers* bitterly, in any place or time,

has ever treated as harshly, as ungratefully, with such hatred and fury and acrimony and so unluckily the next generation as has the generation which preceded us. It bears toward us truly feelings like those of the old ogresses in the fairy tales who always wished to devour the young queen like a string of green peas.[320]

In a sense, Louis de Broglie was part of this movement, this isoemotional wave. Not a Normalien, and in any case, preoccupied with basic problems rather than examinations, he found profound meaning in Bergson's philosophy.

By 1910 Bergson's ideas had begun to permeate the teachings of younger professors. The antagonists of Bergson's philosophy charged, not without grounds, that students attracted to Bergson were left with "a curious state of decision," with a disdain for positive science. Young intellectuals, it was alleged, were led to belittle the sciences as good enough for technicians, doctors, and engineers, but of no importance for one concerned with the real: "Psychological facts must be lived . . . ," they said. Bergsonism seemed to his critics to be abetting a "philosophical insincerity." With some heat, Henri Bergson denied that he had ever "written a line, a word, which could be interpreted in this fashion!"[321] Mathematics was not a game; it came, like physics, into genuine contact with the absolute, he said. But it did eventuate in fragmenting nature in accordance with our practical interests. Young disciples, however, often cherished the passages in which the character of duration was depicted by Bergson as distorted by mathematical method which reduced movement to immobility, and continuities to multiple, separate atoms.

Louis de Broglie was never a member or self-conscious associate of any Bergsonian circle. Indeed, he was not in any sense an "adherent" of any metaphysics. "I read Bergson indeed when I was very young," he has written, "but if certain of his ideas then interested me, I was never truly an adherent."[322] Solitary by nature, he was of those who imbibe their ideas from the intellectual environment without however becoming part of it, who reshape and use these ideas altogether to their own purpose and bent; he was not a Bergsonian but he could sublimate the metaphysical discontent which the young Bergsonians felt into an expression even more enduring and significant. Of himself, de Broglie has written:

> What characterizes me most is that I have been, despite the obligations of my career, a great "solitary," having spent a part of my youth, after the death of my father, with two aged women (my mother and her mother) and having always detested the worldly life which many of my family led. I think it was this which was most important in my life. . . .[323]

Louis de Broglie broke with the way of life customary to his family. He fell back on his own intellectual resources and participated, as it were, vicariously in the stirrings of new ideas among those of his own generation. He would scarcely have fit into the circle of the young Bergsonians, yet in a curious, indirect way, one might say, he was associated with their "pilot wave." Probably Louis de Broglie's relationship to the young Bergsonians was much like that of the novelist Marcel Proust. From Bergson's lectures, Proust derived the notion that duration permeates memory: "A Bergsonian rhythm of change and flux and mutability pulsates through *Remembrance of things past*. . . ."[324] To a friend who was reading Bergson in 1909, Proust wrote: "It is as though we had been together on a great height." He had read enough of Bergson, he said, so that he could sense the direction of *Creative Evolution*, for "the parabola of his thought is already sufficient discernible after only a single generation."[325] As he was dying in 1922, he said that he had written his novel as though his aim were to bring time within the purview of his telescope, to reveal to the conscious mind unconscious phenomena of the forgotten past, thereby concurring with Bergson; nevertheless, he said that "there has never been, insofar as I am aware, any direct influence." It was through a similarly indirect path that de Broglie's scientific work, like Proust's art, knew the Bergsonian influence.[326]

When Bergson published his "Essai sur les Données Immédiates de la Conscience" in 1889, he broke decisively with the intellectual fashions and idols of his own youth—the environmental determinism of Hippolyte Taine and the positivist determinism of Auguste Comte and Herbert Spencer. The scientific intellect, with its spatialized handling of reality, was regarded by Bergson as thrusting its distortion of discontinuities and divisions upon it; intuition alone could recapture the inner nature of continuous movement. The materialistic reduction or the positivist analysis could never avail against the duration of inner consciousness which defied all geometrical categories.

Louis de Broglie could scarcely accept Bergson's radical criticism of mathematical method as belying the nature of ultimate reality. But in his lectures and above all in his book, *Matière et Mémoire* (1896), Bergson gave his philosophical endorsement to those physicists who were trying to replace the atomistic mode of thought with conceptions of a nature that was more continuous, more fluid. Granted that the practical requirements of action led human beings to impose a "primary discontinuity" upon the nature of matter. Still, physicists were beginning to free themselves from these "customary images."

> And, indeed, we see force and matter drawing nearer together the more deeply the physicist has penetrated into their effects. We see force more and more materialized, the atom more and more idealized, the two terms converging towards a common limit and the universe thus recovering its continuity. We may still speak of atoms . . . but the solidity and the inertia of the atom dissolve either into movements or into lines of force whose reciprocal solidarity brings back to us universal continuity.

To such a conclusion, according to Bergson, had come "the two physicists of the last century who have most closely investigated the constitution of matter, Lord Kelvin and Faraday." For Faraday, the individuality of the atom dissolved into indefinite lines of force, extending through all space and interpenetrating with all atoms. For Kelvin, the atom became a vortex ring whirling in a continuous medium. "But on either hypothesis," Bergson affirmed, "the nearer we draw to the ultimate elements of matter the better we note the vanishing of that discontinuity which our senses perceived on the surface." "Every philosophy of nature," according to Bergson, "ends by finding it [discontinuity] incompatible with the general properties of matter." Thus, the evolution of physical science will bring it into accord with the intuitions of psychological analysis.[327]

Such were the bold ideas that intrigued Louis de Broglie when, having completed one degree in history, he undertook in the years from 1911 to 1913 to take the degree of licentiate in the sciences. They were ideas that were out of keeping with the central tenor of the "scientific revolution" that was taking place. Of those developments, Louis' brother Maurice, already an established and noted physicist, investigating X-rays in his private laboratory, kept him abundantly informed. As Louis de Broglie tells:

In October 1911, there took place at Brussels the first Solvay Council of Physics at which were discussed the questions, still then very poorly clarified, relative to the quanta. My brother, who had been one of the secretaries of the Council and was preparing the publication of its record, communicated to me the text of the discussions. With the ardor of my age, I had been filled with enthuasism by the interest of these problems being investigated and I had promised to devote all my efforts toward understanding the mysterious quanta which Max Planck had introduced into theoretical physics ten years previously but whose profound significance was not yet perceived.[328]

The Bergsonian ideas, however, to which Louis was also being drawn were diametrically opposed to the standpoint of discontinuities which Planck, Einstein, and Bohr were introducing into the conceptions of the atom and energy.

The young French nobleman was indeed involved in a confluence of emotional vectors of a most exceptional kind. On the one hand, he had turned from the traditional preoccupation of his family with a monarchical past toward the world of science. At the same time, he was in a curious way a member of a besieged, highly select minority; as the prince de Broglie, whose ancestors had fought and lost their lives on behalf of the monarchy and had espoused its church's doctrines, he retained a loyalty which in a strange way was not unlike the loyalty that bound Einstein to the persecuted Jews. A persecuted elite can exist both among social aristocrats and social pariahs. Thus Louis de Broglie could never give himself wholeheartedly to the republican positivist skepticism of Poincaré. But in Bergson's doctrines, imbued with a protest against the materialist standpoint and appealing to subtle discriminations, intuitions, and nuances against the categories of practical action, he found something that articulated his emotional longings.

Retreat at the Military Wireless

At this juncture, World War I intervened: "The world war of 1914–18 came interrupting brutally for several years thoughts which, subsequent events were to show well, were oriented in a good direction," writes de Broglie.[329] Thus began a period in his life not unlike Albert Einstein's years of gestation in the Berne patent office.[330] Louis entered the French army, and worked in the military wireless service under the command of General Gustave Ferrié and Colonel Brenot. The young physicist regretted that he scarcely had the time to concentrate on his theoretical studies. He was glad, however, to be acquiring a practical knowledge of electrotechnology and to be working in the atmosphere of laboratories. He was especially involved in the projects that developed the use of the Eiffel Tower for military wireless transmission. The character of General Ferrié exerted an immense influence on the young Louis de Broglie. Ferrié was one of those military men so peculiar to France, a professional soldier who was at the same time a scientist and thinker. A graduate of the Ecole Polytechnique, Ferrié had become a pioneer in the development of and experiments in radiotelegraphy; the chief figure in solving the problems of adapting wireless to military communications, he was also an investigator in geophysics and meteorology. In his life's work, above all, was embodied the reality of wave motion, the Hertzian electromagnetic waves whose use and application he succeeded in demonstrating to uncomprehending and inertial bureaucrats. In 1914, at the outbreak of the war, he overcame immense difficulties to secure the manufacture in France of the essential triode vacuum tubes, which had only just appeared in the United States. De Broglie spoke to commemorate his general on November 12, 1949:

> Thus Ferrié, apart from his productive military career, was also a great scientist whose works concerned the most varied domains of science. He knew how to search and find out himself, always engaging in personal work. He knew also how to make others work. Having all the qualities of a chief, he knew admirably how to surround himself with collaborators necessary to him, and to animate them with a constant ardor. Benevolent and skilful, he succeeded in getting to work together savants of different

specialties for great enterprises often of moment both to our Country and the progress of Science.[331]

In Ferrié's personality, Louis de Broglie found a union of the rival dualities of the old order and the new republican France. He recalled from his military experience:

> The post of the Eiffel Tower was part of the zone of the interior. Advancement in it was difficult and the members of the mobilized technical service obtained with difficulty the modest stripes of a corporal or noncommissioned officer. Often in charge of quarters was a simple corporal, and this permitted us, my comrades and I, to say at times: "This evening the service is guaranteed by a general and a corporal." This contrast between the two extreme grades of the military hierarchy made us smile for we were young, but we felt the moral lesson that it conveyed: we knew that General Ferrié was thus giving us a great example.[332]

The Birth of Wave Mechanics

After the war ended and de Broglie was demobilized in 1919, he resumed his studies, working in his brother's laboratory, experimenting on X-rays, carrying on long discussions with his brother, and enjoying the give-and-take with young coworkers. The laboratory's experiments on the wave properties of X-rays, his brother's interest, further led him "to reflect profoundly on the necessity, for associating always the point of view of waves with that of corpuscles."[333] Meditating on Bohr's theory of quantum orbits, de Broglie wondered how the discontinuous orbits, electronic jumps, and liberations of energy could be linked with wave structures and manifestations. A duality of standpoint was accepted with respect to the phenomena of light and X-rays. But no one had ventured to suggest that the atoms of matter themselves were constituted of underlying waves. Could he then show that the corpuscular theory of matter had its counterpart in a wave mechanics? Then, in September and October of 1923 came the weeks of illumination. Louis de Broglie wrote three notes in which he formulated the basic ideas of wave mechanics.[334] It was a revolutionary hypothesis, which had the stamp as well of de Broglie's traditionalism, for it argued the dual primacy of the principles of the French classical physicists Fermat and Maupertuis of the seventeenth and eighteenth centuries. He showed how Maupertuis' principle of least action, as applied to an electronic motion, corresponded analytically to Fermat's principle for the propagation of a wave. His essential originality was in his development of the notion that particles of matter are guided, steered, or piloted by a new type of wave; de Broglie called them "matter waves," but scientists later christened them "de Broglie waves." The hypothesis at once showed its heuristic power. For it made possible the derivation of the preferred, stationary, discontinuous orbits which Niels Bohr had assigned to his electrons in correspondence with the experimental spectral data. In de Broglie's theory, the preferred orbits arose from the cumulative interference of the matter waves; they were analogous to the bright regions in the interference patterns that ordinary light projects as it is diffracted upon a screen. The rejected possible orbits corresponded to the dark regions, where waves out of phase extinguished rather than reinforced each other. A simple mathematical law, according to de Broglie, governed the relationship of the pilot waves to their particles: the wave-lengths were inversely proportional to the momenta of their particles.

Louis assembled his results into a thesis for the doctorate in science which he defended at the Sorbonne on November 25, 1924. His jury was an eminent one including Paul Langevin, the noted atomic physicist, Jean Perrin, and the mathematician Elie Cartan.[335] They pondered de Broglie's unusual hypotheses that associated with every atomic corpuscle are waves, whose lengths, in a one-to-one correspondence with electrons in their quantal orbits, could be derived mathematically. De Broglie believed that experimental evidence would sustain the hypothesis of these pilot waves, but as yet none could be cited. Was it all then a *jeu d'esprit*, a *comédie française*, as some called it? Langevin perceived that the doctoral candidate was proposing a thesis that reconstructed Newtonian mechanics in a way as profound as Einstein's. Although he was a revolutionist in politics, Langevin was taken aback by the young aristocrat's calm daring. As Philipp Frank writes, Langevin "was staggered by the new proposals," which "seemed fairly absurd to him." Nevertheless, Langevin recognized that the absurd and the true strangely commingled in contemporary thought. Langevin sent the thesis to Einstein, who declared that Louis de Broglie "had lifted a corner of the great veil"; he told the philosopher Emile Meyerson that de Broglie's work was a "stroke of genius" (*un veritable coup de génie*); it was a "very remarkable thesis," Einstein wrote.[336] Finally, from an unexpected quarter came the experimental evidence for the existence of the pilot waves. In 1927, two

American physicists, Clinton J. Davisson and Lester H. Germer, working in the laboratories of the Bell Telephone Company unexpectedly discovered the phenomenon of the diffraction of a beam of electrons by a nickel crystal. At first they did not comprehend the significance of their discovery; Davisson, however, sent his results to the theoretical physicist Max Born, who remembered Einstein's enthusiasm over de Broglie's hypothesis. Born directed a coworker to investigate whether Davisson's curious maxima could be interpreted as the interference fringes of de Broglie waves. Thus the American investigators learned that they had confirmed the basic idea of wave mechanics.[337]

There was scarcely any premonition among physicists in the mid-1920s that the situation in their theory demanded a wave theory of the electron. The recollections of Professor Victor F. Lenzen evoke the intellectual climate of that time among young physicists:

> When de Broglie's article on his wave conception was published, Professor W. H. Williams, who then taught the theoretical physics in Berkeley, reported on it at a colloquium. I remember quite well that he said that he would not think that it amounted to anything, except that Einstein thought there was something in it. During the year 1927–28 I was a Guggenheim Fellow in Göttingen, and I attended a seminar at which Max Born reported on the experimental findings of Davisson and Germer.[338]

Underlying de Broglie's conception was a metaphysical vision in which the notion of "wave" was restored to primacy as a character of reality: "One can conceive then that as a consequence of a great law of nature, to every piece of energy of proper mass M_0 there is joined a periodic phenomenon of frequency V_0."[339]

> We have been guided by the idea that the corpuscle and its phase wave are not physically different realities. If one reflects one will see that the following conclusion seems to follow: Our dynamics (its Einsteinian form understood) has lagged behind optics; it still is in the stage of geometrical optics. If it seems to us today probable enough that every wave comports concentrations of energy, by contrast the dynamics of a material point undoubtedly conceals a propagation of waves, and the true sense of the principle of least action is to express a concordance of phase.[340]

Corresponding to the idea of the essential physical identity of the corpuscle with the wave, de Broglie wrote that he was "guided by the idea of the principle of least action and that of Fermat."[341] In the seventeenth century the great French mathematician, Pierre de Fermat, proposed as a fundamental law on metaphysical, Aristotelian grounds that nature always acts by the shortest course. From it he derived the law of reflection of light—that the path taken by light between a point on the incident ray and a point on the reflected ray is the shortest possible consistent with the light's meeting the mirror. Fermat then derived the law of refraction: the paths described by light as it goes from one medium to another are those which take the least time, given the resistances of the different media. The result, wrote Fermat in 1661, was "the most extraordinary, the most unforeseen and the happiest, that ever was. . . ."[342] In 1744 and 1746 Maupertuis modified Fermat's Principle, which Huygens had derived from his wave theory of light, and making it conform to a corpuscular theory, advanced the hypothesis that light follows a path in which the least action is expended; this he generalized into a law for all moving objects, for all nature considered as a dynamical system. But now in de Broglie's formulation, the classical wave standpoint, together with Fermat's Principle, was reinstated as a reality polar, if not ultimately identical (in some metaphysical sense) with Maupertuis' corpuscular conception: "The principle of least action in its Maupertuisian form and the principle of concordance of phase due to Fermat might well be two aspects of a single law."[343] Pierre de Fermat, a contemporary of Corneille and Racine and friend of Father Mersenne and the mystic Pascal, a councilor of the Toulouse parlement who pursued mathematics as an avocation with little thought to publication, modest, content to regard himself as following in the footsteps of the Greek algebraists Diophantus and Apollonius, was a scientist of France's classical age.[344] Maupertuis, on the other hand, was a friend of the *philosophes*, their aggressive chief in the battle for Newtonianism, and the admired intimate of Voltaire.[345]

It was fitting that de Broglie, in whom the classical tradition coexisted with a renovator's impulses, restored Fermat's standpoint to an equality with that of Maupertuis:

> We arrive then at the following affirmation: "The principle of Fermat applied to the phase wave is identical with Maupertuis' principle applied to the moving object; the trajectories dynamically possible for the moving object are identical with the possible rays of the wave."[346]

This became de Broglie's heuristic principle for realizing a synthesis of waves and quanta. This idea, wrote de Broglie, that the movement of a material point always dissimulates the propagation of a wave, must be studied and completed, for if one could render it a satisfactory form, it would represent "a synthesis of a great rational beauty" (*une synthèse d'une grande beauté rationnelle*).[347] And this "rational beauty" was perhaps the universe, as a young Bergsonian aristocrat with the temper of a scientist would have longed to see it.

Bergson has often been criticized as a philosopher of anti-intellectualism, a protagonist of instinct as against reason. It is all the more remarkable then, that Louis de Broglie avowed that he found the most significant analogies between Bergson's ideas and those of quantum and wave mechanics. And we may infer that these were the isomorphemes, the guiding forms of ideas, which impressed him from early youth, answering some deep emotional longing. Above all, they shaped the hypothesis of the wave conceptions, which he then sought to provide with a mathematical formalism consonant with the experimental facts.

In later years, Louis de Broglie wrote a revealing essay on the basic similarities between Bergson's insights and the standpoint of quantum mechanics. "Is there any analogy between Bergson's critique of the idea of motion and the conceptions of contemporary quantum theories? It seems that the reply ought to be in the affirmative," wrote de Broglie.[348] According to Bergson, Zeno of Elea had shown that the notion that an arrow in flight is at any instant in a determinate position would make it impossible to understand movement, for how can the arrow be said to be moving when at every instant it is immobile?

> Passage is a movement and stoppage an immobility. . . . When I see the moving object pass through a point . . . I tend to consider its passage as an arrest. . . . All points of space necessarily being fixed, I must be careful not to attribute to the moving object itself the immobility of the point with which it coincides. . . .[349]

Guided by Zeno of Elea, Bergson had affirmed that trying to locate the position at any instant of a moving object does violence to the fact of its movement. To Louis de Broglie, this was the kernel of the idea of quantum mechanics that when one regards things on a small enough scale there is no trajectory assignable to the moving object, for one can determine through a series of necessarily discontinuous measurements only certain instantaneous positions of the physical entity in motion, and each of these determinations implies a total renunciation of the possibility of grasping at the same time its state of motion.

Bergson had written his first critique of the spatialized distortion of movement in 1889, more than forty years before Heisenberg formulated his uncertainty principle. Today he could have availed himself of the language of quantum mechanics, and said: " 'If one attempts to localize the moving object, through a measurement or an observation, at a point of space, one will obtain only a position and the state of motion will entirely escape.' "[350]

Bergson had indeed been the intuitive prophet of quantum mechanics, according to Louis de Broglie. Where elementary physical entities had formerly been represented by the concept of a particle located in geometrical space, there was now the essential complementary "concept of a *wave*, which in wave mechanics represented motion in a pure state with no spatial location."[351] The images of mobility and localization cannot be simultaneously employed. Prescient in rejecting the classical representation of movement through successive positions on a continuous curve, Bergson had also first renounced the rigorous determinism of classical physics. According to Bergson, the future was not entirely determined because in real duration "there is unceasingly being created in it . . . something unforeseeable and new." Louis de Broglie saw a confirmation of this view in the developments of quantum physics:

If Bergson could have studied quantum theory in detail, he would have observed that in the image of the evolution of the physical world which it offers us, at each instant nature is described as if hesitating between a multiplicity of possibilities, and he could doubtless have repeated as in *The Creative Mind* that "time is this very hesitation or it is nothing."[352]

Quantum theory also seemed to de Broglie to have confirmed Bergson's insight that reality was characterized by interpenetration, by fusion of its components; individualities such as atoms or sensations, related to each other by purely external relations were demarcated for practical reasons, but were not intrinsic to the nature of reality. Reality was more akin to a symphony of music in which the notes modify each other, compenetrating in melody. So de Broglie declared that in wave mechanics, too, it was impossible, in dealing with a group of particles of the same physical nature, to give to each a distinct individuality. For that is possible only when one can distinguish particles through their different spatial positions, whereas

> in wave mechanics, one cannot in general attribute to particles well-defined positions in space. . . . If their regions of possible presence merge or overlap—which will most often happen—how can one follow their individuality? Thus wave mechanics has given up individualizing particles and following the evolution of each separately with the course of time. . . .[353]

The rapprochement, however, of Heisenberg's indeterminacy to Bergson's continuous duration came in de Broglie's later years after he had tried to salvage his wave theory from a probabilist outcome. More significant was the Bergsonian analogy of the wave which imbued the thinking of his most original years.

The central images in Bergson's philosophy were indeed those of the physics of Louis de Broglie. That is why Bergson could claim that in his last writings he had shown "the direct consequences of the theory of Louis de Broglie."[354] Among the philosophies of the early twentieth century, Bergson's was unique for the primacy it attached to the image of the wave as underlying reality. It emerged in the vision of *Creative Evolution.*

> Life as a whole, from the initial impulsion that thrust it into the world, will appear as a wave which rises, and which is opposed by a descending movement of matter. . . . This rising wave is consciousness, and like all consciousness, it includes potentialities without number which interpenetrate. . . . From our point of view, life appears in its entirety as an immense wave which, starting from a center, spreads outwards, and which on almost the whole of its circumference is stopped and converted into oscillation. . . .[355]

The emotional-intellectual valence for the notion of a wave in this French culture circle was conveyed by Bergson's writings: wave theory became an isomorpheme for young French intellectuals at odds with bourgeois, republican atomism and materialism. Then too the curious yet seminal analogy of the continuous musical melody, so much emphasized in Bergson's work, appeared in de Broglie's presentation of his wave mechanics. The stability condition for the electron in its orbit, wrote de Broglie, should be that its "phase wave is tuned with the length of the path."[356] The musical analogy in de Broglie's work seemed to George Gamow so pronounced that he attributed it in part to de Broglie's attainment as a "connoisseur of chamber music"; thus "de Broglie chose to look at an atom as some kind of musical instrument which, depending on the way it is constructed, can emit a certain basic tone and a sequence of overtones."[357] The metaphor of the musical tune had however, made its canonical appearance in Bergson's first book: a succession of pendulum oscillations could be understood in their durational reality only when perceived as "each permeating the other and organizing themselves like the notes of a tune," with "a continuous or qualitative multiplicity"; duration was like "the notes of a tune . . . melting, so to speak, into one another;" movement could be understood only as "a unity resembling that of a phrase . . . in a melody."[358] He praised Whitehead many years later, as a "very profound philosopher who began as a mathematician," and had envisaged a piece of iron as " 'a melodic continuity.' "[359] To be sure, Louis de Broglie as a mathematical physicist could not agree with the polemic which Bergson directed against the idea of number and mathematical method itself. Nor did he think well of Bergson's stance against the theory of relativity.[360] Yet, granting that the "numerous critiques" of Bergson were often well-founded, still, said de Broglie, his "work is powerful":

> We have been struck by the analogy between certain new concepts of contemporary physics and certain brilliant intuitions of the philosophy of duration. And we have been still more surprised by the fact that most of these intuitions are found already expressed in *Time and Free Will*, Bergson's first work and also perhaps the most remarkable. . . . This essay, its author's doctor's thesis, dates from 1889 and consequently antedates by forty years the ideas of Niels Bohr and Werner Heisenberg on the physical interpretation of wave mechanics.[361]

What impressed Albert Einstein most in Louis de Broglie's "unique" book, *Physics and Microphysics*, was its sincere account of a struggle for a logical basis of physics: "I found the consideration of Bergson's and Zeno's philosophy from the point of view of the newly acquired concepts highly fascinating."[362] Previously Einstein had had a low opinion of Bergson's philosophy.[363] Throughout his life, Louis de Broglie has drawn sustenance from Bergsonian intuition. Confronted with the moral crises posed by the atomic bomb and the dawn of the atomic era, he turned to the pages of Bergson's "great work," *The Two Sources of Morality and Religion.*[364] Louis de Broglie's scientific intuition was guided by a generational rebellion, in which the primary wave was overlaid with a secondary wave—the emotions of a scion of the displaced aristocracy in a bourgeois world. In Bergson's intuitions, Louis de Broglie found his "pilot wave."

The personage in the history of French science with whom Louis de Broglie most identified was André-Marie Ampère. Indeed he wrote a short biography of Ampère, "to show through what trials in that tragic epoch the generation of scientists grew up . . ."[365] For the father of André-Marie Ampère had, like de Broglie's own ancestor, "paid with his head for his opposition to the masters of the hour," for his resistance to the dictatorship of the Convention. The father's execution so affected young André-Marie "that his reason was disturbed for a whole year." "Timid, impressionable, unconfident of himself, he was almost never happy." "He was never to rediscover the happy years of his first youth." As a youth he was a member of a "sect of young philosophers" called the mystical school, and later part of a philosophical circle whose leading spirit was Maine de Biran.[366] In this pattern of trauma and philosophical response, Louis de Broglie could perceive something essentially akin to that of his own life. The idealistic Maine de Biran indeed was regarded by Henri Bergson as his intellectual precursor.[367] And Ampère virtually alone had opposed the dominant mechanical caloric hypothesis as de Broglie opposed the pure atomism.

Notes

1. The greater part of this chapter was presented to the Seminar in the Sociology of Ideas at the University of Toronto in the early spring of 1970. In the fall of 1970 the essay by Gerald Holton, "The Roots of Complementarity," appeared in *Daedalus*, vol. 99, no. 4, of the *Proceedings of the American Academy of Arts and Sciences*, pp. 1,015–1,055. This admirable work also emphasizes the role of Kierkegaard and Høffding in the formation of Bohr's thought. I have differed, however, with Professor Holton in specifying the principal ideas in Kierkegaard and Høffding that helped shape Bohr's way of thinking, and also the shift in the character of the Kierkegaardian component from Bohr's youth to his middle and old age. In addition, I have used the indications of Bohr's attachment to Møller's novel, *The Adventures of a Danish Student*, and the Icelandic sagas in my effort to interpret Bohr's unconscious motivations.

2. Josephine Goldmark and A. H. Hollman, *Democracy in Denmark* (Washington, D.C., 1936), pt. 2, p. 149.

3. Ibid., pt. 1, p. v.

4. Frederic C. Howe, *Denmark: A Cooperative Commonwealth* (New York, 1921), p. 139.

5. The German occupation authorities demanded in 1942 that the Danish government dismiss Einar Cohn from his post as Director-General. The Danes refused. In

the fall of 1943, Cohn fled to Sweden where he lectured at the Stockholm School of Economics and Commerce, and served as counselor to the Danish embassy. Cf. Leni Yahil, *The Rescue of Danish Jewry: Test of a Democracy*, trans. Morris Gradel (Philadelphia, 1969), pp. 104–105.

6. Stefan Rozental, ed., *Niels Bohr: His Life and Work As Seen by His Friends and Colleagues* (Amsterdam, 1967), pp. 24, 25, 31. *Dansk Biografisk Leksikon* (Copenhagen, 1934–1942), vol. 4, p. 293; vol. 5, p. 359; vol. 17, pp. 338–341; vol. 20, p. 267; vol. 22, p. 215. Harald Høffding, "Philosophy in the North in the Last Decade," *The Monist*, vol. 36 (1926), pp. 194–195.

7. Svend Dahl, ed., *Danish Theses for the Doctorate and Commemorative Publications of the University of Copenhagen, 1836–1926* (Copenhagen, 1929), p. 251. Theodor Geiger, "An Historical Study of the Origins and Structure of the Danish Intelligentsia," *The British Journal of Sociology*, Vol. 1 (1950), pp. 214, 216, 219.

8. Børge Jensen, "Harald Bohr in Memoriam," *Acta Mathematica*, vol. 86 (1951), pp. 1–2. Harald Bohr, *Collected Mathematical Works* (Copenhagen, 1952), vol. 1, p. xxxiii.

9. Rozental, *Niels Bohr*, p. 24.

10. Høffding, "A Philosophical Confession," *The Journal of Philosophy, Psychology, and Scientific Methods*, vol. 2 (February 16, 1905), p. 86.

11. "'We live forwards,' a Danish scholar has said, 'but we understand backwards'." (William James, *Pragmatism* [New York, 1907], p. 223). "'We live forward, but we understand backward,'—a phrase of Kierkegaard's which Høffding quotes" ("How Two Minds Can Know One Thing?," *The Journal of Philosophy, Psychology and Scientific Methods*, vol. 2 [March 30, 1905], p. 180; reprinted in *Essays in Radical Empiricism*, ed. Ralph Barton Perry [New York, 1912], p. 132). See also, William James, "Is Radical Empiricism Solipsistic?," *The Journal of Philosophy, Psychology and Scientific Methods*, vol. 2 (April 27, 1905), p. 237. "'We live forward, we understand backward,' said a Danish writer; and to understand life by concepts is to arrest its movement . . . (*A Pluralistic Universe* [New York, 1909], p. 244).

12. *Letters of William James*, ed. Henry James (New York, 1920), vol. 2, p. 216. Høffding, however, felt in later years that when James described him to Schiller as "a good Pluralist and Irrationalist," he was at least exaggerating the pluralism (See Høffding, *Erindringer* [Copenhagen, 1928], p. 213).

13. Høffding, "A Philosophical Confession," *The Journal of Philosophy, Psychology, and Scientific Methods*, p. 90.

14. Høffding, *The Problems of Philosophy*, trans. Galen M. Fisher (New York, 1905), p. xi.

15. Ibid., pp. 121, 122, 125.

16. William James, *Some Problems of Philosophy* (New York, 1911), p. 154.

17. Høffding, *The Problems of Philosophy*, p. 2.

18. Ibid., pp. 150, 168, 170.

19. Høffding, "Philosophy and Life," *International Journal of Ethics*, vol. 12 (January 1902), pp. 138, 140, 143.

20. Høffding, *The Philosophy of Religion*, trans. B. E. Meyer (London, 1906), pp. 23, 57, 69, 70, 71, 91, 340.

21. Høffding, "A Philosophical Confession," pp. 86–90.

22. Høffding, *The Problems of Philosophy*, p. 8.

23. Ibid., p. 76. "The deepest foundation for the principle of natural causation is the need for continuity, which lies in the nature of our consciousness . . ." (Høffding, *Philosophy of Religion*, p. 23).

24. Høffding, *Philosophy of Religion*, p. 13.

25. Ibid., p. 263.

26. Høffding, "The Law of Relativity in Ethics," *The International Journal of Ethics*, vol. 1 (1890), pp. 39, 51.

27. Høffding, "The Conflict between the Old and the New: A Retrospect and a Prospect," *The International Journal of Ethics*, vol. 6 (1896), pp. 324, 337.

28. Høffding, "On the Relation between Sociology and Ethics," *The American Journal of Sociology*, vol. 10 (1905) pp. 674–675.

29. Høffding, *Jean Jacques Rousseau and His Philosophy*, trans. William Richards and Leo E. Saidla (New Haven, 1930), pp. 62–63.

30. Høffding, *Spinozas Ethika: Analyse und Charakteristik* (Heidelberg, 1924), pp. 1–2; *A History of Modern Philosophy*, vol. 1, pp. 292–204, 305. Also, see Høffding, in *A History of Psychology in Autobiography*, ed. C. Murchison (Worcester, 1932), vol. 2, pp. 199, 203.

31. L. Rosenfeld, "Niels Bohr's Contribution to Epistemology," *Physics Today*, vol. 16 (1963), p. 48.

32. Rozental, *Niels Bohr*, p. 13. See also J. Rud Nielson, "Memories of Niels Bohr," *Physics Today*, vol. 16, no. 10 (October 1963), p. 22.

33. *Correspondance entre Harald Høffding et Emile Meyerson* (Copenhagen, 1939), p. 149.

34. Ibid., p. 156.

35. Ibid., p. 169.

36. Ibid., p. 123.

37. Ibid., pp. 51, 131.

38. Ibid., p. 70.

39. Ibid., pp. 2, 115, 138, 143.

40. Valdemar Hansen, "Harald Høffding, 1843–1931," *The American-Scandinavian Review*, vol. 19 (October 1931), pp. 599–604. When Høffding celebrated his seventieth birthday in 1913, Henri Bergson joined with John Dewey and other philosophers in sending him a letter of appreciation on his "long and eminent service as a teacher of philosophy as well as of your valuable contributions to the literature of the subject." "Letter to Dr. Harald Høffding," *The Journal of Philosophy, Psychology and Scientific Methods*, vol. 10 [1913], pp. 279–280).

41. Rozental, *Niels Bohr*, p. 270.

42. Høffding, *Erindringer*, pp. 206–207.

43. Ibid., pp. 207–208.

44. To Høffding, Kierkegaard was "one of the most Danish authors who ever lived. . . . (He) loved our country, its nature and its language, with a tender and understanding love" ("Søren Kierkegaard: 5 Mai 1813–5 Mai 1913," *Revue de Metaphysique et de Morale*, vol. 21 [1913], p. 720).

45. Høffding, *Søren Kierkegaard als Philosoph*, 1892, trans. A. Dorner and Chr. Schrempf (Stuttgart, 1922), p. 75.

46. Ibid., p. 75.

47. Ibid., p. 73.

48. Ibid., p. 76.

49. Ibid., p. 77.

50. Ibid., p. 78.

51. Ibid., p. 79.

52. Ibid., p. 96.

53. Ibid., p. 97.

54. Ibid., p. 98.

55. Ibid., p. 129.

56. Ibid., p. 129.

57. Høffding, "Søren Kierkegaard," p. 723.

58. Høffding, *Søren Kierkegaard als Philosoph*, pp. 83–84.

59. Ibid., p. 128.

60. Ibid.

61. Edgar Rubin, "Psychology Regarded as a Positive Science," "Concerning the Soul and the Dualistic Nature of the Individual," *Ninth International Congress of Psychology: Proceedings and Papers*, Yale University, New Haven, Connecticut, September 1st to 7th, 1929 (Princeton, 1930), pp. 370–372. Rubin's experimental researches on "figure" and "ground" in visual perception were published as *Visuell Wahrgenommene Figuren* (Copenhagen, 1921). The review by R. M. Ogden noted their "important bearings upon systematic psychology," and observed that they "have been eagerly welcomed by the 'Gestalt' psychologists . . ." (*The Psychological Bulletin*, vol. 20 [1923], pp. 219–223). See also *The Selected Papers of Wolfgang Köhler*, ed. Mary Henle (New York, 1971), pp. 116, 162. Rubin died after a long

illness partially caused by the hardships he suffered during his flight to Sweden from the Nazis. His "figure-ground" perception has a striking resemblance to the notion of complementarity. David Katz, "Edgar Rubin, 1886–1951," *The Psychological Review*, vol. 58 (1951), pp. 387–388.

62. Gerald Holton, "The Roots of Complementarity," *Daedalus* (Fall 1970), p. 1,034. Part of the transcript of an interview with Bohr by Thomas S. Kuhn and Aage Petersen on November 17, 1962 is reprinted in Holton's study. Holton conjectures that Bohr's term "complementarity" might have been derived from James's use of "complementary" (Ibid., pp. 1,037–1,038).

63. Leon Rosenfeld, "Niels Bohr in the Thirties," in *Niels Bohr*, ed. S. Rozental, p. 121.

64. Mogens Pihl, "Niels Bohr and the Danish Community," in *Niels Bohr*, ed. S. Rozental, pp. 292–293. Hans Bohr, "My Father," in ibid., p. 330.

65. Niels Bohr, *Essays 1958–1962 on Atomic Physics and Human Knowledge* (New York, 1966), p. 13.

66. P. M. Mitchell, *A History of Danish Literature* (Copenhagen, 1957), pp. 123–125. S. Foster Damon and Robert S. Hillyer, *A Book of Danish Verse*, ed. Oluf Friis (New York, 1922), pp. 63–65.

67. Walter Lowrie, *Kierkegaard* (London, 1938), pp. 147–149.

68. Ibid., pp. 143–144.

69. Elias Bredsdorff, Brita Mortensen, and Ronald Popperwell, *An Introduction to Scandinavian Literature* (Copenhagen, 1951), pp. 89–90. Bohr, *Essays 1958–1962 on Atomic Physics and Human Knowledge*, p. 13.

70. Bohr, *Essays 1958–1962 on Atomic Physics and Human Knowledge*, p. 13.

71. Rosenfeld, "Niels Bohr's Contribution to Epistemology," *Physics Today* (October 1963), pp. 48–49.

72. Werner Heisenberg, "Quantum Theory and Its Interpretation," in *Niels Bohr*, ed. S. Rozental, p. 95.

73. Oskar Klein, "Glimpses of Niels Bohr As Scientist and Thinker," in S. Rozental, ed. *Niels Bohr*, p. 75. J. Rud Nielson, "Memories of Niels Bohr," p. 28. L. Rosenfeld, *Niels Bohr: An Essay* (Amsterdam, 1945), pp. 3, 18. According to Heisenberg, Bohr had a similar opinion of the positivist Philipp Frank as "an expert on metaphysics" (*Physics and Beyond: Encounters and Conversations*, trans. Arnold J. Pomerans [New York, 1971], p. 210).

74. Rosenfeld, "Niels Bohr's Contribution to Epistemology," p. 49.

75. Ibid., p. 49.

76. Rozental, *Niels Bohr*, pp. 27–28.

77. Nielson, "Memories of Niels Bohr," pp. 27–28.

78. Aage Henriksen, *Methods and Results of Kierkegaard Studies in Scandinavia: A Historical Critical Survey* (Copenhagen, 1951), pp. 44, 52, 56, 62, 67.

79. Ibid., p. 38.

80. Ibid., p. 12.

81. George Brandes, *Friedrich Nietzsche*, trans. A. G. Chater (New York, 1915), p. 69: *Henrik Ibsen: A Critical Study*, trans. Jessie Muir (London, 1899; reprint ed., New York, 1964), p. 48.

82. Henriksen, *Methods and Results of Kierkegaard Studies*, pp. 23–25.

83. Edmund W. Gosse, *Studies in the Literature of Northern Europe* (London, 1879), pp. 164–165. See also Kemp Malone, "Grundtvig's Philosophy of History," *Journal of the History of Ideas*, vol. 1 (1940), p. 281.

84. Brandes, *Henrik Ibsen: A Critical Study*, p. 48.

85. Ibid., p. 99.

86. Ibid., p. 62.

87. Constantin Stanislavsky, *My Life in Art*, trans. J. J. Robbins (London, 1924), pp. 378–379.

> In that time of political unrest—it was but a little while before the first revolution—the feeling of protest was very strong in all spheres of society. They waited for the hero who could tell the truth strongly and bravely in the very teeth of the government. It is not to be wondered at that the image of Doctor

Stockmann became popular at once in Moscow, and especially so in Petrograd. "The Enemy of the People" became the favorite play of the revolutionists, notwithstanding the fact that Stockmann himself despised the solid majority, and believed in individuals to whom he would entrust the conduct of life. But Stockmann protested, Stockmann told the truth, and that was considered enough.

88. Thomas Henry Croxall, *Kierkegaard Commentary* (London, 1956), p. 11; *The Journals of Søren Kierkegaard*, trans. Alexander Dru (London, 1959), pp. 102, 104; Walter Lowrie, *Kierkegaard* (London, 1938), p. 234.

89. Kierkegaard, *Johannes Climacus*, trans. T. H. Croxall (London, 1958), p. 81.

90. *Kierkegaard's Concluding Unscientific Postscript*, trans. David F. Swenson (Princeton, 1944), p. 306; Croxall, *Kierkegaard Commentary*, p. 134.

91. Bohr, "On the Constitution of Atoms and Molecules, pt. 3 *Philosophical Magazine*, vol. 26, 6th ser. (1913), p. 874.

92. Kierkegaard, *Stages of Life's Way*, trans. Walter Lowrie (Princeton, 1940), p. 430. Also see pp. 400–401.

93. Régis Jolivet, *Introduction to Kierkegaard*, trans. W. H. Barber (London, 1950), pp. 113, 115.

94. Høffding, *A History of Modern Philosophy*, trans. B. E. Meyer, vol. 2 (London, 1924), p. 286.

95. David R. Swenson, *Something about Kierkegaard*, ed. Lillian Marvin Swenson (Minneapolis, 1941), pp. 122–123.

96. Henriksen, *Methods and Results of Kierkegaard Studies* pp. 32–33. Walter Lowrie, a convinced Kierkegaardian, complained that Høffding's "chief fault" was that he even made S. K. dull. Was this, however, the same as saying that Høffding wisely made no effort to imitate Kierkegaard? (*Kierkegaard*, vol. 1, p. 6).

97. Bohr, *Atomic Theory and the Description of Nature*, p. 20.

98. Rozental, *Niels Bohr*, p. 75.

99. Rosenfeld, *Niels Bohr: An Essay*, p. 8.

100. See Barbara Lovett Cline, *Men Who Made a New Physics: Physicists and the Quantum Theory* (Toronto, 1969), p. 85.

101. Rosenfeld, *Niels Bohr: An Essay*, pp. 7–8.

102. George Gamow, *Thirty Years That Shook Physics: The Story of Quantum Theory* (New York, 1966), pp. 72–73. We might observe that such a nonconservative cosmology, which envisaged the accretions of novel droplets of energy in the universe, was a favored idea of William James. Cf. William James, *A Pluralistic Universe* (New York, 1909), pp. 231–232; *Memories and Studies* (New York, 1911), pp. 192–193; *Some Problems of Philosophy* (New York, 1911), pp. 185–187.

In 1924 Bohr once again challenged the strict validity of the law of conservation of energy. To resolve the conflict between the wave and corpuscular aspects of light, Bohr, Kramers, and Slater proposed a theory in which the law of conservation did not hold for individual atomic processes, though it preserved a statistical validity when large numbers of atomic processes were considered. Bohr's anticonservationist theory, however, met with experimental objections; the confirmation of the existence of the neutrino, a particle which had been postulated to safeguard the principle of conservation, was a crucial victory for the conservationists. See P. A. M. Divac, "Does Conservation of Energy Hold in Atomic Processes?" *Nature*, vol. 137 (February 22, 1936), pp. 298–299. Bohr came to regard the surrender of the principle of conservation as "too cheap a way of overcoming the quantum paradox" (Oscar Klein, "Glimpses of Niels Bohr as Scientist and Thinker," in *Niels Bohr*, ed. S. Rozental, p. 77).

103. Schilpp, *Albert Einstein: Philosopher-Scientist*, pp. 233, 235, 238, 202, 204, 206, 211.

104. Bohr, *Atomic Theory and the Description of Nature*, p. 24.

105. Ibid., p. 24. The motif of renunciation runs throughout Bohr's writings (ibid., pp. 56, 77, 84, 90, 109, 114, 115). See also, Bohr, "Causality and Complementarity," *Philosophy of Science*, vol. 4 (1937), pp. 291, 292, 293, 294.

106. Nielson, "Memories of Niels Bohr," p. 26.

107. Robert Bretall, ed., *A Kierkegaard Anthology* (Princeton, 1951), pp. 127–128, 457.

108. Rosenfeld, *Niels Bohr: An Essay*, p. 9.

109. Bohr, *Atomic Theory and the Description of Nature*, pp. 20, 108–109.

110. Rosenfeld, *Niels Bohr: An Essay*, p. 9.

111. Bretall, *A Kierkegaard Anthology*, p. 211.

112. Bohr, *Atomic Theory and the Description of Nature*, p. 15; italics in original.

113. Ibid., p. 96.

114. Ibid., p. 97.

115. Bohr, "Causality and Complementarity," p. 297.

116. Bohr, *Atomic Theory and the Description of Nature*, pp. 100–101.

117. Bohr, *Atomic Theory and the Description of Nature*, pp. 99–100.

118. Peter Rohde, *Søren Kierkegaard: An Introduction to His Life and Philosophy*, trans. Alan Moray Williams (London, 1963), p. 148.

119. Rozental, *Niels Bohr*, p. 328. Also see p. 238. On Kierkegaard's reflection on the 70,000 fathoms beneath him, see Croxall, *A Kierkegaard Commentary*, p. 21.

120. Rozental, *Niels Bohr*, p. 328.

121. "Almost every cardinal idea in this poem is to be found in Kierkegaard, and its hero's life has its prototype in his. It actually seems as if Ibsen had aspired to the honor of being called Kierkegaard's poet" (Brandes, *Henrik Ibsen*, p. 21).

122. Rosenfeld, "Niels Bohr's Contribution in Epistemology," p. 52.

123. Ibid., p. 51.

124. Nielson, "Memories of Niels Bohr," p. 29.

125. Rosenfeld, "Bohr's Contribution to Epistemology," p. 48.

126. Rosenfeld, "Niels Bohr's Contribution to Epistemology," p. 49.

127. Rozental, *Niels Bohr*, p. 13; Nielson, "Memories of Niels Bohr," p. 22.

128. Rosenfeld, "Niels Bohr's Contribution to Epistemology," p. 47.

129. L. Rosenfeld, "Niels Bohr in the Thirties," p. 132.

130. Ibid.

131. Ibid., p. 133; see also Rosenfeld, "Niels Bohr's Contribution to Epistemology," p. 47.

132. William James, *A Pluralistic Universe*, p. 280.

133. Einstein, in *Living Philosophies* (New York, 1931), p. 4.

134. Rozental, *Niels Bohr*, p. 164.

135. K. B. Anderson, "Political and Cultural Development in Nineteenth-Century Denmark," in *Scandinavian Democracy: Development of Democratic Thought and Institutions in Denmark, Norway and Sweden*, ed. J. A. Lawverys (Copenhagen, 1958), p. 150.

136. Ivan Maisky, *Journey into the Past*, trans. Frederick Holt (London, 1962), p. 167.

137. Pihl, "Niels Bohr and the Danish Community," p. 262. Edmund Gosse observed that Grundtvig's folk-schools were "extremely popular, and the spirit of hatred towards the German 'tyrant' is strongly fostered in them, for every Grundtvigian is, above all things, intensely a Dane" (*Studies in the Literature of Northern Europe*, p. 164).

138. Kierkegaard, *Stages on Life's Way*, p. 439.

139. Pihl, "Niels Bohr and the Danish Community," p. 292.

140. Rozental, *Niels Bohr*, p. 262; See also, Hans Bohr "My Father," p. 335.

141. George Paget Thomson, *J. J. Thomson: Discoverer of the Electron* (London, 1964; reprint ed. New York, 1966), p. 181. Max Born writes: "Pauli . . . has used the expression 'styles,' styles of thinking, styles not only in art, but also in science." "Physical theory has its styles . . . relatively a priori with respect to that period" (*Physics in My Generation* [London, 1956], p. 123).

142. *Dansk Biografisk Leksikon*, vol. 3, pp. 371, 379.

143. Rozental, *Niels Bohr*, p. 13.

144. Sir Charles Sherrington, *Goethe on Nature and Science* (Cambridge, 1942), p. 15.

145. See Hermann von Helmholtz, "On Goethe's Scientific Researches," trans.

H. W. Eve, in *Popular Scientific Lectures* (1881; reprint ed., Dover, 1962), p. 11. As Werner Heisenberg describes it:

> Newton's starting point appeared strange and unnatural to Goethe. White light, that is, light in its purest form, is to be downgraded to a composite. Instead the physicist is to accept as the basic form a light tormented and forced through narrow slits, lenses, prisms and all sorts of complicated devices" (*Philosophic Problems of Nuclear Science*, trans. F. C. Hayes, [New York, 1952; reprint ed., 1966], p. 70).

146. Sherrington, *Goethe on Nature and Science*, p. 9.
147. Arthur O. Lovejoy, *The Reason, The Understanding, and Time* (Baltimore 1961), p. 130.
148. Sigmund Freud, *Two Case Histories, The Standard Edition of the Complete Psychological Works of Sigmund Freud*, trans. James Strachey et al. (London, 1955), vol. 10, p. 244.
149. See Freud, "The Disposition to Obsessional Neurosis," *Standard Edition*, vol. 12, p. 324; *Two Case Histories*, pp. 241–243.
150. Freud, *Two Case Histories*, p. 245.
151. Rozental, *Niels Bohr*, p. 133; see also Rosenfeld, "Niels Bohr's Contributions to Epistemology," p. 47.
152. Heisenberg, *Physics and Beyond*, p. 50.
153. Peter Hallberg, *The Icelandic Saga*, trans. Paul Schach (Lincoln, Neb., 1962), p. 126.
154. E. R. Eddison, trans., *Egil's Saga* (Cambridge, 1930), pp. 60, 61, 75, 76.
155. *The Jewish Encyclopaedia* (New York, 1901), vol. 1, p. 194; Rozental, *Niels Bohr*, pp. 14, 290.
156. In 1836 Goldschmidt was prevented by the Danish Church orthodoxy from taking a B.A. degree at Copenhagen. He founded the liberal journal *The Corsair* which Kierkegaard fought, though as for its editor, writes Lowrie, "there was no other man to whom he felt so much drawn" (*Kierkegaard*, p. 438). *The Jewish Encyclopedia*, vol. 6 (New York, 1904), p. 27; *Denmark's Best Stories*, ed. Hanna Astrup Larsen (New York, 1928), p. 70.
157. Yahil, *The Rescue of Danish Jewry*, pp. 10, 11. Also see Julius Moritzen, "Denmark's Jews," *Contemporary Jewish Record*, vol. 3 (May–June 1940), pp. 276–277.
158. Yahil, *The Rescue of Danish Jewry*, p. 18.
159. Ibid., p. 91.
160. Ibid., p. 201.
161. Ibid., p. 330.
162. Chr. Buur, "Hanna Adler," *Dansk Biografiak Leksikon*, vol. 1, pp. 126–127.
163. Yahil, *The Rescue of Danish Jewry*, p. 490.
164. Heisenberg, *Physics and Beyond*, p. 211.
165. Freud, *Inhibitions, Symptoms and Anxiety, Standard Edition*, vol. 20, p. 220, 165–166.
166. Erwin Schrödinger, *Science, Theory, and Man*, trans. James Murphy (1935; reprint ed., New York, 1957), pp. 100, 112–113.
167. Werner Heisenberg, *Philosophic Problems of Nuclear Science*, trans. F. C. Hayes (New York, 1966), p. 124.
168. Werner Ohnsorge, "August Heisenberg," *Neue Deutsche Biographie*, (Berlin, 1969), vol. 8, pp. 455–456. Alexander Alexandrovich Vasiliev, *History of the Byzantine Empire: 324–1453* (Madison, Wis., 1952), pp. 552, 554–55, 30, 550, 713.
169. Heisenberg, *Physics and Beyond: Encounters and Conversations*, trans. Arnold J. Pomerans (New York, 1971), pp. 42, 47.
170. Ibid., p. 28.
171. Heisenberg, *The Physicist's Conception of Nature*, trans. Arnold J. Pomerans (London, 1958), pp. 53–54.
172. Heisenberg, *Der Teil und das Ganze: Gespräche im Umkreis der Atomphysik* (Munich, 1969), pp. 19–20.

173. Heisenberg, *A Physicist's Conception of Nature*, pp. 59–60.
174. Ibid., p. 60.
175. Ibid., p. 58.
176. Ibid., p. 58.
177. Ibid., pp. 63–64.
178. Ibid., p. 59.
179. Heisenberg, *Der Teil und das Ganze*, p. 11.
180. Ibid., p. 11 f.
181. Ibid., p. 12.
182. Ibid., p. 13.
183. Ibid., p. 20.
184. Ibid., p. 16.
185. Heisenberg, *Physics and Philosophy: The Revolution in Modern Science* (New York, 1962), p. 69.
186. Ibid., p. 67.
187. Ibid., p. 71.
188. Ibid., pp. 71–72.
189. Walter Z. Laqueur, *Young Germany: A History of the German Youth Movement* (New York, 1962), p. 141. Cf. Theodore Abel, *Why Hitler Came Into Power: An Answer Based on the Original Life Stories of Six Hundred of His Followers* (New York, 1938), pp. 45, 51; William K. Pfeiler, *War and the German Mind* (New York, 1941), p. 83 f. Max Weinreich, *Hitler's Professors* (New York, 1946), p. 11 ff.
190. Laqueur, *Young Germany*, pp. 142–143.
191. Robert Jungk, *Brighter than a Thousand Suns: A Personal History of the Atomic Scientists*, trans. James Cleogh (New York, 1958), p. 26.
192. Laqueur, *Young Germany*, pp. 138–141.
193. Alan Mitchell, *Revolution in Bavaria, 1918–1919: The Eisner Regime and the Soviet Republic* (Princeton, 1965), pp. 99–101. Kurt Eisner declared in December, 1918: "A politician who is not at the same time a poet is also not a politician. The poet is the prophet of the future world." Political poetry seemed indeed about to rule the world. Charles B. Maurer, *Call to Revolution: The Mystical Anarchism of Gustav Landauer* (Detroit, 1971), p. 186.
194. Georg Franz, "Munich: Birthplace and Center of the National Socialist German Workers' Party," *The Journal of Modern History*, vol. 29 (December 1957), pp. 319–334.
195. A. J. Ryder, *The German Revolution of 1918: A Study of German Socialism in War and Revolt* (Cambridge, Mass., 1967), p. 213; Mitchell, *Revolution in Bavaria*, p. 272.
196. Georg Franz, "Munich," p. 319. For contemporary anarchist views of Mühsam and Landauer, see C. W., "Gustav Landauer," *Anarchy*, vol. 5, no. 54 (August 1965), pp. 244–251; J. F., "Erich Mühsam," ibid., pp. 255–256.
197. Ernst Toller, *I Was a German: An Autobiography*, trans. Edward Crankshaw (London, 1934). Pp. 98, 90. See also Margarete Turnowsky-Pinner, "A Student's Friendship with Ernst Toller," *Publications of the Leo Baeck Institute: Year Book XV* (London, 1970), pp. 214–217.
198. It is noteworthy that the English philosopher and mathematician, Alfred North Whitehead, also found in Plato's *Timaeus* a classical model for the refutation of scientific materialism. The distinguished expositor of Plato, A. E. Taylor, felt that Whitehead's philosophy of nature had indeed recaptured the genuine meaning of Plato's cosmology. See A. E. Taylor, *A Commentary on Plato's "Timaeus"* (Oxford, 1926), pp. 71, 313, 326–327.
199. Heisenberg, *Physics and Beyond*, p. 22.
200. Ibid., p. 17.
201. Ibid., p. 43.
202. Ibid., p. 45.
203. Ibid., p. 475.
204. Ibid., p. 53.
205. Ibid., p. 54.
206. Ibid., p. 83.

207. Ibid., p. 245.
208. Ibid.
209. Ibid., p. 114.
210. Ibid., p. 186.
211. Ibid., p. 166.
212. Ibid., p. 78.
213. Werner Heisenberg, "Über den anschaulichen Inhalt der quantentheoretischen Kinematik und Mechanik," [On the Intuitive Content of Quantum Theoretical Kinematics and Mechanics]," *Zeitschrift für Physik*, vol. 43 (1927), pp. 174–175.
214. Ibid., p. 197.
215. Heisenberg, "Schwabing versöhnlicher geist," in *Geliebtes Schwabing: Verse und Prosa*, ed. Werner Rukwid, (Munich, 1970), pp. 152–153.
216. Schrödinger, *Science, Theory, and Man*, pp. xvii, 147.
217. Max Toperczer, "Franz Exner," *Neue Deutsche Biographie*, vol. 4, (Berlin, 1959), p. 699.
218. William T. Scott, *Ervin Schrödinger: An Introduction to his Writings* (Amherst, 1967), pp. 2, 30–31; Martin J. Klein, *Paul Ehrenfest: Volume I: The Making of a Theoretical Physicist* (Amsterdam, 1970), p. 175; Ernst Cassirer, *Determinism and Indeterminism in Modern Physics*, trans. O. Theodore Benfey (New Haven, 1956), pp. 84–87.
219. "The first mention published in print and referring to many-valued logics is to be found in *The Farewell Lecture Delivered by Professor Jan Lukasiewicz in the Great Hall of Warsaw University on March 7, 1918*. This proves that his system of three-valued logic was constructed by Lukasiewicz as early as 1917." See Jan Lukasiewicz, *Elements of Mathematical Logic*, trans. Olgierd Wojtasiewicz (Warsaw, 1963), p. 119.
220. Lukasiewicz, "On Determinism," in *Polish Logic*, ed. Storrs McCall (Oxford, 1967), pp. 22–23.
221. Ibid., p. 19.
222. Ibid., pp. 37, 4.
223. Ibid., pp. 38–39.
224. Heisenberg, *The Physical Principles of the Quantum Theory*, trans. C. Eckart and F. C. Hoyt (Chicago, 1930), p. 20.
225. Heisenberg, *Physics and Philosophy: The Revolution in Modern Science*, p. 181.
226. Jungk, *Brighter than a Thousand Suns*, p. 43.
227. Bohr, *Essays 1958–1962 on Atomic Physics and Human Knowledge*, pp. 5–6.
228. Max Born, *Natural Philosophy of Cause and Chance* (Oxford, 1949), p. 107.
229. Wolfgang Pauli, "Hans Reichenbach, Philosophic Foundations of Quantum Mechanics." *Dialectica*, vol. 1, no. 2 (1947), pp. 177–178. Max Born, "Einstein's Statistical Theories," in *Albert Einstein*, ed. P. A. Schilpp, p. 177.
230. Ludwig Wittgenstein, *Tractatus Logico-Philosophicus*, trans. D. F. Pears and B. F. McGuinness (London, 1961). He also wrote: "But it is clear that the causal nexus is not a nexus at all" (*Notebooks, 1914–1916*, trans. G. E. M. Anscombe [Oxford, 1961], p. 84e).
231. Paul Engelmann, *Letters from Ludwig Wittgenstein, with a Memoir*, trans. L. Furtmuller (New York, 1967), pp. 94, 73, 121.
232. Wittgenstein, *Tractatus Logico-Philosophicus*, pp. 145–147.
233. Wittgenstein, *Notebooks, 1914–1916*, p. 82e.
234. Ibid., p. 75e.
235. Wittgenstein, *Notebooks, 1914–1916*, p. 73e.
236. Ibid., p. 78e.
237. Engelmann, *Letters from Wittgenstein*, p. 107.
238. Wittgenstein, *Notebooks, 1914–1916*, p. 82e.
239. Engelmann, *Letters from Wittgenstein*, pp. 53, 55. John Moran, "Wittgenstein and Russia," *New Left Review*, no. 73 (May–June 1972)pp. 85–89.
240. Engelmann, p. 100.
241. Ibid., p. 32.
242. Ibid., pp. 23, 27, 37, 41.

243. Robert Motherwell, ed., *The Dada Painters and Poets: An Anthology* (New York, 1951), p. xviii. See also Etienne Gilson, *Painting and Reality* (Princeton, 1957), p. 269.

244. Hans Richter, *Dada: Art and Anti-Art* (New York, 1965), p. 16. See also Miklavz Prosenc, *Die Dadaisten in Zurich* (Bonn, 1967), p. 58.

245. Richter, *Dada*, p. 16.

246. Ibid., pp. 56, 58.

247. Ibid., p. 60.

248. Ibid., p. 34.

249. Ibid., p. 219.

250. Georges Ribemont-Dessaignes, "History of Dada," in Motherwell, *The Dada Painters and Poets*, p. 102.

251. Richard Huelsenbeck, "En Avant Dada: A History of Dadaism," (1920), in Motherwell, *The Dada Painters and Poets*, p. 36.

252. Huelsenbeck, "En Avant Dada," pp. 23–24.

253. Ibid., pp. 39, 42, 44–45.

254. Walter Mehring, a leading Berlin Dadaist, wrote how he had already responded in 1911 to the futurist wave. Aroused to contradiction of the "authority of my father," he was enflamed by a leaflet which called for action against the fall show: "Destroy the Museums. Burn the libraries down" (Prosenc, *Die Dadaisten in Zurich*, p. 18).

255. Huelsenbeck, *Dada Almanach* (Berlin, 1920; reprint ed., New York, 1966), p. 125. Tzara later recognized explicitly his similarity in spirit with Heisenberg's indeterminacy. Cf. Elmer Peterson, *Tristan Tzara: Dada and Surrational Theorist* (New Brunswick, 1971), pp. 133, 211.

256. Richter, *Dada*, p. 49.

257. Ibid., pp. 51–52.

258. Ibid., p. 55.

259. Ibid., p. 59.

260. Matthew Josephson, *Life among the Surrealists: A Memoir* (New York, 1962), p. 337.

261. Ibid., p. 220.

262. Ibid., pp. 114, 22.

263. Ibid., p. 226.

264. Ibid., p. 103.

265. Marie-Antoinette Tonnelat, "La Notion de Réalité Physique et L'Oeuvre de M. Louis de Broglie," *Louis de Broglie: Physicien et Penseur* (Paris, 1953), p. 150. [Auhor's translation.]

266. Peter Blake, *The Master Builders* (London, 1960), p. 41. Le Corbusier, *Creation Is a Patient Speech*, trans. James Palmer (New York, 1960), p. 49.

267. Blake, *The Master Builders*, p. 29.

268. Ibid., p. 39. See also Le Corbusier and Pierre Jeanneret, *Oeuvre Complète 1910–1929* (Zurich, 1964), pp. 29, 45.

269. Blake, *The Master Builders*, p. 56.

270. Ibid., p. 42.

271. Ibid., p. 33.

272. Ibid., p. 36.

273. Ibid., p. 37.

274. Ibid., p. 119.

275. Charles S. Peirce, *Chance, Love and Logic: Philosophical Essays*, ed. Morris R. Cohen (New York, 1923), pp. 157–158.

276. Bertrand Russell, *Mysticism and Logic* (London, 1927), p. 38.

277. Russell, "The Philosophy of Logical Atomism," *The Monist*, vol. 29 (1919), p. 367; vol. 28 (1918), p. 497.

278. Russell, *The Analysis of Matter* (London, 1927), p. 38.

279. Russell, *The Analysis of Matter*, p. 74. Russell, *The ABC of Relativity* (New York, 1925), p. 196. Also cf. Vivian Harper, "Bertrand Russell and the Anarchists," *Anarchy*, no. 109, Vol. 10, (March 1970), pp. 68–77. Bertrand Russell, *Proposed Roads to Freedom* (New York, 1919), pp. 183, 192, 211.

280. Russell, *The Scientific Outlook* (New York, 1931), p. 95.

281. Russell, *The Conquest of Happiness* (1930; reprint ed., New York, 1951), p. 104.

282. Russell, *Philosophy* (New York, 1927), pp. 98, 106.

283. *The Autobiography of Bertrand Russell: The Middle Years: 1918–1944* (1968; reprint ed., Bantam ed. 1969), pp. 6, 95, 110, 166, 140, 225.

284. Russell, *Religion and Science* (New York, 1935), pp. 161–162. "Communism," wrote Russell in 1932,

> creates an economic system which appears to be the only practicable alternative to one of masters and slaves. It destroys that separation of the school from life which the school owes to its monkish origin, and owing to which the intellectual in the West is becoming an increasingly useless member of society. It offers to young men and women hope which is not chimerical and an activity in the usefulness of which they feel no doubt. And if it conquers the world, as it may do, it will solve most of the major evils of our time. On these grounds, in spite of reservations, it deserves support (cf. Russell, *Education and the Modern World* [New York, 1932], pp. 189–190).

285. Louis de Broglie, *Physics and Microphysics*, trans. Martin Davidson (New York, 1955), p. 200.

286. Rosenfeld, "L'Evidence de la Complémentarité," in *Louis de Broglie: Physicien et Penseur*, ed. André George (Paris, 1953), p. 63.

287. Auguste Comte, *Cours de Philosophie Positive* (1830; 5th ed., Paris, 1892), vol. 1, pp. 29–30.

288. Høffding, *A History of Modern Philosophy*, vol. 2, p. 343.

289. William James, *The Principles of Psychology* (New York, 1890), vol. 1, p. 188.

290. In a noteworthy essay, which came to my hand belatedly, Paul Forman has adduced much evidence to show that German physicists at the end of World War I were, as a consequence of their sociointellectual environment, predisposed toward "acausal laws of nature." Professor Forman also argues that the work of Oswald Spengler published in 1919 was an especial source of the revolt against determinism. In this regard, I differ from Professor Forman's interpretation. For, as far as history was concerned, Spengler was actually the strictest determinist. To be sure, he distinguished between Destiny and Causality, but both indeed were forms of determinism. Destiny was what we might call a "meteorological determinism," making use of the analogy of the seasons, but subjecting history to a fate that was inevitable and unavoidable. The direction of Destiny was to be grasped intuitively, whereas Causality was apprehended mathematically, as the sum total of the functional equations describing the relations of events. The contrast between Destiny and Causality was not one between freedom and determinism. Thus Richard von Mises justly regarded Spengler as a historian in the determinist tradition who mistakenly thought he was the "first" to try "to predetermine history" (Cf. Richard von Mises, *Positivism: A Study in Human Understanding*, trans. Jerry Bernstein and Roger G. Newton [Cambridge, 1951], p. 361, and Paul Forman, "Weimar Culture, Causality and Quantum Theory, 1918–1927: Adaptation by German Physicists and Mathematicians to a Hostile Intellectual Environment," in *Historical Studies in the Physical Sciences*, vol. 3 [1971]).

In the case of von Mises, an attachment to the Roman Catholic Church and the poetry of Rainer Maria Rilke were the most potent emotional-intellectual influences away from determinism. Von Mises had "the greatest privately-owned Rilke Collection in the world." At the same time, his closest friend Philipp Frank, who had no theological or poetical propensity whatsoever, was equally drawn to the notions of indeterminacy. A wavelike movement toward a new "generational postulate" was taking place which was primary, and the postwar world was going to be regarded from the standpoint of these novel emotions (Sydney Goldstein, "Biographical Note" in *Selected Papers of Richard von Mises*, ed. G. Birkhoff et al. [Providence, 1963], pp. ix, xi, xii, xiii).

291. Louis de Broglie, *Continu et Discontinu en Physique Moderne* (Paris, 1941), p. 123.

292. P. W. Bridgman, *The Logic of Modern Physics* (New York, 1927), pp. 93–94.

293. F. S. C. Schiller, *Studies in Humanism*, 2d ed. (London, 1912), p. 85. Schiller felt that adherence to the classical law of identity was part and parcel of the "bureaucratic spirit" which has always tried to suppress "the revolts of the men of action" (cf. *Formal Logic: A Scientific and Social Problem* [London, 1912], p. 407).

294. In films, wrote Sergei Eisenstein: "We brought collective and mass action onto the screen, in contrast to individualism and the 'triangle' drama of the bourgeois cinema." Discarding the individualist conception of the bourgeois hero, his films presented "the mass ás hero" (*Film Form: Essays in Film Theory*, trans. Jay Leyda [London, 1963], p. 16.

295. Cf. John Dewey, *Individualism Old and New* (New York, 1929), pp. 51 ff; José Ortega y Gasset, *The Revolt of the Masses, trans.* (New York, 1932), chaps. 5, 6; Lewis Mumford, *The Story of Utopias* (New York, 1922), p. 289.

296. Charlotte Touzalin Muret, *French Royalist Doctrines since the Revolution* (New York, 1933), pp. 105–118.

297. Albert, Duc de Broglie, *The King's Secret: Being the Secret Correspondence of Louis XV, with his Diplomatic Agents from 1752–1774* (London, n.d.), vol. 1, p. v.

298. Ibid., pp. vi, vii, ix.

299. Comtesse Jean de Pange (Pauline de Broglie), "Comment J'ai Vu 1900: V", *La Revue des Deux Mondes* (April 15, 1962), p. 553.

300. De Pange, "Comment J'ai Vu 1900: L'Age Ingrat", *La Revue des Deux Mondes* (August 15, 1964), pp. 569, 570, 575, 576.

301. De Pange, "Deuil 1900", *La Revue des Deux Mondes* (February 1, 1967), pp. 359–360.

302. de Pange, "Souvenirs de Jeunesse", *La Revue des Deux Mondes* (March 15, 1965), pp. 188–191. See also *Dictionary of Scientific Biography* (New York, 1970), vol. 2, p. 487.

303. M. de Broglie, "La Jeunesse et les Orientations Intellectuelles de Louis de Broglie," p. 427.

304. Maurice de Broglie, "La Jeunesse et les Orientations Intellectuelles de Louis de Broglie," in *Louis de Broglie: Physicien et Penseur*, ed. André George, (Paris, 1953), pp. 424–425.

305. Louis de Broglie, *Savants et Découvertes* (Paris, 1951), p. 64.

306. Ibid., p. 55.

307. Ibid.

308. Louis de Broglie, "The Concepts of Contemporary Physics and Bergson's Ideas on Time and Motion," in *Bergson and the Evolution of Physics*, ed. and trans., P.A.Y. Gunter (Knoxville, 1969), p. 46. Louis de Broglie, *Physique et Microphysique* (Paris, 1947), p. 192 [this passage was among those omitted in the translation, *Physics and Microphysics*, trans. Martin Davidson (New York, 1955)].

309. *Louis de Broglie: Physicien et Penseur*, p. 429.

310. Arthur O. Lovejoy, "The Metaphysician of the Life Force," *The Nation*, vol. 89 (1909) pp. 298–301.

311. Daniel Halévy, *Péguy and "Les Cahiers de la Quinzaine,"* trans. Ruth Bethell (London, 1946), p. 65. Péguy said of Bergson: "He will never be forgiven for setting us free" (ibid., p. 29).

312. André Henry, *Bergson: Maître de Péguy* (Paris, 1948), p. 13. Bergson was denied a post at the Sorbonne because he was "not of the party" (Halévy, *Péguy and "Les Cahiers de la Quinzaine,"* p. 93).

313. Etienne Gilson, *God and Philosophy* (New Haven, 1941), pp. xii, xiii. Bergson also deeply affected the novelists Marcel Proust and André Gide; see George D. Painter, *Marcel Proust: A Biography* (London, 1959), vol. 1, p. 68, and Wallace Fowlie, *André Gide: His Life and Art* (New York, 1965), pp. 7–8.

314. Enid Starkie, "Bergson and Literature," in *The Bergsonian Heritage*, ed. Thomas Hanna (New York, 1962), p. 76. Isaac Benrubi, *Souvenirs sur Henri Bergson* (Neuchâtel, Paris, 1942), pp. 11, 32; *Contemporary Thought of France*, trans. Ernest B. Dicker (New York, 1926), p. 169. Pierre Andreu, "Bergson et Sorel," in *Les Etudes Bergsoniennes* (Paris, 1952), vol. 3, pp. 41–78, 170–180.

315. Jacques Maritain, *Antimoderne*, cited in Raisaa Maritain, *We Have Been Friends Together: Memoirs*, trans. Julie Kernan (New York, 1942), pp. 61–62.

316. Ibid., p. 79.

317. Julien Benda, *Sur le Succès du Bergsonisme* (Paris, 1914), p. 165.

318. Romain Rolland, *Péguy* (Paris, 1944), pp. 25–40. See also Robert J. Niess, *Julien Benda* (Ann Arbor, 1956), pp. 124–128.

319. Charles Péguy, *Pensées*, p. 51.

320. Joseph Lotte, "Le Sorbonne Moderne," in *Cahiers de la Quinzaine*, (September 19, 1911), pp. 140–142.

321. Alfred Binet, "Une Enquête sur l'Evolution de l'Enseignement de la Philosophie," *L'Année Psychologique* (Paris, 1908), 14th yearly issue, p. 19 f, pp. 229–230.

322. Letter of de Broglie to the author, October 1971, translated.

323. Ibid.

324. *Letters of Marcel Proust*, trans. Mina Curtiss, introd. Harry Levin (New York, 1949), p. xxvii.

325. Ibid., p. 197.

326. Curiously, Proust knew the Princess Albertine de Broglie; he used to see her at the Salon d'Haussonville which he frequented. The Salon was the headquarters of the nobly-born members of the Academy, the "party of the dukes." Its master was the grandson of Albertine de Staël, who had married Achille-Charles-Victor de Broglie, later prime minister of France. It has been conjectured that Proust named one of his characters "Albertine" partially under the influence of the princess (George D. Painter, *Proust: The Early Years* [Boston, 1959], pp. 202–203; *Proust: The Later Years* [Boston, 1965], p. 76).

327. Henri Bergson, *Matter and Memory*, trans. Nancy Margaret Paul and W. Scott Palmer (London, 1911), pp. 261–266. In a lecture at Oxford in 1920, Bergson described how the atomic, discontinuous mode of thought evolved into its opposite; "masses are pulverized into molecules; molecules into atoms, atoms into electrons or corpuscles," and lastly, "a series of extremely rapid vibrations," "a movement of movements" (*The Creative Mind*, trans. Mabelle L. Andison [New York, 1946], p. 175).

328. De Broglie, "Vue d'Ensemble sur mes Travaux Scientifiques," *Louis de Broglie: Physicien et Penseur*, p. 458.

329. *Louis de Broglie: Physicien et Penseur*, p. 458.

330. As André George remarks: "the passage through the Wireless—military though it was—would finally no more harm his scientific youth than Einstein's was when he had to examine, in the Berne Office, the purely technical procedures of inventors seeking patents" (ibid., p. 449).

331. De Broglie, "L'Oeuvre Scientifique du Général Ferrié," *Savants et Découvertes*, p. 87.

332. Ibid., p. 88.

333. *Louis de Broglie: Physicien et Penseur*, p. 459.

334. Ibid., pp. 461–462.

335. *Louis de Broglie: Physicien et Penseur*, pp. 430, 462.

336. Frank, *Einstein: His Life and Times*, p. 210; André George, "Louis de Broglie," *Laffont-Bompiani Dictionnaire Bibliographique des Auteurs*, 2d ed. (Paris, 1964), p. 218. Citation from Walter Hertler, "The Departure from Classical Thought in Modern Physics," in *Albert Einstein: Philosopher-Scientist*, ed. Schilpp, p. 186. Emile Meyerson, *Réel et Determinisme dans la Physique Quantique* (Paris, 1933), p. 20.

337. Louis de Broglie, *Sur les Sentiers de la Science* (Paris, 1960), p. 367; Max Born, "Einstein's Statistical Theories," in *Albert Einstein: Philosopher-Scientist*, p. 174. De Broglie curiously misdescribes Davisson and Germer as "research engineers" and as principally preoccupied with applied electronics. Both men were physicists. Also Max Born's account is somewhat at variance with that in the *Dictionary of Scientific Biography*. Davisson, according to the latter, discussed his experiments with Born and James Franck in 1926 at Oxford at the meeting of the British Association for the Advancement of Science. There he was persuaded that his results were due to de Broglie waves. He returned home and in 1927 performed the systematic experiments

on the diffraction of electrons that confirmed the existence of de Broglie waves. Cf. Kenkichiro Koizumi, "Clinton Joseph Davisson," *Dictionary of Scientific Biography*, vol. 3 (New York, 1971), pp. 597–598.

338. Letter of Professor Victor F. Lenzen to the author, January 19, 1972.

339. Louis de Broglie, "Recherches sur la Théorie des Quanta," *Annales de Physique*, vol. 3 (1925), p. 33.

340. Ibid., p. 81.

341. De Broglie, "Recherches sur la Théorie des Quanta," p. 45.

342. Philip E. B. Jourdain, "Maupertuis and the Principle of Least Action," *The Monist*, vol. 22 (1912), p. 286; E. T. Whittaker, *A History of the Theorie of Aether and Electricity from the Age of Descartes to the Close of the Nineteenth Century* (London, 1910), pp. 9–10, 102–103; A. E. Bell, *Christian Huygens and the Development of Science in the Seventeenth Century* (London, 1947), pp. 81, 182–183.

343. De Broglie, "Recherches sur la Théorie des Quanta," p. 126.

344. E. T. Bell, *The Development of Mathematics*, 2d ed. (New York, 1945), pp. 142–143; R. Huron, "L'Aventure Mathématique de Fermat," in Fédération des Sociétés Académiques et Savantes de Languedoc, Pyrénées, Gascogne, *Pierre de Fermat: Toulouse et Sa Région* (Toulouse, 1966), pp. 16–17, 33–34.

345. Léon Velluz, *Maupertuis* (Paris, 1969), pp. 43–45, 59–63; Pierre Brunet, *Maupertuis: Etude Biographique* (Paris, 1929), vol. 1, pp. 33 ff.

346. De Broglie, "Recherches sur la Théorie des Quanta," p. 56.

347. Ibid., p. 126.

348. De Broglie, "The Concepts of Contemporary Physics and Bergson's Ideas on Time and Motion," p. 52.

349. Ibid., cited from Bergson, *Matter and Memory*, p. 247.

350. Ibid., p. 54.

351. Ibid.

352. Ibid., p. 57.

353. Ibid., p. 59.

354. Here is what I said about it in *The Thought and the Moving*. . . . I couldn't take account of it in my preceding works because the new physics didn't exist. But I cleared this point up progressively in my last volumes until the last where I showed the direct consequences of the theory of Louis de Broglie.

(Jacques Chevalier, *Entretiens avec Bergson* [Paris, 1959], p. 292. This conversation with Bergson is reported as having taken place on March 27, 1939.)

355. Bergson, *Creative Evolution*, pp. 280, 293.

356. De Broglie, "A Tentative Theory of Light Quanta," *The London, Edinburgh, and Dublin Philosophical Magazine and Journal of Science*, vol. 47 (1924), p. 446.

357. George Gamow, *Thirty Years That Shook Physics*, p. 81.

358. Bergson, *Time and Free Will: An Essay on the Immediate Data of Consciousness*, trans. F. L. Pogson (London, 1910), pp. 100, 105, 111.

359. Bergson, *The Creative Mind*, p 45.

360. De Broglie, "The Concepts of Contemporary Physics and Bergson's Ideas on Time and Motion," p. 50.

361. Ibid., pp. 46–47.

362. De Broglie, *Physics and Microphysics*, p. 7.

363. "Lettre de M. Albert Einstein à M. André Metz," *Revue de Philosophie*, vol. 31 (1924), p. 440. See also P.A.Y. Gunter, ed. *Bergson and the Evolution of Physics*, p. 190. To Solovine, Einstein wrote in 1923: "Bergson, in the book on the theory of relativity has committed some bad blunders. God will forgive him" (Einstein, *Lettres à Maurice Solovine*, p. 45).

364. De Broglie, *Physics and Microphysics*, p. 25 f.

365. De Broglie, "Un Glorieux Moment de la Pensée Scientifique Française," *Physique et Microphysique*, p. 246.

366. Ibid., p. 247.

367. Bergson regarded Maine de Biran as the greatest metaphysician France had produced since Descartes and Malebranche, and as having shown that the knowledge of the depths of our inner life takes us to reality "in itself" (Jacques Chevalier, *Henri Bergson*, trans. Lilian A. Clare [New York, 1928], pp. 16–17).

3

Generational Movements and "Scientific Revolutions"

The Idea of Scientific Revolution

Our previous inquiries have shown how crucial for the young scientist is that step in his life which Ignazio Silone calls "the choice of comrades." Though Silone was writing of political choices, what he says bears on the scientist's choices as well:

> The revolt of a young man against tradition is a frequent occurrence in all times and all countries, and it rarely happens without at least some ambiguity. Depending on circumstances, the revolt may lead into the Foreign Legion, gangsterism, the film world or political extremism. What defined my revolt was the choice of companions.[1]

Einstein made his choice of comrades among Friedrich Adler and the Olympia circle; Niels Bohr made his among the *Ekliptika.* Werner Heisenberg identified himself with the White Knights of the German Youth Movement; Louis de Broglie was spiritually one of the young Bergsonians. With the choice of comrades went a choice of philosophy: the Zurich-Berne circle was Machian; the Copenhagen group was Kierkegaardian; the Munich activists were romantic idealists. Each circle had its distinctive coordinates in the social world: the Zurich-Berne circle was international in its composition, its members mostly unemployed Jews (the son of an Austrian socialist leader, a down-at-the-heel Rumanian, an Italian engineer, some future high school teachers); the Copenhagen circle was comprised of children of the well-to-do Danish intellectual elite; members of the Munich group were children of the anti-semitic, nationalistic

German bourgeoisie; an aristocratic and military circle nurtured the more isolated de Broglie. The choice of comrades also involved a stand with respect to the contemporary political events: for Einstein, it signified a sympathy with socialism and internationalism; for Bohr, a loyalty to liberalism, moderation, and a mistrust of any total standpoint; for Heisenberg, an aversion to Marxism, materialism, and determinism, a longing for a world in which ideal forms were supreme; for Louis de Broglie, an idealistic conservatism critical of the Third Republic and its official positivism. The circle, its philosophy, its social composition, and its outlook on the contemporary world-historical scene—all these were social conditions important to the formation of scientific standpoints.

No social movement, class, or philosophy can be said to be privileged in the apprehension of physical reality. Each standpoint, under certain world-historical conditions, when it intersects favorably with a specific stage in the experimental and theoretical work of a science, has its moment of privileged insight. No great work in science, art, or politics can be accomplished, however, unless some definite philosophy, or sense of the underlying pattern of things, moves its proponent. But the creative usefulness of a philosophy depends on a coincidence with a situation, set by the contemporary experimental technology and the problems of physical science, to which its temper of mind is uniquely suited. The young English absolutists at the beginning of the twentieth century were not in a frame of mind or emotion to conceive the theory of relativity. Einstein, longing in the late 1920s for an ordered Spinozist universe, was in no mood to perceive complementarities or indeterminacies.

The creative talent at turning points in the history of science has generally the traits of a generational revolutionist. This is especially true, as we have seen, in the cases of Einstein and Heisenberg. Moreover, during the last two centuries when some basic theoretical controversy has divided scientists with respect to their conceptions as to the underlying mode of thought, the division has been primarily along generational lines. The most creative original scientists have looked to the support of the young rather than the old. When, for instance, the now classical paper of Hermann von Helmholtz on the conservation of energy was rejected by the editor of the *Annalen*, "all the younger physicists and physiologists" were enthusiastically behind Helmholtz, interceding and pressing for its publication. Then, when it was published independently, "the memoir was enthusiastically welcomed by the younger physicists and physiologists of Berlin," while "the older scientists with hardly an exception rejected the ideas which the work expressed, fearing, strangely enough, that such speculations would revive the phantasm of Hegel's nature-philosophy. . . ."[2] The older generation had in its youth fought for liberation from the *Naturphilosophie*; to them, the spectacle of the new generation affirming a novel, comprehensive world-generalization, aroused memories of a traumatic experience. A few years later, the theory of natural selection evoked a similar generational conflict. Charles Darwin was not a young man when he published *The Origin of Species* in 1859, but he could count on no allies among the old: "Nearly all men past a moderate age, either in actual years or in mind, are, I am fully convinced, incapable of looking at facts under a new point of view," he wrote to Joseph Hooker in 1860.[3] Then to Thomas Henry Huxley he wrote later that same year: "If my view is ever to be generally adopted, it will be by young men growing up and replacing the old workers. . . ."[4] The cleavage of generations determined allegiances within science: "The younger and middle-aged geologists are coming round. . . . Not one of the older geologists (except Lyell) has been ever shaken in his views of the eternal immutability of species," wrote Darwin in 1861.[5] Darwin jibed in his letters at the "old fogies."

This experience of generational insurgence, deeply felt in the consciousness of scientists, more than any other reason accounts for the appeal that the notion of scientific revolution has had. The language of scientific revolution entered the human vocabulary during the period of social and generational restlessness that preceded the French Revolution.

It was Denis Diderot who in 1755 introduced the concept of scientific revolution. As the editor of the epoch-making *Encyclopaedia*, he had labored through the years with an unparalleled courage; sustained by a conviction in the redeeming power of scientific revolution, he had withstood the ordeals of censorship, imprisonment, and financial desperation. No one, in Diderot's view, as yet understood the heights which the human race might attain with the help of revolutions: "Revolutions are necessary; there have always been revolutions, and there always will be; the maximum interval between one revolution and another is a fixed quantity. . . ." The labors of extraordinary individuals acquired significance when their resultant was a scientific revolution.

Add to the labors of this extraordinary individual those of another like him, and of still others, until you have filled up the whole interval of time between one scientific revolution and the revolution most remote from it in time, and you will be able to form some notion of the greatest perfection attainable by the whole human race. . . .

Those who doubted the reality of scientific revolutions were at odds with the linguistic testimony provided by dictionaries:

Revolutionary changes may be less abrupt and less obvious in the sciences and liberal arts than in the mechanical arts, but change has nonetheless occurred. One need only open the dictionaries of the last century. One will not find under the word "aberration" the slightest hint of what our astronomers understand by this term; on "electricity," that extremely promising phenomenon, there will be found a few lines which contain nothing but false notions and ancient prejudices. How many terms are there relating to mineralogy or natural history of which the same could be said![6]

Thus, during the years preceding the French Revolution it became fashionable to speak of revolution in the sciences. Sometimes the anxiety arose that the revolution would be in a negative direction.[7] Usually, however, to speak of revolution was to speak of a joyful experience of progress. Kant spoke of the "beneficent revolution" in the point of view of physics.[8] D'Alembert, the gifted mathematical physicist and Diderot's coeditor, praised Descartes as "a chief of conspirators, who had the courage to rise up first against an arbitrary and despotic power, and preparing a signal revolution, laid the foundations of a government juster and happier such as he had not seen."[9] To D'Alembert revolution in science was also tied to the creative energies of the young. When Newtonian ideas supplanted the Cartesian, "it was indeed the young geometers, as much in France as in foreign countries, who decided the fate of the two philosophies," he wrote. For "young people who are ordinarily regarded as having poor judgment, are perhaps the best in philosophical matters and many others, when they are not deprived of light; for since everything is equally new to them, they have no other interest but to choose well."[10] A few years later, the noble-spirited Condorcet, in extolling D'Alembert before the Academy of Sciences after his death in 1783, spoke of the latter's discovery at the age of twenty-six of the principle of equilibrium as the basis for all the laws of mechanics: this principle, he said, introduced "the epoch of a great revolution in the physico-mathematical sciences."[11] Condorcet linked the attraction that a branch of study had for a youth like D'Alembert with the possibilities it offered for participation in a revolution. D'Alembert, he said, had first studied law but had dropped it, for Montesquieu had not yet done his work, and "one could not foresee the revolution which it would accomplish in our minds. . . ."[12] In his *Pensées*, reflecting on the intellectual climate of the eighteenth century, D'Alembert indeed affirmed that the time was one of "very remarkable change in our ideas," in manners, works, conversations, and of such rapidity that it heralded "an even greater change." "It was for time to fix the object, the nature and the limits of this revolution whose inconveniences and advantages our posterity will know better than we."[13]

Most influential of the intellectual movements which gave a vogue to the idea of scientific revolution was the socialist. Above all, it was the forerunner of the socialist wave, Count Claude-Henri de Saint-Simon (1760–1825), exalted by the vision of a new scientific, industrial society, to be governed by an intellectual elite, who wrote in grandiose, apocalyptic terms of the role of scientific revolutions. He addressed a "Letter to the Physiologists" (found in "Memoire sur la Science de l'Homme," *Oeuvres*, vol. 5 [Paris, 1966] whose help he felt was most essential for the making of a "great and useful scientific revolution":

> History shows that the scientific revolutions and the political ones have alternated, that they have been successively, with respect to one another, as causes and effects. Let us recall those which have been the most striking since the fifteenth century. This review will prove to you that the very next revolution must be a scientific revolution, even as my work will demonstrate, . . . that it is principally you who must make this revolution. . . .

To Saint-Simon, who said he had been educated by d'Alembert, political revolutions were the corollaries of the scientific ones of Copernicus, Bacon, Newton, and Locke; the French Revolution was the consequence (through the Encyclopaedists) of Newton's work, as the English Revolution was of Bacon's. Behind Saint-Simon's conception of scientific revolution was the longing to see the scientist as also the bearer of political revolution. The scientist in other words was the only professional revolutionist that the world's history had known.

Among French scientists and intellectuals bred upon this tradition, the Revolution acquired the connotation of a personified god. "Revolution" became a sacred, ritualistic word bringing their scientific work into communion with their political religion; all wished to be immersed in its holy fluids. François Arago, the secretary of the Academy, eulogized "the contribution of France to these admirable revolutions in astronomical science."[14] Antoine Augustin Cournot, the founder of mathematical economics in France, and the foremost exponent during the nineteenth century of the role of chance in history, had witnessed three revolutions in his lifetime—those of 1830, 1848, and the Paris Commune of 1871. It was natural that his panoramic book published in 1872, *Considérations sur la Marche des Idées et des Évènements dans les Temps Modernes*, gave primacy to the role of revolutions in the progress of science. He wrote of the "revolution of the sciences in the seventeenth century," of "the revolution in mathematics" provoked by the invention of the calculus and method of fluxions, and the "revolution" in chemistry initiated by Lavoisier.[15] Then, toward the end of the nineteenth century, the notion of a revolution in science appeared in the title of a book, *La Révolution Chimique: Lavoisier*, by M. Berthelot, published in 1890. The Revolution at this time, as Romain Rolland said, "was the official cult, definitely established in the seats of power; its consecration took place in 1889, the apotheosis of the Bastille, taken a hundred years before by the big bourgeoisie, who," in his opinion, "had rebuilt it as a cashbox."[16] Berthelot was not only a distinguished chemist but was an example of a type not uncommon in France—the politician-savant. He later served as minister of foreign affairs in a short-lived radical cabinet; he cherished the tradition of the Revolution, and said that any effort at royalist restoration would be met by the peasants rising to defend their revolutionary gains.[17] In his book, Berthelot associated Lavoisier's chemical revolution to the spirit of the French Revolution:

France has just celebrated the hundredth anniversary of the great Revolution which changed its institutions, reconstituted the society among us on new foundations, and demarcated an era basically new in the history of humanity.

This anniversary is also that of one of the memorable moments in science and natural philosophy. At that time, indeed, science was transformed by a considerable revolution in the ideas reigning until then, not only in chemistry, but in all the ensemble of the physical natural sciences.

These discoveries and scientific transformations present, in the way that they were effected, a character remarkably similar to that of the Social Revolution with which they coincided: they were not effected gradually, through the slow evolution of years and the accumulated work of several generations of thinkers and experimenters. No! They were on the contrary produced suddenly: fifteen years were enough to accomplish them.[18]

Meanwhile, spurred by the French example, English, German, and Danish scientists took to speaking of revolutions in science. Hans Christian Oersted, whose patience and imagination rendered him in 1820 the discoverer of the relationship between magnetism and electricity, imbibed as a youth the enthusiasm of the French Revolution and the German Romantic movement; his formative years, as well as those of his friends, wrote his biographer P. L. Möller, "happened during the greatest mental fermentation which has been exhibited in modern times." In 1806, as he lectured on the history of chemistry, Oersted declared that "nothing is better adapted to form a mind which is capable of great development, than living and participating in great scientific revolutions."[19] John Herschel, in his pathfinding *Preliminary Discourse on the Study of Natural Philosophy*, told how Humphrey Davy "conceived the happy idea of bringing to bear the enormous batteries" of the Royal Institution for decomposing alkalis and earths: "a total revolution was thus effected in chemistry. . . ."[20] The founder of agricultural chemistry, Justus von Liebig, wrote how as a youngster returning to Germany, he found that "a great revolution in inorganic chemistry had already begun," while his own biographer, W. A. Shenstone, told how von Liebig's friend, Wöhler, "revolutionized" the view of chemists on organic chemistry by producing urea from purely inorganic materials.[21]

In Britain, however, the concept of revolution in science found little favor. The two ablest and most scholarly British historians of science, William Whewell and John Theodore Merz, made virtually no use of the concept of scientific revolution. In his classical *History of the Inductive Sciences* Whewell used the concept of revolution only once, and that was in reference to the philosophical work of Francis Bacon: "He announced a New Method, not merely a correction of special current errors; he thus converted the Insurrection into a Revolution, and established a new philosophical Dynasty."[22] Whewell used the analogy of revolution only in reference to the struggle for experimental scientific method in general as against medieval scholasticism. He did not use it to characterize any of the specific landmarks of scientific discovery. Though he wrote of Newton's law of gravitation as "the greatest scientific discovery ever made," he did not describe a Newtonian revolution.[23] Likewise, though John Merz in his magisterial *History of European Thought in the Nineteenth Century* wrote of the "radicalism which lays bare the roots of our ideas, which delves deep into the ground of our opinions and principles, or which points out new methods by which we may test the correctness and consistency of our axioms," the notion of revolution never obtruded into his work, so filled with love for the sciences.[24] And John Stuart Mill, for all his enthusiasm for political revolutions, began his *System of Logic*, the most influential work on the subject in the nineteenth century, with the statement that "there would be a very strong presumption against any one who should imagine that he had effected a revolution in the theory of the investigation of truth. . . ." Mill disclaimed such presumption.

The notion of scientific revolution, on the other hand, became central to the Marxist standpoint. According to Engels, the dialectical conception applied not only to the laws of nature but to the history of human thinking about them:

> Amid the welter of innumerable changes taking place in nature, the same dialectical laws of motion are in operation as those which in history govern the apparent fortuitousness of events; the same laws as those which similarly form the thread running through the history of the development of human thought. . . .

The empirical findings of the sciences were themselves, according to Engels, forcing a revolution upon them:

> For the revolution which is being forced on theoretical natural science by the mere need to set in order the purely empirical discoveries, great masses of which are now being piled up, is of such a kind that it must bring the dialectical character of natural events more and more to the consciousness even of those empiricists who are most opposed to it.[25]

Engels spent most of his time from 1872 to 1882 writing a manuscript on the dialectic of nature and the history of science. Natural science, he declared was "developed in the midst of the general revolution and was itself thoroughly revolutionary." The sciences were, in his view, doubly "dialectical." For apart from their dialectical relationship to social circumstances, their internal pattern of development was also by way of a dialectic of contradictions:

> The form of development of natural science, in so far as it thinks, is the *hypothesis.* A new fact is observed which makes impossible the previous mode of explanation of the facts belonging to the same group. From this fact onwards new modes of explanation are required—at first based on only a limited number of facts and observations. Further observational material weeds out these hypotheses. . . .[26]

All human knowledge, wrote Engels, "develops in a curve which twists many times";[27] these "twists" were the critical points of dialectical transition, of scientific revolution. Mathematics too was conceived as having developed through contradictions engendered in its specific stages. The English Marxist scientist, J. B. S. Haldane, later added that contradictions always arise, that they can never be eliminated; therefore, mathematicians who claim today to have done so "have only pushed the contradictions into the background where they remain in the field of mathematical logic."[28]

At the turn of the century, the psychological undercurrent of a will to revolution in the sciences again began to be felt. The discovery of radioactivity undermined the classical doctrine of immutable atoms. Henri Poincaré, the most respected philosopher of science of the period, wrote of "radium, that grand revolutionist of the present time," which in his view imperiled such classical principles as the conservation of energy.[29] But Poincaré's aim was to salvage the menaced principles; he had little sympathy for avowed revolutionary aims. When such logicians as Peano, Couturat, and Russell professed an aim to make a revolution in thought, Poincaré wrote that some of their symbols might be convenient, "but that they are destined to revolutionize all philosophy is a different matter." He even ridiculed the name they had chosen for their revolutionary discipline, that is, "logistic"—a word employed "at the Military Academy, to designate the art of the quartermaster of cavalry, the art of marching and cantoning troops. . . ." But, added Poincaré, "it is at once seen that the new name implies the design of revolutionizing logic."[30]

Among Poincaré's most careful readers was the Russian revolutionist V. I. Lenin, then living in exile. He was not drawn to Poincaré's philosophy, but was stirred by the picture Poincaré drew of the so-called bankruptcy of science and the atmosphere of crisis among scientific theories: "The laity are struck to see how ephemeral scientific theories are. After some years of prosperity, they see them successively abandoned; they see ruins accumulate upon ruins; they foresee that the theories fashionable today will shortly succumb in their turn. . . . This is what they call the bankruptcy of science." To Lenin it seemed that Poincaré's words were a corollary of the Marxist world-view. Just as the capitalist system first enjoyed a period of prosperity, but then, beleaguered by contradictions, became ever more insolvent as its rate of profit declined and enterprises failed, so similarly were the scientific theories, conceived in the bourgeois world, entrapped in the contradictions of bourgeois philosophy. A scientific revolution, coeval with the social revolution, would restore clarity to science with the help of theories conceived in the spirit of dialectical materialism, the ideological matrix of isomorphemes emanating from the proletarian revolution. Lenin quoted Poincaré's phrase about "radium, that grand revolutionist," but took him to task for not responding to the "crisis" in physics, the "revolution" and the "general destruction of principles" with recourse to the powers of dialectical materialism.[31]

To support his sociological conception of revolutions in science, Lenin turned to the writings of Abel Rey, the Comtist historian of science:

> In the history of physics as in any other history one can distinguish great periods which are characterised by a certain aspect of general form of its theories. . . . When a discovery is made which affects all fields of physics, which establishes some cardinal fact hitherto unknown or unappreciated, the entire aspect of physics is changed; a new period commences. So it was after Newton's discovery, so it was after the discoveries of Joule-Meyer and Carnot-Clausius. . . . The same result, apparently, has followed upon the discovery of radioactivity. . . . This is a typical crisis in its natural growth occasioned by great discoveries.[32]

Science advanced, according to Rey, through the occurrence of "contradictions" and "conflicts"; it was indeed, psychologically speaking, a process describable as "dialectical" in the Hegelian sense. A historian, however, Rey noted, might nonetheless perceive the discontinuous transitions as moments of an underlying continuous process: "The historian who can observe events with the requisite aloofness, will without difficulty discern a continuity of evolution where our contemporaries see only conflicts, contradictions, and divisions into various schools. . . ."[33] But it was the emphasis on the dialectical structure of scientific revolutions that was especially imparted through the Leninist influence not only to Marxist-minded scientists but to a legion of scientists and expositors of science during the years of economic depression beginning in 1929. Always the emphasis was on the synchronic isomorphism between social and scientific revolutions.

In 1931, at the outset of the depression, a powerful delegation of Soviet thinkers attended the International Congress of the History of Science and Technology in London. Led by the last Soviet Marxist of independent mind, Nikolai Bukharin, their volume of published papers *Science at the Crossroads* began: "In the capitalist world the profound economic decline is reflected in the paralysing crisis of scientific thought and philosophy generally." One Soviet scholar argued that the "crisis" in mathematics was connected with "the entire crisis within capitalism as a whole." In the apparently gloomy prognosis of the end of the world, pronounced as a consequence of the second law of thermodynamics by many Western scientists, the Soviet thinker saw the manifestation of the cowed bourgeois ideologist, unable to participate in the scientific revolution: "To us Marxists-Leninists it is obvious that this physical theory merely reflects the general tendency of bourgeois ideology, which interprets the approaching and inevitable end of the capitalist system as the approach of anarchy."[34]

To Marxists, a process of scientific revolutions was a corollary of their view of human history; the doctrines of historical and dialectical materialism entailed the notion that man's scientific and philosophic ideas, reflecting somehow the underlying economic foundation, must change dialectically as his economic relations do.

During the 1930s, the notion of scientific revolution became a commonplace among Marxian philosophers of science. The British Marxist, Christopher Caudwell, wrote:

> This crisis is different from the previous crises of physics. It is a revolutionary crisis, [one which] occurs when the contradictions discovered in practice cannot be met by a rearrangement of content within the categories of the domain of ideology concerned . . . [one which cannot be solved] unless the most basic and fundamental of categories, those common to all domains of ideology, are more or less rapidly transformed.

The "integrations" achieved by Einstein and Planck, had, according to Caudwell, given rise to a "contradiction" which had "burst asunder the much-patched fabric of physics." The " 'younger men'—Jeans, Eddington, Heisenberg, and Schrödinger," were not, according to the Marxist, "revolutionary in a real sense." "For it is the essence of a revolution that such a transformation can only take place as part of the transformation of the fabric in all fields of ideology, and this in turn is part of a still deeper transformation." Einstein's philosophy, according to Caudwell, had failed to produce a synthesis of "the whole complexity of modern physics, whose anarchy it has indeed helped to produce." A "pending revolution" was in the making, which would eventuate in a larger synthesis. Thus each revolution resolved contradictions, but each synthesis generated novel contradictions. "At every stage contradictions already latent have become open as a result of extended observation of Nature; and at every emergence they have been resolved by means of a new theory which lifted physics to a higher plane."[35]

Furthermore, the Marxist affirmed that the crises in scientific categories and systems arose from the underlying economic crises. So Caudwell wrote: "It is no accident therefore that the crisis in physics occurs at the same time as an unprecedented economic crisis, which has become worldwide."[36]

The concept of scientific revolution also attracted thinkers who were far from Marxist in their outlooks. The noted astronomer, Arthur Eddington, probably contributed more than any other individual toward its diffusion. Speaking of "the Bolshevism of modern science," Eddington revealed his feeling that revolutionary emotion in science had a certain kinship with that in politics. In his Gifford Lectures of 1927, published the following year as *The Nature of the Physical World* (a work which had phenomenal success), Eddington began his account of the recent scientific revolution: "The new ideas of space and time were regarded on all sides as revolutionary." Nevertheless, "when the evidence became overwhelmingly convincing it [the new idea of matter] quickly supplanted previous theories. No great shock was felt." Still, "protests against the Bolshevism of modern science and regrets for the old-established order" were occasionally heard. "The epithet 'revolutionary,' " observed Eddington, "is usually reserved for two great modern developments—the Relativity Theory and the Quantum Theory." For these theories involved "changes in our mode of thought about the world"; they were not merely additive discoveries of fact. Eddington wondered whether the next thirty years would see another "revolution" or "perhaps even a complete reaction." He speculated as to how religion would fare under "the next scientific revolution." He wondered, moreover, whether perhaps the underlying development of science might somehow transcend the revolutionary analogy: "But in each revolution of scientific thought new words are set to the old music, and that which has gone before is not destroyed but refocussed . . . the kernel of scientific truth steadily grows. . . ."[37]

By 1970, forty years after Eddington wrote, no word was as dear, precious, and meaningful to intellectuals as "revolution"; it occurred in the titles of their articles and books even more often than another favorite catchword, "alienation."[38] An era that began with Harold J. Laski's academic *Reflections on the Revolution of Our Era* terminated with such works as *The Erotic Revolution* and *Revolution for the Hell of It.* When George Gamow undertook to write a popular account of the history of atomic physics in the high tradition of such scientific writers as Thomas Henry Huxley, he entitled his book *Thirty Years That Shook Physics.* Though he himself had fled the Soviet scene, the temptation to follow the literary model of *Ten Days That Shook the World* was too strong to resist. Crane Brinton, the only eminent American historian much influenced by Pareto's sociological theories, declared in 1965 that most Americans would respond with unpleasant feelings to the word "revolution"; this hardly seemed the case among the intellectual community. Many intellectuals now feel themselves possessed of a revolutionary vocation. Their forebears three centuries ago might have wished to look at things *sub specie aeternitatis.* Not so the modern intellectual, who is emotionally compelled to look at history, past and present, *sub specie revolutionis.* Such is his emotional a priori.

It is natural therefore, given the contemporary intellectual climate, that the history of science, as of societies, economics, and the arts, would be regarded as driven by revolutionary processes. The isomorpheme of revolution is the "unit-idea" (Lovejoy's phrase) to which intellectuals above all feel themselves attracted. (See Chapter 4 for a discussion of the notion of isomorpheme.) Yet the questions arise: Is the history of science misperceived when it is viewed through the categories imposed by revolutionary emotions? Does a distortion occur when the processes of basic scientific advancement are subsumed under the isomorpheme of revolution? Has the notion of scientific revolution given rise to a perspective on the work and interrelations of scientists that is ideological rather than scientific?

The metaphor of revolutions in science has not been without its rivals in modern times. During the 1920s, it was fashionable to use the metaphor of the frontier. At that time, one spoke of *frontiers of knowledge*, and the scientist was compared to the pioneer penetrating the unknown wilderness. During World War II and since then, it has also been fashionable to regard the scientific enterprise as a military operation. A discovery became a *breakthrough*, a hypothesis was a *beachhead*, scientific method was a *strategy*, and the *tactics* were the specific techniques to be used in particular situations. A discussion with a fellow scientist became a *briefing session*; there were *operational reports*, *reconnaissances*, *attacks in depth.* The militarized language of scientific research still retains a certain vogue, but it has never evoked generational emotions to the degree associated with *revolution.*

Romain Rolland observed many years ago that "the word 'revolution' always exercises a prestige [over] the younger generation."[39] Precisely because any theory fashioned on the revolutionary model exerts such a strong emotional attraction for the young, and perhaps even expresses an underlying psychological motive of its most creative workers, we must be doubly careful to inquire whether from a sociological standpoint the model of revolution accurately characterizes the progress of science.

The Disanalogy of Scientific Revolution: The Absence of Revolutionary Situations

In several respects the structural analogy, or isomorpheme, of revolution is misleading when applied to the development of science. First, every social revolution is preceded by what theorists and practitioners of revolution call a *revolutionary situation.* Thus, the Bolshevik Revolution was preceded by social disorganization characterized by food shortages, high prices, mass desertions from the army, roving mobs, unemployment, and the breakdown of the government's authority. The French Revolution was preceded by a similar period of intensified crisis and the collapse of the political regime. In a revolutionary situation all classes of society have a sense of the instability of things. Every social relation—in fields, factories, families—is affected by anxiety and uncertainty. Many people, worried whether food will be available to them, and troubled by threats to their personal physical safety, lose confidence in their social moorings; they grasp for new options, new solutions. The psychological conditions thus arise by which, according to Lenin, one can gauge that the time for revolution is ripe. At such a time, Lenin wrote: (1) "All the class forces hostile to us have become sufficiently confused, are sufficiently at loggerheads with each other, have sufficiently weakened themselves . . .;" (2) "All the vacillating, wavering, unstable, intermediate elements . . . have sufficiently disgraced themselves . . ."; and (3) "Among the proletariat a mass mood in favor of supporting the most determined, unreservedly bold, revolutionary action against the bourgeoisie has arisen. . . ."[40]

The scientific revolutions, by contrast, have shown a remarkable absence of a preceding sense of crisis, or revolutionary situation. Instead, the scientific community, on the eve of revolution, has generally been placid, confident, and content with itself. The most striking example of this unrevolutionary calm was found in the writings and remarks of Albert A. Michelson, whose name was attached to the celebrated experiment upon which Einstein's relativist theory later relied. Michelson was wholly unaware that any crisis in physical science was in the making; he would have scoffed at any suggestion that the experiment he performed twenty years earlier was the harbinger of a revolutionary overturn in theory. In 1903, shortly before the publication of Einstein's paper on his special theory of relativity, Michelson wrote in his book, *Light Waves and Their Uses*: "The more important fundamental laws and facts of physical science have all been discovered and these are now so firmly established that the possibility of their ever being supplanted in consequence of new discoveries is exceedingly remote." There were, to be sure, Michelson added, "apparent exceptions to most of these laws, and this is particularly true when the observations are pushed to a limit. . . ." The discovery of such exceptions, he said, was "in most cases due to the increasing order of accuracy made possible by improvements in measuring instruments. . . ." But Michelson adhered serenely to the view that the system of known physical laws would be adequate to cope with the "apparent exceptions": "Many other instances might be cited, but these will suffice to justify the statement that our future discoveries must be looked for in the sixth place of decimals."

Physical science, Michelson indeed believed, offered no opportunity for the young man who longed to be a daring innovator, let alone revolutionist, in physical theory. All that was left to do was to refine measurements and to explain "apparent" exceptions to known laws. The opinion was widely held among graduate students at that time that physics was a "dead subject." Robert A. Millikan, later awarded a Nobel Prize for his work in determining the charge of an electron, tells how in 1894 when he was a graduate student living in New York and sharing rooms with four other Columbia graduate students (three of whom were in sociology and political science),

> I was ragged continuously by all of them for sticking to a "finished," yes, a "dead subject," like physics, when the new, "live" field of the social sciences was just being opened up. But here in Roentgen's discovery only one year later was a door opening into a new, theretofore undreamed of field of physics, a big qualitative field which had nothing to do with great refinement of measurements. . . .[41]

When Millikan went to Chicago that year, he heard Michelson talk of physics in much the same way as the sociology graduate students —the system of physical theory was for the foreseeable future in a state of stable equilibrium:

> My first view of Michelson was at the convocation [of the Ryerson Laboratory]. He gave the address on the place of very refined measurement in the progress of physics—an address in which he quoted someone else, I think it was Kelvin, as saying that it was probable that the great discoveries in physics had all been made, and that future progress was likely to be found in the sixth place of decimals. Later, in conversation with me, he was to upbraid himself roundly for this remark.[42]

Michelson was in a good position to observe the state of mind that prevailed among physicists at the end of the nineteenth century; he found they generally believed that the system of physical theory was in a state of stable equilibrium. As his pupil, Robert Millikan tells:

> The transition from the old to the new mode of thought in physics was probably made as dramatically in my case as in that of anyone in the world; for I was in the fortunate position of having entered the field just

three years before the end of the complete dominance of nineteenth century modes of thought. In those three years I had the privilege of personally meeting and hearing lectures by the most outstanding inventors of nineteenth century physics—Kelvin, Helmholtz, Boltzmann, Poincaré, Rayleigh, Van't Hoff, Michelson, Ostwald, Lorentz—every one of whom I met and heard between 1892 and 1896. In one of these lectures I listened with rapt attention to the expression of a point of view which was undoubtedly held by most of them—indeed, by practically all the physicists of that epoch; for it had been given expression to more than once by the most distinguished men of the nineteenth century.

The speaker reviewed the triumphant progress of modern physics; first, the establishment during the seventeenth and eighteenth centuries of the principles of mechanics, culminating in Laplace's *Mécanique Celeste*; then from 1800 to 1830 the verification of the wave theory of light; the development in the middle of the century of the principle of conservation of energy; then during the first two decades of the century's second half the establishment of the principle of entropy; and finally, the formation of Maxwell's theory of electromagnetism, its experimental confirmation in 1886 by Hertz, and the dominance of the physics of the ether.

Then, summarizing this wonderfully complete, well-verified, and apparently all-inclusive set of laws and principles into which it seemed that all physical phenomena must forever fit, the speaker concluded that it was probable that all the great discoveries in physics had already been made and that future progress was to be looked for, not in bringing to light qualitatively new phenomena, but rather in making more exact measurements upon old phenomena.

Just a little more than one year later, and before I had ceased pondering over the aforementioned lecture, I was present in Berlin on Christmas Eve, 1895, when Professor Roentgen presented to the German Physical Society his first x-ray photographs.

Here was a completely new phenomenon—a qualitatively new discovery and one having nothing to do with the principles of exact measurement. As I listened and as the world listened, we all began to see that the nineteenth century physicists had taken themselves a little too seriously, that we had not come quite as near sounding the depths of the universe, even in the matter of fundamental physical principles, as we thought we had.[43]

Other young scientists encountered the view that the theoretical labor of physics was virtually closed. At McGill University, shortly before the youthful Ernest Rutherford arrived to undertake his epoch-making experiments on the disintegration of the atom, the chairman of his department and his later protector, Cox, "said one day that he was feeling rather dispirited because there seemed nothing new going on in Physics. The main things, he said, had all been found out and the work which remained was to carry on a great number of experiments and researches into relatively minor matters." It has been said that "when Rutherford got going, Cox was ready and glad to sing another tune."[44]

There was, on the other hand, the distinguished but exceptional figure of Henri Poincaré, who was at this time writing with verve of his feeling that a "crisis" obtained in the foundations of mathematical physics; Poincaré was grappling with the interpretation of the experimental arguments for the earth's rotation in absolute space. He treated this theme in his book *La Science et l'Hypothèse.* Nevertheless, when its English version was published in 1905, its eminent reviewers showed no sense of sharing his apprehension. The young Bertrand Russell found Poincaré's chapter on absolute motion "unsatisfactory." Poincaré, he noted, argued, on the one hand, that " 'the earth turns round,' has no meaning, since it cannot be verified by experiment," but he also argued, on the other, that " 'it is more convenient to suppose that the earth turns round.' " To which Russell replied: "But if 'the earth turns round' has no meaning, it has the same meaning as 'Abracadabra,' and therefore if M. Poincaré is right, it has the same meaning as 'it is more convenient to suppose that Abracadabra.' " It was difficult to see, Russell declared, what convenience resulted from supposing "Abracadabra." Russell concluded that "it would seem, the facts which make it more convenient to suppose that the earth turns round prove that there is such a thing as absolute rotation."[45] The physicist Arthur Schuster was even more unimpressed by Poincaré's misgivings about absolute motion; in his long notice in *Nature* he gave but one clause to the subject. Poincaré's discussions of modern theories of physics, he wrote, "will be read with interest, though many of us will not agree that 'a day will come when the ether will be rejected as useless.' " Certainly Schuster's review in *Nature* showed no sense of anxiety, or awareness of a pending revolutionary situation in physics.[46]

The literature of science just prior to the publication of Einstein's first paper on the theory of relativity in 1905 showed a strange serenity. In 1904, for example, the distinguuhed Cambridge scientist and historian of science, William Cecil Dampier Whetham, published his book, *The Recent Development of Physical Science.* Within that same year, it quickly went through three editions. His book can be taken as an accurate report on the state of mind of British science at the time. Whetham, a fellow of Trinity College, remarked in his preface: "The writer has been fortunate in his surroundings, where the knowledge and insight of one worker are placed freely and ungrudgingly at the service of another in the day of need." Such distinguished colleagues as J. J. Thomson and Joseph Larmor read the various chapters. The chapter entitled "Atoms and Aether" conveyed an atmosphere of tranquillity, untouched by the portent of a pending storm:

> As long as the aether was invoked only to explain the phenomena of light, the difficulties of interpretation might well suggest doubts about the fundamental hypothesis, but when Clerk Maxwell showed that it was possible to explain the phenomena of the electro-magnetic field by an aether having properties identical with those of the luminiferous medium, the evidence for both theories was strengthened almost indefinitely.[47]

Dampier's entire book looked forward to the day when a theory of aether would provide the explanation for both the phenomena of gravitation and electricity. Though adequate mechanical models for the aether were no longer expected, the hypothetical aether still seemed a fruitful notion for atomic theory: "Aether, being now regarded as a sub-material medium, is not necessarily described by the experimental laws to which the facts of ordinary mechanics conform. In dealing with the aether, we are on an entirely different plane, and have no right to assume that a mechanical model of its properties is possible." The long-term crucial test for the theory, he felt, would be "the problem of the nature of gravitation, which must some day be explained in terms of the universal medium, if that medium is to survive as a permanent conception in physical science." The Michelson-Morley experiment received a passing reference, though not by name:

> The passage of light over the surface of the earth is not affected by a change in direction relative to the earth's total motion, the velocity of light is the same whether it is passing with or against the motion of the earth. . . . It is possible to reconcile these results by certain suppositions as to the effect of moving matter on the absolute velocity of light within it. . . .[48]

When he revised his book in 1924 for a fifth edition, Dampier had to rewrite it extensively. "The theory of relativity, founded by Einstein and Minkowski," he then added "leads to another revolution in scientific thought." He no longer referred only parenthetically to the Michelson-Morley experiment; rather he described it in some detail, and went on to narrate how Einstein, "impressed by this experimental result," had in 1905 "accepted the constancy of the measured velocity of light as an ultimate fact of Nature," and then shown that "the ideas of absolute space and time are mere figments of the imagination."[49]

In the theoretical calm at the end of the nineteenth century, only a few scientists or philosophers would have conjectured that the whole relationship of geometry and physics might soon be reconstructed, and that physical science might avail itself of a non-Euclidean space. Henri Poincaré, no doubt, spoke for the overwhelming majority when he wrote that the scientific spirit would always cling to the Euclidean geometry as the most convenient and prefer any solution rather than introduce a non-Euclidean geometry. The philosopher Emile Meyerson, who endorsed Poincaré's view, had to add a footnote in 1926 to a later edition of his classical treatise *Identity and Reality*: "We are obliged to recognize that Poincaré's prevision which we had followed in this matter has not been confirmed by the later progress of science."[50] The general opinion, "even among the mathematicians and above all among the physicists," had certainly not followed those who like Helmholtz speculated on the use of non-Euclidean geometries in physics. But this view, Meyerson ruminated in a later book, "was given the lie by events." The scientific world, "after a period of hesitation and struggle which, given the immense extent of the revolution being accomplished, was remarkably short," abandoned the Newtonian axioms of space and time.[51] Evidently, however, this was a revolution that was not preceded by the growing consciousness of a revolutionary situation, a revolution that did not conform to the isomorpheme of social revolution.

The current of opinion in the Harvard University scientific community was likewise remarkably placid before and during the first stage of the "revolution" in science. Almost every American campus has a scientist who is regarded as its sensitive recorder of the latest trends in thought. Such was the Harvard biochemist Lawrence J. Henderson who was always alert to current philosophies and conjectures and prided himself on being in the scientific vanguard. There can be little doubt that Henderson reflected the prevalent calm and trust in the received scientific scheme when, in 1912, he wrote:

In spite of all assaults of philosophers and mathematicians space remains for practical purposes more certainly than ever the Euclidian space of the ancients. . . . And time is now and forever that which flows equably, wholly independent of all else, though almost all else is dependent upon time. It is Euclidian space in which the earth moves and describes its ellipse, parallel rays of light never do meet in our practical experience and our crystals are in form the figures of Euclidian geometry.[52]

Was the quantum revolution preceded by the analogue of a revolutionary situation, by a "period of pronounced professional insecurity," which (in Thomas Kuhn's view) "has generally preceded" the emergence of new theories?[53] Max Born, in his recollections of that period, tells of no such interval of tension. "It has been generally acknowledged," he writes, "that the year 1900 of Planck's discovery marks indeed the beginning of a new epoch in physics. It was the time of my own student days, and I remember that Planck's idea was hardly mentioned in our lectures, and if so as a kind of preliminary 'working hypothesis' which ought of course to be eliminated."[54] Again, one does not get the sense that a profound malaise obtained among the young physicists, a feeling that the foundations of thermodynamical theory were in sore straits, that a systematic structure was in danger of collapse.

Eleven years after Planck's discovery, on October 30, 1911, the first Solvay Congress convened in Brussels, under the leadership of the physical chemist Walther Nernst, to discuss the problems posed by quantum theory. Twenty-one noted scientists attended this symposium which lasted until November 3rd. It was indeed a momentous meeting. Yet it had hardly been preceded by a sense of crisis on the part of the scientific community. When the list of invitees was being drawn up, Max Planck "could think of only four other relatively well-known physicists who had given evidence of truly caring."[55] And the number four was apparently an overestimate, for of that group, composed of Einstein, Lorentz, Wien, and Larmor, the last declined the invitation on the ground that he had not had time to inform himself of quantum theory. Moreover, the celebrated Poincaré, with his vast knowledge of the developments in physics and mathematics, had ignored quantum theory, and encountered its problems for the first time at the congress in 1911; eleven years after its inception, the theory was new to him:

This unfamiliarity with the quantum theory in one who was *au courant* with important developments in theoretical physics points up in a striking way the obscurity of the subject at that time. If the Brussels meeting had brought together only those scientists who had already proven themselves to be deeply preoccupied with quanta, it would have amounted to a very small group indeed.

The cautious Planck did not want to undertake to make a scientific revolution when there was no consciousness on the part of the scientific community of a revolutionary situation. He had therefore wished to postpone the discussion for a few years until the situation became "*intolerable* for every true theorist."[56] Lenin's October Revolution was the culmination of a collapse in the social system which was widely experienced; the Solvay Congress, largely because of its impact on one man, Henri Poincaré, promoted the reception of the quantum revolution in a way which could not have been extrapolated from the preceding state of the scientific community.

The propounding of the greatest physical generalization of the nineteenth century, the law of the conservation of energy, likewise did not follow a revolutionary situation. No diffused sense of insecurity was provoked or existed when James P. Joule, a young brewer from Manchester, presented his research in 1843 at the meeting of the British Association for the Advancement of Science at Cork, Ireland. Joule presented an experimental determination of the mechanical equivalent of heat: 838 foot-pounds of work was the equivalent of the heat necessary to raise one pound of water by one degree (Fahrenheit) of temperature. His paper "was received with general silence." As Joule tells, with some exceptions, "the subject did not excite much general attention. . . ." His conclusion, all too sweeping in its generality of the 'convertibility of heat and mechanical power into one another, according to the above numerical relations' was received with entire incredulity."[57] Undaunted, Joule returned in 1845 to read to the British Association at Cambridge his complementary paper, "On the Mechanical Equivalent of Heat" with its report on his paddle-wheel experiments for raising the temperature of water; still no discussion arose. Evidently an unruffled placidity reigned among British scientists on both the eve and morrow of the presentation of the greatest generalization of the nineteenth century. Again, Joule came forward with a paper for the 1847 meeting at Oxford; the chairman, not wishing to see time unduly misallocated to the subject, suggested that Joule not read the paper, but simply give a verbal report on his experiments. "The chairman suggested that, as the business of the section pressed, I should not read my paper, but confine myself to a short verbal description of my experiments." This Joule agreed to do. Once more it seemed as though his work would be courteously passed over.

A new protagonist was, however, at hand. As Joule tells the story:

> Discussion not being invited, the communication would have passed without comment if a young man had not risen in the section, and by his intelligent observations created a lively interest in the new theory. The young man was William Thomson, who had two years previously passed the University of Cambridge, with the highest honor, and is now [Lord Kelvin] probably the foremost scientific authority of the age.[58]

Years later Thomson said that distinguished scientists "were for many years quite incredulous as to Joule's results, because they all depended on fractions of a degree of temperature, sometimes very small fractions," and that there was a reluctance "in making such large conclusions from such very small observational effects."[59] Thomson's reiterated public support, though with some suspended judgment, began to turn the tide for Joule. It is clear however that there had been no feeling previously among British scientists that their current set of ideas involved them in contradictions requiring a new approach.

The Darwinian revolution was also notable for the absence of any general feeling among biologists prior to its inception that their theoretical standpoints were in a critical stage. As Charles Darwin wrote in his *Autobiography*:

> It has sometimes been said that the success of the *Origin* proved "that the subject was in the air," or "that men's minds were prepared for it." I do not think that this is strictly true, for I occasionally sounded not a few naturalists, and never happened to come across a single one who seemed to doubt about the permanence of species. Even Lyell and Hooker, though they would listen with interest to me, never seemed to agree. I tried once or twice to explain to able men what I meant by Natural Selection, but signally failed. What I believe was strictly true is that

innumerable well-observed facts were stored in the minds of naturalists ready to take their proper places as soon as any theory which would receive them was sufficiently explained.[60]

According to Leonard Huxley's account of his father's life, Thomas Henry Huxley felt that most of his contemporary fellow-scientists "were very much in my own state of mind—inclined to say to both Mosaists and Evolutionists, 'a plague on both your houses,' and disposed to turn aside from an interminable and apparently fruitless discussion, to labor in the fertile fields of ascertainable fact." Quite clearly, most naturalists were in a placid state of mind insofar as the notion of the fixity of species was concerned; they were not suffused with anxiety or apprehension. The "well-observed facts," which Darwin perceived as the foundation for a new theory, were not regarded by the naturalists as posing a crucial challenge to the received theory; their feelings were much like those of Michelson who perceived in the null result of his experiment no basic challenge to the received physical theory.

In the eighteenth century, the chemical revolution, initiated and largely accomplished by Lavoisier, was similarly not preceded by a revolutionary situation of widely felt insecurity and tension arising from the existing system of ideas. The generous letter that Joseph Black wrote to Lavoisier in 1791 best illustrates the state of contentment in which chemists had rested with the old ideas:

The numerous experiments which you have made on a large scale, and which you have so well devised, have been pursued with so much care and with such scrupulous attention to details, that nothing can be more satisfactory than the proofs you have obtained. The system which you have based on the facts is so simple and so intelligible, that it must become more and more generally approved and adopted by a great number of chemists who have long been accustomed to the old system. . . . Having for thirty years believed and taught the doctrine of phlogiston as it was understood before the discovery of your system, I, for a long time, felt inimical to the new system, which represented as absurd that which I had hitherto regarded as sound doctrine; but this enmity, which springs only from the force of habit, has gradually diminished, subdued by the clearness of your proofs and the soundness of your plan.[61]

Prior to Lavoisier's work there was evidently no sense among chemists that their phlogiston theory had reached an impasse from which they could extricate themselves only through heroic measures.

The great landmark in the history of chemistry in the nineteenth century was, by common consent, the formulation of the periodic table of the elements by the Siberian chemist of the town of Tobolsk, Dmitry Mendeleev. Grandson of a church official, Dmitry had, however, experienced the stir of revolutionary ideas from the man who gave him his first lessons in science. This was Nikolai Basargin, formerly a lieutenant, now a political exile; Basargin had been a Decembrist, one of those involved in the brave, ill-fated attempt in 1825 to win for Russia a constitution. Basargin had been among the most moderate of the Decembrists; when he learned that he was sentenced to Siberian exile, he considered himself no longer an inhabitant of the world. But he found a vocation in the Siberian settlements, and wrote a survey of their social, economic, and educational problems.[62] Basargin was married to Dmitry's elder sister and thus something of the restless spirit of questioning that pervaded the European youth of the mid-century was transmitted to the young Mendeleev.[63] Yet in 1869, when Mendeleev presented his synthesis to the Russian Chemical Society, it could hardly be said that there existed a sense of crisis in the consciousness of chemists. "The law of periodicity," said Mendeleev, "may be expressed as follows: the properties of the elements, and thus the properties of simple or compound bodes of these elements, are dependent in a periodic way [recur regularly] on the magnitude of the atomic weights of the elements."[64] Only three years earlier, in 1866, the English chemist, John A. R. Newlands, had tried to show that elements could be arranged in groups of eight along horizontal lines in such a fashion that those in vertical columns would then have similar properties. But his listeners at the English Chemical Society responded with ridicule; one of them even asked whether he had thought of arranging the elements according to the first letters of their names. Similarly, Mendeleev in 1869 did not find the ground prepared for a warm reception to his theory; chemists evidently felt they were doing satisfactorily without an all-embracing system. As he later wrote: "The idea of seeking for a relation between the atomic weights of all elements was foreign to the ideas then current, so that neither the *vis tellurique* of De Chancourtois, nor the *law of octaves* of Newlands, could secure anybody's attention."[65] Moreover, Mendeleev had dealt rather boldly with those accepted atomic weights of elements that didn't fit into his scheme. In such cases, he said, it was not because his scheme was wrong but rather that the existing determinations were in error; thus, he held that the atomic weight of gold, for instance, was greater than those of platinum, iridium, and osmium, not less. Such an attitude was irritating to scientists who generally prefer to see innovations in theory conform to the data rather than to have a theory call into question a collection of existing experimental facts.

"The great majority of scientists in Mendeleyev's time," writes a Russian biographer, "were firmly convinced" that the atoms of different elements were in no way connected with each other in a general system. And no novel experimental fact, anomaly, or contradiction moved Mendeleev to the discovery of the periodic law. It was simply that he had undertaken to write a textbook of chemistry, and no textbook writer likes disorder; he seeks to order the facts in a systematic way. As Mendeleev explained:

> Undertaking the compilation of the manual of chemistry called the *Foundations of Chemistry*, I had to draw up some kind of system of simple bodies so that in placing them I had some definite point to start from and was not guided by pure chance as if following my instinct. . . . Any system that is based on accurately observed numbers is, of course, for that alone preferable to systems that have no numerical basis.[66]

What won the broad acceptance of Mendeleev's periodic table was not the aesthetic beauty or simplicity of his classification of the elements. Most crucial were his predictions concerning elements yet to be found that would fill in the empty boxes of his table. The importance of the periodic law, wrote its historian,

> does not seem to have been generally recognized at first, nor was it widely accepted as a law. In fact for several years it nearly dropped out of sight and it was only the lucky discovery of some new elements, thus fulfilling certain predictions of Mendeleeff, that brought it prominently before the chemical world. How long it would otherwise have laid unnoticed can only be guessed at.[67]

The discovery of two new elements, gallium and scandium, whose observed properties virtually coincided with those which Mendeleev had foretold for the missing *eka-aluminium* and *eka-boron* brought the periodic law to the fore. Mendeleev then affirmed that it "was now certain that the Periodic Law offers consequences that the old systems had scarcely ventured to foresee." Seven years had passed, he said, since he had devoted himself to the problems of the periodic law, but now its "logical consequences" confirmed, it was becoming "popular."[68] Then in 1886, Winkler discovered germanium which had properties almost identical with those of Mendeleev's predicted *eka-silicon.* "These wonderful predictions," as the present writer's professor of chemistry, Benjamin Harrow, related, "did more to convince scientists of the validity of the law than anything else could have done. The soundness of a theory is best exemplified by the use to which it can be put. Does it explain anomalies? Does it guide along future paths of investigation?"[69] Thus, a basic change in chemical theory was accomplished, but without the sociological traits of experienced tensions and crisis that characterize revolution.

The pattern of acceptance of the periodic law was similar to the patterns of acceptance characterizing the theory of relativity and the principle of conservation of energy. George Sarton, who observed the reception of Einstein's ideas, wrote in 1929:

> We have witnessed recently a similar sequence of events with regard to the theory of relativity. As long as it remained a mathematical abstraction, it was considered with general indifference or scepticism, but as soon as the experimental confirmations were provided, the whole scientific world, with but few exceptions, hastened to accept it.[70]

Indeed, we may assert it as a sociological law of the scientific community that no new theory—whatever its appeals of elegance or simplicity, will generally supersede the old unless it leads to new experimental discoveries of fact; such resultant discoveries may, with the help of auxiliary hypotheses, be rendered consistent with the older doctrines, but would not have been foretold by the latter's adherents. Heuristicity, the fruitful direction of scientific thinking and scientific labor toward novel consequences is, pragmatically speaking, the primary criterion by which the scientific community judges a novel idea. Again, this is the common platform of the scientific community, something that overrides the conflicts of schools, so that there is not a sociological situation analogous to the struggle of classes preceding a social revolution, in which all sense of community is destroyed.

Neither have the so-called revolutions in the social sciences been preceded by situations in which the communities of social scientists felt that their categories and generalizations had simply broken down under the impact of novel experiential fact. This is most significantly seen in economics, the only social science in which there is a developed body of mathematical theory. Until the advent of John Maynard Keynes, the one revolution in economic theory recognized by English-speaking economists was the so-called Jevonian revolution. W. Stanley Jevons first conceived and formulated in England the concept of marginal utility; soon afterward the Ricardian labor theory of value fell into disuse. A generation earlier, John Stuart Mill had spoken for economists generally when he wrote that the theory of value was henceforth a closed chapter, a concluded subject: "Happily, there is nothing in the laws of Value which remains for the present or any future writer to clear up; the theory of the subject is complete."[71] Certainly, when Jevons published his work there was no consciousness among economists that some impasse had arisen in economic theory.[72] Jevons adduced no logical objections to the classical theory that hadn't already been thought of by the classical economists;[73] his marginal analysis won acceptance because it was a far more heuristic mode of thought for a multitude of problems than the clumsy labor theory. The sociologist Pareto regarded Marx's labor theory as the logical analogue in economics of the phlogiston theory in chemistry. It was more a language for dialectical arguments about facts than an instrument for discovery.[74]

Jevons, though devoid of the superficial traits of the revolutionist's personality, was a restless, original spirit, one might say, a generational rebel. He closed his book, *The Theory of Political Economy*, with an appeal for political economy to emulate the example of the physical sciences and to make its revolution:

> In matters of philosophy and science authority has ever been the great opponent of truth. A despotic calm is the triumph of error; in the republic of the sciences sedition and even anarchy are beneficial in the long run to the greatest happiness of the greatest number.
>
> In the physical sciences authority has greatly lost its noxious influence. Chemistry, in its brief existence of a century, has undergone three or four complete revolutions of theory. . . .
>
> I have added these words because I think there is some fear of the too great influence of authoritative writers in Political Economy. . . . Our science is becoming far too much a stagnant one. . . .
>
> Under these circumstances it is a positive service to break the monotonous repetition of current questionable doctrines, even at the risk of new error.[75]

Jevons had a strong emotional dislike for Mill's radicalism in politics as well as his inductive empiricism in logic. "My attack on Mill is as much a matter of the heart as head," he wrote in April 1878. Judged by Joseph A. Schumpeter to be "one of the most genuinely original economists who ever lived," Jevons was moved in his generational rebellion less for goals of social reconstruction than those of social realism and clearer analysis.[76] "Whether wisely or not, I declared war against Mill's crotchets some years ago," wrote Jevons in 1875; Mill's socialistic interventionism was one of the crotchets. In that period gestative of socialist intellectuals, Jevons, by contrast, as Bernard Shaw observed, considered it as an open question whether the government should run the post office.[77]

The "Jevonian revolution" did meet with some resistance, but scarcely of the magnitude associated with the word "revolution". Jevons wrote to the French economist Leon Walras in 1875: "I have no doubt whatever about the ultimate success of our efforts, but it will take some fighting; the disciples of J. S. Mill being bitterly opposed to any innovation upon his doctrine." Jevons had already run afoul of the Millite orthodoxy when in 1868 he wrote several articles criticizing the inconsistencies in Mill's logical theories; the articles were rejected by the editor of a leading magazine. The Millites were a determined and self-righteous academic sect: "The Mill faction never scrupled at putting their lecturers and examiners wherever they could, but I believe it only requires a little logic and a little time to overthrow them," wrote Jevons in 1879. And indeed he was right.[78]

The Keynesian revolution in our time, like the Jevonian revolution, triumphed less by virtue of any crucial experiment than by the fact that it was far more heuristic than classical economic theory, especially with regard to proposals for governmental action and intervention. As Keynes's biographer, Roy Harrod, wrote:

> In the physical sciences some crucial test is usually available to decide between conflicting theories. If Keynes was really to be successful, he should have been able, it is argued, to refute, say, Mr. D. H. Robertson, by showing a set of facts which the Keynesian doctrine would fit, while the other would not. Unhappily the state of economics is not so advanced. It is true to say that the Keynesian scheme consisted in essence of new definitions and a re-classification. He asked us to look upon the multifarious phenomena of business life, and order them in our minds in a

different way. . . . One can only judge the alternative systems by using them in relation to various problems and situations. It is by actual use and application, not by logic, that Keynes has been, and will, I am confident, continue to be, triumphantly vindicated.[79]

Such classical economic theorists as Schumpeter could provide explanations for the advent of the depression that were equally convincing. They did not, however, fulfill the emotional need for intervention in economic processes. The classical theorists were like social seismologists who could analyze the equation for social upheaval but could offer no program for action. A new generation stirred by a new philosophy took up the Keynesian standpoint. But, as Harrod observes, it was not an immanently developing impasse in logic or experiment that led to the discarding of classical theory.

Generational Collaboration in Scientific Revolutions

We have seen that the scientific revolutions have been characterized by the absence of genuine revolutionary situations in which established systems are viewed as practically unworkable. There is yet a second respect in which the development of science fails to exhibit the sociological characteristics of revolutionary change. In a social or political revolution, a struggle for power takes place and has its outcome in the removal, physical destruction, or "liquidation" of a social class. The revolutionary party, in its challenge to the status quo, invariably finds that to spread and propagate its views, it must found its own journals, magazines, and newspapers, for its ideas are so contrary to those of the ruling class, governing elite, or establishment that it cannot expect that the latter's resources of publication will be placed at its service. The Russian revolutionists before 1917 were always founding newspapers, organizing illegal presses, and devising channels for underground circulation in Russia.[80] The political factions of Lenin, Plekhanov, and the Mensheviks therefore engaged in bitter struggles for the control of the party newspaper *Iskra*. The French Revolution was preceded by a period in which the *philosophes*, such as Diderot and

D'Alembert, were hard-pressed by the agencies of censorship and the law; the *philosophes* showed remarkable ingenuity in circumventing the authorities not always successfully, however.

But the scientific revolutions have almost never had to undergo what we might call *the ordeal of closed publication.* Let us consider, for instance, the scientific revolutions of the last century, taking as our examples those so denominated in the titles of books: the Darwinian, Einsteinian, Freudian, Keynesian, and quantum revolutions. With only one exception (Freud), there were no difficulties in securing access to publication. In every other instance, the leaders of scientific revolutions published their works in the recognized, established journals. The first publications of the theory of natural selection by Charles Darwin and Alfred Wallace appeared in the *Journal of the Linnaean Society* (1859) under the auspices of the respected geologist Sir Charles Lyell and the botanist Joseph D. Hooker. The young Albert Einstein published his most innovative paper in the thoroughly reputable *Annalen der Physik* (1905), chiefly edited by Max Planck. Keynes was himself for many years the editor of the most esteemed organ of the Royal Economic Society, *The Economic Journal*, and published his *General Theory* during that time.

For all the furor that Darwin's theory evoked, its acceptance and his personal recognition by the establishment were exceedingly rapid. The novelist Charles Kingsley, remarking in 1863 on the "most curious state of the scientific world," noted that Darwin was "conquering everywhere and rushing in like a flood." The next year, England's highest honor, the Copley Medal of the Royal Society, was awarded to Darwin. It was a strange revolution in which the archrevolutionist was honored by the establishment within five years of the revolutionary act. The reluctant president of the Royal Society did indeed try to mollify the routed opposition with the suggestion that the award was conferred upon Darwin the working naturalist rather than Darwin the theorist of natural selection. However, Darwin's friends were influential enough to have the offending passage stricken from the printed version of the president's address.[81] Similarly, the Swedish Academy of Sciences in 1922 sought to by-pass troubled waters when it awarded the Nobel Prize for 1921 to Einstein; it announced that the honor was conferred for his merits in mathematical physics, especially the "discovery of the law of the photoelectric effect" without daring to mention the theory of relativity.[82]

The psychoanalytic revolution was the one case where the innovations in standpoint and the discoveries of fact required a new set of journals as well as societies for their publication and diffusion. As Freud writes: "Official science had pronounced its great ban and boycott against physicians and institutions practising psychoanalysis."[83] The first investigations of Breuer and Freud had been published in 1893 in the official *Neurologisches Zentralblatt*, but Freud then encountered increasing antagonism among his colleagues.

My publications, for which I found shelter despite some difficulty, could safely remain far behind my state of knowledge. . . . The analysis of "Dora" was finished at the end of 1899 . . . but was not published until 1905. In the meantime, my writings were not reviewed in the medical periodicals, or if an exception was made, they were always treated with scornful or pitying condescension. Now and then, a colleague would refer to me in one of his publications, in very short and unflattering terms, such as "unbalanced," "extreme," or "very odd."[84]

The behavior of his opponents in public discussions was such ("only very few persons are capable of remaining polite"), that Freud lost whatever faith he had had in "the value of majorities at congresses or societies."[85] Freud and his followers began to establish their own societies and journals. The Vienna, Berlin, and Zurich circles coalesced to form the International Psychoanalytic Association; henceforth they could train fellow physicians in their methods, and encourage each other in friendly discussion. The *Zentralblatt für Psychoanalyse* was founded with Freud as its editor; its first number appeared in September 1910. It was a regular journal unlike the monographs of the *Schriften zur angewandten Seelenkunden* which had been published irregularly beginning in 1907. A yearbook, the *Jahrbuch für psychoanalytische und psychopathologische Forschungen*, also commenced publication. Still other organs emerged as editorial controversies ensued, and the domain of psychoanalysis expanded. *The Internationale Zeitschrift für Ärztliche Psychoanalyse* became an accredited organ in 1913, while *Imago* was founded in 1912 as a forum for work in which the psychoanalytic method was applied to the arts, literature, and philosophy.[86]

The psychoanalytic revolution was, however, the only scientific-intellectual movement in modern times which has had to undergo an ordeal of closed publication. And this was, as Freud noted, because it was confronted by an "official science." What is meant by an *official science*? The notion deserves a brief digression. An official science is not to be confused with the *normal science* of Thomas Kuhn's usage. Official science connotes something other than the usual activities and dominant mode of thought among the scientists of a period. An official science arises whenever a mode of thought is regarded as functionally necessary to the stable working of the social order. An official science therefore is allied with and entrenched behind guardian social institutions; it is fortified by extrascientific, administrative, arbitrary measures. In the Soviet Union, for instance, Pavlov's model of conditioned reflexes was long enforced as psychologists' only acceptable isomorpheme; other modes of investigation ranging from psychoanalysis to intelligence testing were placed under an official ban. In biology, the authority of an ill-informed Stalinist henchman, T. D. Lysenko, ("a real fanatic, a Savonarola of science," as Julian Huxley described him) was reinforced to throttle work that made use of Mendelian genetic modes of explanation.[87] Freud narrated incidents that indicate how an official science can function under more enlightened circumstances. "At a time when psycho-analysis was solemnly put on its trial before the learned societies of Germany," Freud wrote, one scientist at a congress declared that if a patient of his began to talk about sexual matters, he shut him up: "The learned society applauded the speaker to the echo instead of feeling suitably ashamed on his account."[88] In the case of the psychoanalytic revolution, the dominant, established, medico-psychological practitioners felt threatened by more than a new mode of thought, or a novel set of experimental facts. They felt that the established values of religion, ethics, family life, and political authority were endangered by the proposed new standpoint. Under the impact of such a threat or anxiety the prevalent mode of scientific thought congeals into an official science; intellectual rigidities replace open-mindedness, and agencies of social control are invoked to repress disagreeable facts and uncongenial theories. By reaction, the revolutionary mode of thought then attracts to itself all the more the alienated, or those drawn to the revolutionary experience for its own sake. Whenever a dominant mode of thought hardens into an official science, the opposing mode of thought is infused with the emotions of a revolutionary science. The latter attracts personalities who share to a great degree the characteristics of social revolutionists. As Marthe Robert writes:

> No one should be surprised that the first generation of psychoanalysis included so many who were unbalanced, socially maladjusted or even

suffering from serious psychical disturbances. Who but men with a personal experience of suffering could have felt the urgency and foreseen the extent of the Freudian revolution . . . ? Psychoanalysis offered at once an outlet and intellectual sustenance to gifted minds, who, through ignorance or lack of opportunity had not been able to use their talents, or to highly strung unstable people who saw a chance of escaping their fate.[89]

The psychoanalytical revolution was exceptional in having had to resort to journals specially founded to secure a hearing for the novel ideas. There were unusual instances, however, in the mid-nineteenth century when publication was denied to papers enunciating the law of conservation of energy; but the resistance to these papers can hardly be regarded as having had counter-revolutionary dimensions.[90] When Julius Robert Mayer in June 1841 wrote a paper that first formulated the principle of conservation of energy, he sent it to the *Annalen der Physik und Chemie*, edited by Johann Christian Poggendorff. The latter, however, rejected it, evidently not without substantial grounds. George Sarton writes that the rejection "was very fortunate, for that paper contained gross errors."[91] But the following year, an improved paper by Mayer on the same subject was accepted for publication by von Liebig of the *Annalen der Chemie und Pharmazie*. The paper entitled "Bemerkungen über die Kräfte der unbelebten Natur" (Observations on the Forces of Inorganic Nature), which was printed in 1842, was the first not only to state clearly the law of the conservation of energy but to make a determination of the mechanical equivalent of heat. A resistance eventuating in a postponement of publication for one year scarcely indicates a serious degree of repression by an official science.

Hermann von Helmholtz ran into similar difficulties with the editor Poggendorff. In his twenty-sixth year, von Helmholtz wrote the paper *Uber die Erhaltung der Kraft* (On the Conservation of Energy), and sent it to Poggendorff. The editor felt it was an important essay, but had misgivings as to whether its character was sufficiently experimental for his journal; he urged instead its publication in independent form. This was a time, indeed, when older physicists were especially suspicious of speculative ideas evocative of the "phantasm of Hegel's "nature-philosophy.' " They had fought hard in their youth to liberate science from apriorist, metaphysical *Naturphilosophie*, and they were determined to preserve the hard-earned respect for experimental truth. In 1847, von Helmholtz's work was indeed published by an independent printer; the young author received not only an honorarium and the acclaim of the younger scientists but also the high praise of the authorities of the Prussian army for his "splendid" work.[92] Clearly, the values of established society were not called into question by von Helmholtz's generalization of the conservation of energy. A merger of the dominant scientific mode of thought with the reigning sociopolitical values which is characteristic of official science had not taken place.

The third scientist who participated in the discovery of the law of conservation of energy, James Prescott Joule, had also (as we have seen encountered resistance to his views when in 1843 he presented his paper "On the Calorific Effects of Magneto-Electricity, and on the Mechanical Value of Heat" to the British Association for the Advancement of Science at Cork, Ireland. Another paper of Joule's was indeed subsequently rejected by the Royal Society, but by 1850, Joule was elected to the Royal Society, and two years later was honored with a royal medal.[93]

The introduction and publication of the law of conservation of energy thus involved no revolutionary conflict with the scientific establishment. No new journals, organizations, or societies, were required for the novel, basic idea to win acceptance within a few years. The resistance to the principle of conservation of energy arose from a suspicion of metaphysical speculation that actually extended to all philosophizing, even of an empiricist variety, when presented in the form of scientific contributions. The editor Poggendorff thus not only turned down papers by Helmholtz and Mayer; twenty years later, he rejected an essay by the young Ernst Mach entitled "On the Definition of Mass." In later years Mach still smarted under this decision: "My note on this subject was returned as unusable by Poggendorff, the then editor of the *Annalen der Physik und Chemie*, after he had had it about a year. . . ." Shortly afterwards, however, in 1868, Mach's essay was published in Carl's *Reportorium.* The first rejection depressed Mach so much that he delayed publishing his investigations on the law of inertia: "If I ran up against the physics of the schools in so simple and clear a matter, what could I expect in a more difficult question?" He wrote ironically of the editor's reluctance to publish an article not written in the language of the dominant orthodoxy: "The inclusion of a short note which is not wholly written in that jargon would obviously greatly lower the value of the *Annalen* in the eyes of the public."[94] Poggendorff, who had been vexed by the speculative character of Helmholtz's principle, was probably appalled by Mach's daring argument that the law of conservation of energy was a purely logical consequence of the law of causality, that it followed "immediately from the supposition of the dependence of phenomena on one another." Nevertheless, Mach did not have to found a new journal to get his ideas considered. He published his original conceptions in journals through the years 1863, 1864, 1865, and 1866. Indeed, his standpoint won so many influential followers that by the mid-eighties the young Max Planck, who was drawn toward an atomistic standpoint, felt that a Machian orthodoxy was weighing down the young generation of physicists.

Johann Christian Poggendorff, the most noted editor in the annals of science, for more than a half-century supervised the publication of the *Annalen der Physik und Chemie*. His editorship was coeval with political turmoil, working-class revolutions, nationalist movements, and the apocalyptic years of 1830, 1848, and 1871. What marked his editorial career was the welcome he extended to scientific conceptions that have been regarded as inaugurating revolutions. During his fifty years as editor 2,167 authors published in the *Annalen*. Michael Faraday, inaugurating the "electromagnetic revolution," had seventy-six contributions; Liebig with his "revolution" in agricultural chemistry had fifty-six; Woehler, the "revolutionist" of organic chemistry, had sixty-five.[95] The works of such investigators as Becquerel and Fresnel were translated from the French and made available in the *Annalen*. This journal transcended national as well as class antagonisms; the community founded on the supremacy of experimental fact transcended political divisions.

Poggendorff might have had reason to nurture political animosities. His father had lost his entire fortune during the occupation of Hamburg by the French army, and Johann born in 1796, was soon obliged to become an apothecary's apprentice. Yet he persisted in his researches and became a meteorological observer for the Berlin Academy of Sciences. The idea of founding a chemical-physical journal occupied his imagination, and in 1824, when the editor of the *Annalen der Physik* died, Poggendorff persuaded its publisher to allow him to found an expanded *Annalen der Physik und Chemie*. Editing that journal became his lifework. To help him in his editing, he carefully collected the writings of his authors and compiled a biographical dictionary of scientists containing thousands of names. His catalogue of scientific literature was a diagram of the extraordinary mosaic to which they all had contributed. "Willing and non-partisan, he at all times let every viewpoint come to expression, worked unflaggingly to bring recognition to every merit, even the least, warded off unjust claims, and assured everyone of what was justly his." The time of Poggendorff's *Annals*, noted his biographer, saw "the beginning and development of electromagnetism and diamagnetism, of induction, electrical telegraphy, galvanoplastics, thermochrosis, the mechanical theory of heat, fluorescence, spectrum analysis and photography."[96] Novel phenomena and novel types of laws and modes of analysis came into being. But Poggendorff, the apothecary's apprentice who loved natural science, always retained an overriding commitment to experimental fact. The half-century of continuous scientific editorship during the time of great political upheaval contravenes the application of the revolutionary analogy to the evolution of science.

The reception of the theory of relativity in Britain is perhaps the most dramatic example of the collaboration of a scientific establishment in the validation of a revolutionary theory. Britain was still at war with Germany in 1918 when plans began to be made for testing Einstein's theory of gravitation. The scheme of thought of Isaac Newton, the greatest English scientist, was being challenged by a professor at Berlin. Nevertheless, the Royal Observatory, the Astronomer Royal, and the British government provided the direction and finances for the expeditions to Principe (led by Arthur Eddington) and Sobral to observe the deflection of light during the total solar eclipse on May 29, 1919. The observations were replicated in 1922 by the Lick expedition during the eclipse that year. Its leader, W. W. Campbell, was asked what he hoped would be the outcome from the photographic plates. "I hoped it would not be true," he replied. Many eminent scientists shared this reluctance to accept the theory of relativity.[97] Yet Newtonianism had not congealed into an official science. The younger scientists, of course, were the ones who embraced relativity. "With the close of the war many scientists and students returned to the universities and were caught up in this new spirit of questioning and speculation about startlingly unfamiliar ideas." To this new generation especially, Eddington's *Report on the Relativity Theory of Gravitation*, already published in 1918, became a kind of manifesto: "Old ideas were in the melting pot; an exciting spirit of adventure was vaguely felt even by those not mathematically equipped to read the book critically."[98]

Subsequently, the results of Eddington's expedition, published in the *Transactions of the Royal Society*, provided the basis for the intergenerational acceptance of the theory of relativity. This was an unusual revolution mediated and partially directed from above, a variety not found in the annals of social revolutions, though used as a rhetorical phrase.

The turning point in the quantum revolution is another remarkable example of generational collaboration in the history of science. We have seen how on October 30, 1911, twenty-one scientists convened in Brussels to discuss the problems of quantum theory at the first Solvay Congress. Only a handful of scientists were then advocates of the quantum hypothesis. Of the six British scientists who had been invited, Lord Rayleigh, Schuster, Larmor, J. J. Thomson, Jeans, and Rutherford, only the last was not unsympathetic to the quantum ideas.[99] The work of the younger generation, taken by itself, failed to convince the elders. Einstein's paper on the photoelectric law, with its light-quantum theory, was regarded as a speculation which in Planck's judgment had "missed the mark"; it was not even reported upon at the congress.

The greater number of physicists, as Louis de Broglie wrote, rejected Einstein's hypothesis of light-quanta, which was at odds with wave optics, as "too simplistic."[100] A few days before the congress, Paul Ehrenfest, a young, unemployed physicist from St. Petersburg, had published a paper showing that a quantum law was a necessary and sufficient condition for what he called the "Avoidance of the Rayleigh-Jeans Catastrophe in the Ultraviolet."[101] But Ehrenfest's paper was ignored in the Solvay deliberations. It was the conversion of Henri Poincaré, the most distinguished mathematical-physical thinker in the classical tradition, to the ideas of the quantum hypothesis that proved decisive for its acceptance. Poincaré came to the congress uninformed about quantum theory, but he joined spiritedly in the discussions. Scarcely a month later he demonstrated the theorem that Planck's radiation formula "can be derived from only one set of physical assumptions, namely, those of the quantum theory."[102] In Poincaré's words: *L'hypothèse des quanta est la seule qui conduise á la loi de Planck.*[103] (The hypothesis of the quanta is the only one which leads to Planck's law.) Poincaré's paper was not an affair of mere mathematical elegance in the formulation of a theory. Its basic theorem was a substantive one of immense importance which showed that the quantization of energy is both a necessary and sufficient condition for the truth of Planck's law. Moreover, besides demonstrating that Planck's law required the discontinuities of the theory of quanta, Poincaré showed that no small or even incremental departures from Planck's law would obviate the necessity for discontinuities in energy. The sheer fact that the total radiation of a black body at a finite temperature is finite requires that the ultimate motions should in some way be discontinuous.[104]

Poincaré's paper made a profound impression on his contemporaries. "Whereas Ehrenfest could be ignored and Einstein not accepted, Poincaré's authority could hardly be questioned. Even James Jeans, a holdout against the quantum for a decade and still an unbeliever at Solvay Congress, yielded to Poincaré's mathematics."[105] Jeans spoke to the British Association for the Advancement of Science at Birmingham on September 12, 1913: "This result is so complete, so definite, and above all so revolutionary, that it will naturally be most closely scrutinised. . . ." He adduced the objections to Poincaré's arguments which "some of the more conservative of us" had felt; they were fully answered. He told of his own experience:

> Speaking for myself, I may perhaps be allowed to say that I have devoted several years of work to an attempt, quite unfruitful as it turned out, to reconcile the laws of radiation with the classical mechanics. . . . [He had been] forced to realise that all the facts are against us . . . [whereas] the new mechanics, based on the quantum hypothesis, show just that power of explaining and predicting facts which is to be expected of a new truth in its infancy.[106]

At the Solvay Congress, Poincaré had been cool toward Jeans whom he admonished for not recognizing that the scientist must choose the theory which is logically the simplest, and should not multiply auxiliary constants or hypotheses.[107] Jeans, however, now unequivocally acknowledged the power of Poincaré's demonstration. But purely mathematical arguments were not so persuasive to physicists who did not share Jeans' mathematical enthusiasm. To such physicists Jeans emphasized the heuristic strength of the quantum theory: "Logical necessity of this kind is made mentally more palatable if we can discover direct evidence, or phenomena other than those from which it is derived, to bear witness to its truth." First, there was the photoelectric effect, "a phenomenon totally unlike anything which can be explained by the classical mechanics," and second, the "most ingenious and suggestive, and I think we must add convincing explanation of the laws of spectral series [which Niels Bohr had just provided]." The 1911 edition of the *Encyclopaedia Britannica* made no mention of the quantum theory. "It was Bohr's work," as Andrade has written, "that forced physicists and chemists to realize that quantum theory was of paramount importance."[108]

In 1914 Jeans published his famous *Report on Radiation and the Quantum Theory* for the Physical Society of London. Because of the outbreak of World War I, writes E. A. Milne, "it did not at first reach a large circle of readers, but it was eagerly read when students of mathematics and physics returned from the war in 1919."[109] Jean's *Report* became a generational manifesto of the new physics, and in Britain shared this role with Eddington's *Report on the Relativity Theory of Gravitation.* But it was the adherence of Henri Poincaré that first transformed an almost pariah hypothesis to an accepted one. "The keynote of the old mechanics," wrote Jeans at the end of his *Report*, "was continuity, *natura non facit saltus.*" The keynote of the new mechanics is discontinuity; in Poincaré's words: "A physical system is only susceptible of a finite number of distinct states; it jumps from one of these states to another without passing through a continuous series of intermediate states."[110]

When ten years later Jeans published a second edition of his *Report*, he described the earlier one as having been written at a time when the quantum theory was "an object of suspicion to most physicists"; his writing therefore had been permeated by a "defensive" note. By 1924 the accumulation of novel experimental evidence by a younger generation had brought hegemony to the quantum ideas. Unlike any social revolution, however, this was a strange revolution in which the last writing of the greatest figure of the scientific establishment conferred the mantle of succession upon the revolutionists.

The introduction of novel ideas on genetics into the field of biology at the beginning of the twentieth century evoked sharp generational conflict, but nonetheless, the scientific establishment was sufficiently pluralistic to open to them the avenues of publication, thereby assisting the experimental vindication of the new ideas. According to L. C. Dunn, genetics was the biological science that developed most rapidly in our century and has "most profoundly affected the development of biology as a whole."[111] The expression "genetic revolution" never gained currency but young geneticists at the time were imbued with a revolutionary fervor. One of the pioneers, R. C. Punnett, recalled that in Britain "the older generation of biologists" was hostile to the new approach, "and the pens of Alfred Russel Wallace, Professor Poulton and Professor J. Arthur Thomson were soon engaged in attempting its belittlement. In this they were supported by [the journal] *Nature*. . . ." Mendelian genetics, however, became newsworthy: "the columns of secular periodicals were opened to us," and the establishment had no monopoly of scientific publications and meetings.[112] William Bateson in 1894 published *Materials for the Study of Variation*, which argued for the notion of discontinuity in variation. Then in 1899 and 1900, three botanists, de Vries, Correns, and Tschermak rediscovered the work of the Abbé Gregor Mendel which had been done thirty-four years before. The first experimental laws of heredity were "confirmed by the unusual process of rediscovery not once but three times and independently."[113] Bateson read the news of the rediscovery of Mendel's work, and, while riding on a train, revised a lecture he was to give to the Royal Horticultural Society. The lines were drawn for a battle of generational methodologies. The dominant biometrical school, founded by Karl Pearson, adhered to the Darwinian assumption of continuity of variation, and enjoyed the staunch support of Weldon, the Linacre Professor at Oxford. Together, Pearson and Weldon published their journal *Biometrika*. As the controversy became more intense, *Biometrika* rejected a paper of Bateson's. In 1903 the editor of *Nature* wrote that he was "not prepared to continue the discussion on Mendel's Principles and therefore returns herewith the papers recently sent him by Mr. Bateson."

Yet the establishment was not monolithic; it was pluralistic enough so that a recessive idea could become dominant.

> It was a difficult time for struggling geneticists when the leading journals refused to punish their contributions to knowledge, and we had to get along as best we could with the more friendly aid of the Cambridge Philosophical Society and the Reports to the Evolution Committee of the Royal Society. . . . The Evolution Committee proved of great service both by providing small grants towards the experiments, and by affording a means of publication.[114]

Moreover, the Cambridge University Press arranged for a private printing of one of Bateson's rejected papers, using the format of *Biometrika*.

William Bateson knew the sociological truth that in the battle of ideas he who has the younger generation on his side will celebrate a congress of victors. As Punnett writes: "Bateson knew very well that if this branch of enquiry was to live and expand it must capture young men and women. . . ."[115] Around himself Bateson gathered a "little band," none of whose members earned his livelihood through research or teaching genetics. Bateson himself looked after the kitchens of St. John's College. In 1899, however, he was appointed deputy to the professor of zoology, and had a more secure base of operations. The debate in 1904 at the session of the zoology section of the British Association marked the decisive turning point for Bateson. Apart from presenting his experimental results, he took the theoretical offensive against the Biometricians: "The imposing Correlation Table into which the biometrical Procrustes fits his arrays of unanalysed data is still no substitute for the common sieve of a trained judgment."[116] The session ended with Pearson's proposal for a three-year truce.

Theological and scientific ideas are generally described as being antithetical, and for the greater part of intellectual history this has been so. The scientific thinking, however, of the great Catholic founder of genetics, Gregor Mendel, may well have had a theological impress. In the opinion of L. C. Dunn, Mendel's work was guided by some "preconceived idea"; the "fit" of the observations to theory was simply "improbably good," but Dunn added, "we are left in the dark as to the time or manner of its conception." But recalling certain biographical facts seems to shed some light. Mendel chose to become an Augustinian monk at the monastery of St. Thomas. Curiously, Mendel's monastery during his youth was a center for daring, subversive ideas; according to one of its pupils in 1843, Theodor Gomperz, later an eminent historian of philosophy, the free-thinking tendencies of the Young Hegelian movement reigned unchecked in its cells and cloisters[117] under the leadership of its canon, Thomas Bratranek, himself the author of *Aesthetics of the Plant World*. Mendel as a young novitiate was for eight years in close intellectual companionship with Bratranek. Mendel became an ordained priest. Elected abbot of his monastery in 1868, he struggled for ten years (1874–1883) against the Austrian political anticlericals who were imposing taxes on church organizations. During this time his papers were published in the journal of a local natural history society.[118]

Mendel's decision to become an Augustinian monk probably signified certain emotional-intellectual attitudes. The Augustinians placed great emphasis on the doctrine of original sin—the inborn nature of man, the character transmitted from generation to generation. At a time when biological theories of evolution were suggesting vistas of progress and novel emergence, the Augustinian monk Mendel conceived a theory in which character traits, like original sin, were all present, dominantly or recessively, in the succession of generations. One indeed wonders whether the operative "preconceived idea" in Mendel's mind was Augustine's theory of original sin, translated into the determinism advocated by the Young Hegelians.

Mendel's life had been cast in a gloom that seemed predestined. As a schoolboy, he had felt hunger gnawing within him; he learned early the harsh realities of life. When his academic hopes seemed to have failed, his distress and prostration were complete: "the gloomy outlook upon the future had so marked an effect upon him that he fell sick, and was compelled to spend a year with his parents for the restoration of his health." He could not put a cheerful face to men's miseries; he found himself utterly unable to cheer the sick, an "infirmity of his" that "made him dangerously ill."[119] He read the deterministic materialist Ludwig Büchner despite the *Index Librorum Prohibitorum.* Büchner, author of the widely read *Force and Matter*, was the militant chief in the *Materialismusstreit* (materialism debate). In the spirit of a disciple of Ludwig Feuerbach, Büchner assailed idealism, speculative philosophy, and vital force, and urged the immutability and universality of laws of nature in all phenomena; he hailed "the new science statistics, which exhibits fixed laws in a mass of phenomena. . . ." Büchner's materialist view of the world may well have been Mendel's guiding standpoint as a biologist. His friend Bratranek suffered penalties for his outspokenness; Mendel preserved a silence on ultimate issues. But Mendel evidently felt that the anticlericals and radical reformers were quarreling not with the social system but with existence itself. Scholars have been puzzled by the failure of Mendel's contemporaries to see the tremendous importance of his work.[120] Perhaps it was because the tenor of his work failed to fit the isoemotional lines of his time. None of the evolutionary radicals associated with the Darwinist, Ernst Haeckel, would have found congenial a model of the immutable gene, or an experimentation which stressed the conservation of characteristics recessive as well as dominant.

Moreover, Haeckel who had been staunchly Christian in his

youth, ceased being so when he embraced Darwinism; instead he became a militant foe of churchmen. It was mainly through Haeckel's "promulgation of it that Darwinism became a watchword for all supporters of the idea of a liberal-minded development in the sphere of social and cultural life, and obviously an abomination to its opponents, the clerical and conservative elements in the community."[121] Haeckel's personal history was "highly characteristic of the generation to which he belonged and therefore explains in some degree how it was that he acquired such an influence over his age."[122] The radical generation that combined natural-scientific radicalism with political radicalism (beginning in the 1870s) was even more dogmatic than its master. Later, the German socialists found in Haeckel a natural ally.

Thus, an amalgam of Darwinism, social radicalism, mechanism, and romantic natural philosophy was the regnant generational doctrine at the time when Mendel conducted his painstaking experiments. He could not expect a friendly reception from his "advanced" scientific contemporaries. Karl Nägeli, the eminent Swiss botanist, did read Mendel's work. But Nägeli held to a grandiose, speculative vision of a creative force in nature, a *Vervollkommungskraft* ("perfecting force"), a *nisus formativus*, which generated the variations of species in a particular direction.[123] That is probably why Nägeli felt that Mendel's experimental studies were "empirical rather than rational." Others evidently felt that Mendel's mathematical laws simply could not encompass living phenomena, especially those of reproduction. Darwin, for all his Malthusian standpoint was a nonmathematical naturalist, who allowed variation to remain fortuitous. Mendel, by contrast, was a mathematical determinist. According to his biographer Hugo Iltis, Mendel never allowed Darwin's name to cross his lips. In a sense, he cut himself off from the ideology of his own generational scientific community. In return, they neglected him, and his work was virtually ignored from 1866 until 1900.[124]

Classical Systems as Limiting Cases of Revolutionary Systems: The Absence of Qualitative Negation in Scientific Revolutions

We have seen that scientific revolutions are not preceded by revolutionary situations or by crises of closed publication. In addition revolutionary theories advanced in science most often do not involve the outright removal of the older system of laws. Indeed, we might define an advanced science as one in which a superseded classical theory is not so much falsified as delimited in its domain of validity. In this respect, we note another basic difference between social revolutions and scientific ones. When a social revolution takes place, the revolutionaries claim that novel class relations have been instituted. With the overthrow of a feudal order, for instance, the emancipated serfs can leave the land, and the social law of the immobility of labor is thus abrogated. The new landowner, moreover, may sell his land, for bourgeois society annuls the law of the immobility of capital. Similarly, in a socialist revolution, private ownership of fields and factories, private investment, and private employment of labor are terminated. No one, unless he were speaking in a highly metaphorical fashion, would affirm that the laws of feudalism were a special case of the more general laws of capitalism; no one would say that the laws of capitalist economy were special cases of more general, underlying laws of socialist economy. The different economies—feudal, capitalist, and socialist—have their specific forms or laws, but none is more general in a logical sense than the others.

However, when a scientific revolution takes place, the laws of the superseded classical system are typically not discarded. When the theory of relativity was accepted, the formulations of the classical laws could still be used provided that it was stipulated that their domain of applicability was restricted to everyday objects moving at usual velocities. The classical laws are regarded as *limiting cases* of the new generalizations. Einstein wrote, for instance, that the theory of relativity

> made it clear that mass is not a constant quantity but depends on (indeed it is equivalent to) the amount of energy content. It also showed that Newton's law of motion was only to be regarded as a limiting law valid for small velocities; in its place it set-up a new law of motion in which the speed of light in vacuo figures as the critical velocity.[125]

The logical expositors of the theory of relativity have all emphasized this relationship of the classical theory as a limiting case to that of the relativist laws. Philipp Frank writes, for example, that Einstein's theory shows that Newton's laws of motion "can only be valid for small velocities, where 'small' means 'small relative to the speed of light,'" and that Newton's laws cannot be universally valid, for they fail to hold for bodies moving with relative speeds comparable to the velocity of light.[126]

In social revolutions the past is something to be undone. "No more tradition's chain shall bind us," sings the *Internationale*. But in scientific work, generalizations remain valid for that segment of experience in which they were tested. Planck's discovery of quanta had the most profound influence upon the fundamentals of physics at the end of the nineteenth century, writes Robert Millikan. Then he adds: "It was not revolutionary, however, in undoing the past; the old laws still held *in the field in which they had been experimentally tested*."[127] That is why the more general revolutionary formulation must always have among its logical consequences corollaries that coincide in extensional content and logical form with the laws of the classical system. Nothing analogous to this pattern obtains in social revolutions.

To reinforce the thesis that all basic changes in the history of science have a revolutionary character, those who are revolution-minded are therefore obliged to assert that the novel system of laws really has nothing in common with those of the classical system, that in no logical sense can the latter be regarded as a limiting, or special case of the former. Curiously, this model is exactly what Marxists regarded as intrinsic to the dialectical conception of change. Engels, for instance, held that the bourgeois economic laws of value would be superseded in a socialist society whose laws would have virtually nothing in common (except for some banalities and tautologies) with those of classical economics. This standpoint we might call the *dialectical model* for the development of science. During the thirties, Marxist theoreticians argued that a proletarian science, whether in physics, genetics, or psychology, must perforce have its unique perspective and laws "qualitatively" different from those of bourgeois science.[128]

To support the notion that the basic changes in scientific theory have had *logically*, not merely psychologically, the character of revolutions, the revolution-minded theorist is therefore compelled to introduce a conception of the logic of scientific theories that is quite at variance with that expressed by Einstein, Frank, and Russell. Where Einstein, for instance, says that Newton's law of motion can be derived as a limiting law for masses moving at low velocities from the theory of relativity itself, Professor Thomas Kuhn affirms that "the derivation is spurious." The equations thus derived as special cases of relativistic mechanics, he says, are not really the same as Newton's laws, unless the latter are reinterpreted in a sense which they could only have received after Einstein's work; in that case, such variables as mass and time would have the intrinsic meanings given to them by Einstein, not by Newton. As Professor Kuhn writes:

> But the physical referents of these Einsteinian concepts are by no means identical with those of the Newtonian concepts that bear the same name (Newtonian mass is conserved; Einsteinian is convertible with energy. Only at low relative velocities may the two be measured in the same way, and even then they must not be conceived to be the same).[129]

What Professor Kuhn's argument overlooks is the logical truth that two generalizations that differ partially in their *intension*, or *connotation*, can still have an overlapping, partially common *extension*, or denotation. The actual instances, the events that are examples in existence of a defining characteristic, are said to be the *extension* of a class; the *intension* is simply the defining characteristic considered in abstraction. According to Bertrand Russell, a class can be defined either by its extension or intension; it is defined extensionally by enumerating its members; it is defined intensionally when we state its defining property.[130] That generationalizations with partially different intensions can still have overlapping extensions is exemplified as follows: the two propositions *all wars have economic causes* and *only wars of capitalist societies have economic causes* have different intensions, yet they would overlap for those wars occurring in capitalist societies. It is this logical truth that the revolution-minded analyst overlooks. Einstein's laws have a different intension from Newton's laws, but for a certain segment of events, they will share a common extension. Einstein's equation for the variation of a mass in accordance with its uniform velocity does coincide with the extension of Newton's law when the velocity is very small relative to the velocity of light. Moreover, Einstein's equation itself, when values of exceedingly small quantities are omitted, takes the form of Newton's equation, reflecting formally the coincidence of extensions in the limiting law. Thus, when m does become virtually equal to m_0, that is, when the numerical value of an object's mass is approximately equal to its rest-mass, then Einstein's law is indistinguishable operationally, and for the limited case formally, from Newton's.

Unless we recognize that such derivations of classical laws from their revolutionary successors are valid, we shall not be able to give an intelligible account of, for instance, the derivation of Kepler's laws of planetary motion from Newton's laws, or the derivation of the laws of gases from the kinetic theory. The revolution-minded theorist will be constrained to say that Kepler's and Newton's laws simply define different types of entities with altogether different structural properties. He will have to say that in the "derivation" of Kepler's third law, for instance, the novel Newtonian meanings of terms are simply not those of Kepler. Kepler affirmed that the squares of the periods of revolution of the different planets were proportional to the cubes of their mean distances from the sun. In the Newtonian derivation, the analogue is the law, $4\pi^2 R^3 = GMT^2$, where R is the planet's mean distance from the sun, and T its period of revolution. The Newtonian equation, however, multiplies the square of the period of the revolution by M, the mass of the sun, and G, the gravitational constant. Moreover, the planets could cease at some future time to move in elliptical orbits; yet the law of gravitation would still be valid. There are thus divergences in intension between the Newtonian and Keplerian laws; nonetheless, in the sense we have described, the latter are derived as limiting laws of the former under specially defined initial conditions.

The quantum revolution was likewise mediated with the classical physics by means of what Niels Bohr called the correspondence principle. Although, as Oskar Klein writes in "Glimpses of Niels Bohr as Scientist and Thinker," Bohr never ceased to emphasize the depth of the abyss which separated the quantum approach from the classical, nevertheless Bohr made a cardinal principle of their correspondence; their descriptions of physical reality coincided in all limiting cases where Planck's quantum of action is very small compared with the actions to be described. The quantum physics, in the words of Bohr's *Essays 1958–62 on Atomic Physics*, "fulfilled the obvious demand of embracing the classical physical description in the limit where the action involved is sufficiently large to permit the neglect of the individual quantum." The correspondence principle is another instance of the principle of the limiting case; the preceding system of classical physics is found to be a special case of the revolutionary physics. But of course all this belies the metaphor of revolution.

In economics, too, in the measure that it has become a mature, cumulative discipline rather than a realm for ideological forays, the conception has gained ground that so-called revolutions in that field do not involve the utter negation of classical theories; the latter are rather assigned the status of special cases of the more general, revolutionary theories. When Keynes was writing his *General Theory* he felt himself subjectively to be revolutionizing economic theory. As he wrote to George Bernard Shaw on January 1, 1935:

> To understand my state of mind, however, you have to know that I believe myself to be writing a book on economic theory which will largely revolutionize—not, I suppose, at once but in the course of the next ten years, the way the world thinks about economic problems. When my new theory has been duly assimilated and mixed with politics and feelings and passions, I can't predict what the final upshot will be in its effect on action and affairs. But there will be a great change, and, in particular, the Ricardian foundations of Marxism will be knocked away.
>
> I can't expect you, or anyone else, to believe this at the present stage. But for myself I don't merely hope what I say,—in my own mind I'm quite sure.[131]

Though psychologically he felt himself to be a revolutionist, Keynes emphasized from the very first page of his book that classical theory still retained its truth as a limiting case under special boundary conditions of his own general theory: "I shall argue that the postulates of the classical theory are applicable to a special case only and not to the general case, the situation which it assumed being a limiting point of the possible positions of equilibrium."

Keynes felt, however, that this *General Theory* would have an importance for everyday economic life because "the characteristics of the special case assumed by the classical theory happen not to be those of the economic society in which we actually live, with the result that its teaching is misleading and disastrous if we attempt to apply it to the facts of experience."[132]

Scientific revolutions have thus involved during the twentieth century elements of continuity in consciousness and logic wholly lacking in social revolutions.

The Non-Immanent, Epigenetic Development of Novel Sciences

The development of science has departed in still a fourth respect from the revolutionary analogy. According to the revolutionary model or isomorpheme, the immanent development of science, the cumulative accretion of its everyday researches, finally generate a situation of strain, tension, and conflict (Marxists would call it "contradiction"), that can be resolved solely by a revolutionary reconstruction of the categories and system. Thus, says Professor Kuhn, out of the development of "normal science" itself arises the discovery of anomalous facts which cannot be assimilated within the traditional system. "The traditional pursuit prepares the way for its own change. . . ." With the perception of the anomaly of a phenomenon, that, in other words, for which the traditional paradigm had not prepared the investigator, the progress of "normal science" itself prepares the way "to a breakthrough quite thoroughly."[133]

Most of the experimental revolutions that have taken place in the history of science have not, however, been the outcome of the cumulation of evidence within the working bounds of a normal science and its "mopping-up operations."[134] Rather, some novel technology has brought to light novel facts; or a social taboo on the report of certain sorts of observations has been overcome; or a new class of workers, insufficiently educated to pursue classical science, has persevered along novel, less highly mathematical, lines of research.[135] In some cases an accident has intervened, or a chance observation, occurring at a certain moment in the researcher's personal history, has been endowed with a challenging significance which it otherwise would not have had. The psychoanalytic revolution, in effect, began when Freud overcame the social taboo against scientific observation of sexual facts. The academic psychologists of the conventional schools had simply omitted sexuality from the inventory of human experience. The classic work of William James, *The Principles of Psychology*, published in 1890, is rich with introspective detail and experiential fact; sex, however, is missing from its index. James, in later years, welcomed Freud's work, but he and academic psychologists generally had a resistance to studying or writing about sexual facts. It was not a developing crisis of observation within academic psychology that led to Freud's work and theories. The academic psychologists were content to continue with the status quo. In the summation of their work in Edwin G. Boring's authoritative *A History of Experimental Psychology*, the recorded studies on sensation by the nineteenth-century scientists are noteworthy for the absence of any work on sexual sensation.[136]

The Newtonian classical system was outflanked by the accumulation of the observations and theories of investigators who lacked the necessary academic education to contribute to Newtonian ideas. As George Sarton once remarked, it was self-taught men, who knew little mathematics, and who had had no university education who, from Franklin to Faraday, were the pioneers in electromagnetic science. The great mathematical physicists of the eighteenth century, d'Alembert, Lagrange, Laplace, were perfecting the Newtonian system. Beginning with the "electrician" Francis Hauksbee, a humbler, less mathematically educated group led the way in their observations of electrical phenomena. The mathematical stage came later. One might say that the nonmathematicians turned their shortcoming into an asset by seeking out for observation fields that had not yet been developed to a stage ripe for mathematical analysis. Joseph Priestley, the self-taught scientist who regarded the study of electricity as "my own favorite amusement," put the matter plainly in his *History of Electricity*:

> In electricity, in particular, there is the greatest room to make discoveries. It is a field but just opened, and requires no great stock of particular preparatory knowledge: so that any person who is tolerably well versed in experimental philosophy, may presently be upon a level with the most experienced electricians. Nay, this history shows, that several raw adventurers have made themselves as considerable, as some who have been, in

other respects, the greatest philosophers. I need not tell my reader of how great weight this consideration is, to induce him to provide himself with an electrical apparatus. The pleasure arising from the most trifling discoveries of one's own, far exceeds what we receive from understanding the much more important discoveries of others. . . .[137]

While on the European continent mathematical physicists followed the leadership of Gauss in trying to derive electrical phenomena from some generalization of Newton's law of gravitation, in England the development was quite different. There, writes Planck, "Faraday impressed upon it the mark of his genius, in that he studied electrical phenomena directly without being in any way influenced by mathematics or astronomy. . . ."[138]

Observers are people conditioned by their societies; the directions, emphases, and repressions of their observations depend in part on the social character of the observers. Depressed Asian societies, for instance, have seen the world under the aspect of a masochist mode of perception. In Western Europe, an activist mode of perception began to prevail in the seventeenth century. The pioneers of psychoanalysis removed the repressions that either blotted out or distorted the observation of sexual phenomena. The pioneers of electromagnetic observation were not restricted in their perceptions by the Newtonian mathematical model. The purely "personal equation" in perception has long been known. In 1796 Nevil Maskelyne, the Astronomer Royal at the Greenwich Observatory, dismissed his assistant because the latter was observing the times of stellar transits as taking place almost a second later than his superior saw them. More than twenty years later in 1819, Fredrich Wilhelm Bessel, the astronomer at Königsberg, studying the records, and proceeding to further comparative observations came to the conclusion that the assistant had been unjustly dismissed, that there was indeed a "personal equation" in observation; it proved later to be dependent not only on movements in the objective stimulus but on the "expectations" of the observer.[139] There is also, however, a social equation in the history of observation which *varies with the repressions and overdeterminations of a culture*. The great observers have been the unsocialized personalities of their time, who were still able to observe with naiveté, with an innocence that was miraculously unconditioned. The psychology of the revolutionary observer has been studied even less than that of the revolutionary theorist. What takes place in the unconscious of the revolutionary experimentalist causing observations previously made by others and discarded as trivial to be suddenly endowed with a significance they had never possessed? The observation, from having been passive and contemplative, suddenly becomes active. Was the observation made at a moment in the investigator's life when his psyche was restless with long repressed revolutionary emotions? Does he see things which the others had not because his feelings are seeking to make his observations into vehicles for rebellious challenge? Perception, then, as it is conditioned by a revolutionary unconscious, might at a fortuitous moment lend to its object a significance it lacked for the ordinary observer. Are emotions of generational revolt, sometimes long delayed, at work in the unconscious of such scientific observers?

The discovery of X-rays in 1895 by Wilhelm Conrad Roentgen, for instance, was "the consequence of his grasping the significance of an unimportant effect of an unknown phenomenon and tracing this effect to its real cause."[140] Roentgen's discovery has often been described as a chance event. He was replicating phenomena which Lenard had observed in experiments with Crookes tubes. It had been found that when a high tension discharge from an induction coil is passed through a vacuum tube, cathode rays are generated which can penetrate for a few centimeters on the outside, when thin aluminum windows were used. Lenard had discovered that the cathode rays darkened the photographic plate. Roentgen wondered whether such effects might exist even when heavier walls were used on the tubes. He found such a phenomenon but also noted a weak light on a bench a yard away which was like a reflection. Suddenly he became very excited; he found that the light emanated from a small barium platinocyanide screen on the bench. What kind of ray was this which could reach so far? If it could penetrate air, could it penetrate other substances? There was a peculiar shadow on the green fluorescent screen; was it caused by a wire running across the tube? He wondered whether he was a victim of deception in working with the green screen. The color green aroused anxieties in him, for Roentgen's eyes did not see green easily, and "he could not distinguish the colors on the caps of the student fraternities."[141] He tested a piece of paper, then a playing card, then a book, placing them between the tube and the screen, and found the rays penetrated them to produce fluorescent effects. Then, as he tested the capacity of lead to stop the rays, he observed on the screen the bones of his hand. He was filled with fear. Were these observations correct? Or would colleagues ostracize him as a charlatan? He was "shaken considerably. It was a horrible and grisly sight to him—his own skeleton."[142] To his wife, it brought a premonition of death.[143] Seeking to reassure himself as to what he saw, Roentgen used photographic plates and took shadow pictures of opaque objects. "During those trying days I was as if in a state of shock," Roentgen wrote.[144] There had been two previous known cases where photographs with X-rays had been accidentally made.[145] But as the psychologist Hugo Münsterberg said: "Suppose chance helped. There were many galvanic effects in the world before Galvani saw by chance the contraction of a frog's leg on an iron gate. The world is always full of such chances, and only the Galvanis and Roentgens are few."[146] It is acknowledged that no such "chance" discovery can be made unless the investigator's mind is prepared by knowledge and work on related problems; there are intellectual preconditions for realizing the significance of an observation that departs from the rule. Yet, it may well be that in the absence of certain emotional preconditions as well, an observation may be ignored or repressed rather than be turned to revolutionary purposes. The anxiety and fear that Roentgen felt were those of a fifty-year-old man who suddenly found a resurgence of rebellious emotions within him. Was it ostracism he feared so much as these forgotten feelings that welled within him once again?

Wilhelm Roentgen had had a troubled youth; his gay, spontaneous rebelliousness had almost cost him his academic, scientific career. He had been expelled as a student from the Utrecht Technical School not long after he registered there in December 1862. His crime was his unrestrained and pleasurable laughter at a caricature of a teacher he disliked which a fellow student had drawn on the firescreen of the classroom. He then refused to divulge the name of the artist. The entire faculty, with one exception, joined in approving Wilhelm's expulsion. His father was desolate at the shattering of his own ambitions for his son. Then Wilhelm studied privately for a year to pass examinations for entrance into a college, but his examiner turned out to be one of the teachers who had participated in expelling him from the technical school and Roentgen failed his examinations. In later years, he retained a hatred for examinations; they evoked the ordeals of his adolescence imposed by the elder authorities. Forty years later, in his renown, he wrote: "Student examinations generally give no clues whatever for the judgment of ability in a special field . . . and they cause repeatedly bad dreams! The real test of ability of any chosen profession or occupation comes actually much later in life."[147] Once more defeated, Wilhelm enrolled as an auditor in science courses at the University of Utrecht. But then he heard from a young Swiss that at the Federal Polytechnic School in Zurich a student might hope to be admitted provided that he passed their strict examinations. To Zurich went Roentgen bearing a letter from a physician saying that he had been under treatment for an eye condition, for he was now two years older than the average student. There Wilhelm began to distinguish himself in his mechanical engineering studies. He spent time at his favorite inn whose keeper was a refugee German revolutionist who had been ousted from Jena in the 1830s for his student activism. They became close friends, and Roentgen later married his daughter. It was difficult for Roentgen to secure an appointment as lecturer at a German university, for he lacked a certified minimal training in classical languages. Fortunately the university at Strassburg in the newly acquired Alsace-Lorraine was less tradition-bound, and there Roentgen began his academic career in 1874.

It is remarkable how the lives of the revolutionary observer Roentgen and the revolutionary theorist Einstein have such psychological similarities. Both men as youngsters were expelled from their high schools, both were filled with dislike for examinations and the classical curricula, and both encountered the active hostility of some of their teachers. Both were the source of extreme concern to their fathers because of their inability to adjust to the world's ways, found their way to the Federal Polytechnic at Zurich, and discovered friendship in circles where a student revolutionary culture prevailed in act or spirit. Both men were impeded when they sought entry to the lowest rung of university work. Roentgen also retained an animus against the trappings of authority; after his famous discovery he refused an offer of nobility, thereby enraging his colleagues at Munich who had achieved the coveted privilege of prefacing their names with "von."[148] He never acquired confidence as a public speaker; his one attempt to do so in 1903 ended in a fiasco of stammering which the local newspapers kindly ignored. Both Einstein and Roentgen aroused the animosity of the famed German reactionary physicist Lenard.[150] And Roentgen, like Einstein, welcomed the advent of a social democratic government in Germany in 1918, while fearing the portent of Hitler.[151] Inevitably, the hypothesis is urged upon us that there is a similarity in the underlying emotions of both the revolutionary observer and theorist. The latter sublimates his rebellious emotions in seeking to undermine the received, dominant system of scientific theory. The observer, likewise, is unconsciously moved to attend to such observations chancing to come his way that possess latent powers to upset the received catalogue of reported data. What characterizes the experimentalist who dismisses irregular phenomena is that his perceptual reporting is more fully conditioned by the categories and expectations of the scientific establishment. What characterizes the revolutionary observer is that his perception has preserved an element of recalcitrance; he's unsocialized with respect to the categories of the establishment, and may be nurtured by a revolutionary culture. "I did not think; I investigated," said Roentgen of that moment when he observed the black line on a piece of barium platinocyanide paper. It was as if he had allowed his unconscious to guide him. Scientists' modes of perception at critical junctures in experimentation and observation may thus bear the impress of the social strivings and torments of their time.[152]

The unknown, unforeseen nature of the physical world indeed has of course, a role in scientific revolutions that is absent in social revolutions. Physicists in 1895 thought their science was a "dead subject" largely because they could have no idea that a whole new subatomic universe was to come into view. Of the eight epoch-making discoveries between 1895 and 1905 enumerated by Robert Millikan, the three that yielded experimental knowledge of the atom could not have been anticipated by theoretical analysis of existing experimental data. Roentgen's X-rays (1895), Becquerel's radioactivity (1896), and Thomson's corpuscles, later known as electrons, (1897), were simply new qualitative facts, whose advents were like chance occurrences.[153]

Thus experimental revolutions, like theoretical revolutions, do not share the pattern of social revolutions. A social revolution, like the French and Russian revolutions, is preceded by a long period of growing class tensions, economic dislocations, crises of taxation, extensive propaganda, riots, disobedience of armed forces, and the breakdown of governmental authority. The experimental revolutions at the end of the nineteenth century were, on the contrary, unexpected, unfamiliar interlopers that burst in upon a scene of calm. They were strange but welcome arrivals to the scientific community.[154]

The Complementarity of Subjective Generational Revolution and Objective, Cumulative Evolution

The pattern of social revolution clearly does not apply to the history of scientific ideas insofar as the latter is the record of a community that carries on investigations, discussions, theorizings, and publishes its conclusions. It is equally clear that creative scientists, seen from the standpoint of their individual psychological motivations, are spurred by revolutionary impulses to their scientific achievements. Not infrequently, these feelings are associated with conscious political or social motives; always however there is the desire to undermine the philosophical hegemony of the mode of thought of the older generation. The individual creative scientist is always a generational revolutionist. The remarkable fact, however, about science as a social institution, is that it transmutes the energies of generational rebellion in constitutional, intergenerational directions. For this reason the same momentous event in the history of science may be regarded as revolutionary from the personal standpoint of the scientist involved and from a social, impersonal standpoint, as latent, inevitable and evolving from the pre-existing state of the science. Thus, in the emotional excitement of conceiving his theories, Einstein felt himself to be a revolutionary, but looking back on his work in later years, he wrote: "The theory of Clerk Maxwell and Lorentz led inevitably to the special theory of relativity. . . ." He also said: "We have here no revolutionary act but the natural continuation of a line that can be traced through centuries. The abandonment of a certain concept connected with space, time and motion hitherto treated as fundamentals must not be regarded as arbitrary, but only as conditioned by observed facts."[155] Werner Heisenberg as an activist of the German Youth Movement emotionally rejected a materialistic older generation and its atomic science. In later years, however, when he witnessed the seething emotions of a later generation of student revolutionists, it was the liberal, cooperative constitution of the scientific community as the agency for the evolution of scientific thought that was pre-eminent in his mind. He recalled telling a Nazi student:

> If we are discussing revolutions in science, we ought to look more closely at what has been happening. Take Planck's quantum theory. No doubt, you know that when Planck first tackled the subject he had no desire to change classical physics in any serious way. He simply wanted to solve a particular problem, namely, the distribution of energy in the spectrum of a black body. He tried to do so in conformity with all the established physical laws, and it took him many years to realize that this was impossible. Only at that stage did he put forward a hypothesis that did not fit into the framework of classical physics, and even then he tried to fill the breach he had made in the old physics with additional assumptions. That proved impossible, and the consequences of Planck's hypothesis finally led to a radical reconstruction of all physics. But even after that those realms of physics that can be described with the concepts of classical physics remained quite unchanged. In other words, only those revolutions in science will prove fruitful and beneficial whose instigators try to change as little as possible and limit themselves to the solution of a particular and clearly defined problem. Any attempt to make a clean sweep of everything or to change things arbitrarily leads to utter confusion. . . . True, I don't know whether scientific revolutions can be compared with social revolutions, but I suspect that even historically the most durable and beneficial revolutions have been the ones designed to serve clearly defined problems and which left the rest strictly alone.[156]

What Heisenberg was urging is what is known among political scientists as the *doctrine of the conservative revolution*—that the only fruitful revolutions are those carried out with a minimum of revolutionary intent and desire for the revolutionary experience for its own sake, and with a minimum of overturning the status quo for the joy of overturning it. The American Revolution, for example, is regarded as a conservative revolution because it affected social relations to only a small degree, since the colonies had all been virtually self-governing through their legislatures long before the Revolution and took to arms to prevent encroachments against liberties they already held.[157] Heisenberg has probably, however, considerably underestimated the psychological component of generational rebellion that characterizes the scientific creator. Even Max Planck, who so often is cited as the reluctant revolutionist in science, as the man who initiated the quantum revolution against his own wish, would from time to time indicate a strain of the revolutionary emotion disciplined but surviving beneath his Kantian rigidity. Planck's son recalled that one day in 1900 when his father was walking with him in the Grunewald, near Berlin, he said: "Today I have made a discovery as important as that of Newton." Planck, of course, notes Max Born, "never said anything like that in public," and indeed, turned for several years to other fields of work. Then "the opinion spread, especially outside Germany, that Planck 'did not seem to know what he had done when he did it,' that he did not realize the range of his discovery."[158]

Planck, as Born writes, "had nothing of the revolutionary, and was thoroughly sceptical about speculations." Withal, "he did not flinch from announcing the most revolutionary idea which ever has shaken physics."[159] The revolutionary strain, however, indicated in the brief aside to his son was on another occasion linked by Planck to the feelings he had had as a young student when he felt himself stifled by the orthodoxy of the elders:

> I myself experienced during the '80's and '90's of the last century what the feelings of a student are who is convinced that he is in possession of an idea which is in fact superior, and who discovers that all the excellent arguments advanced by him are disregarded simply because his voice is not powerful enough to draw the attention of the scientific world. Men having the authority of Wilhelm Ostwald, Georg Helm, and Ernst Mach were simply above argument.[160]

The polemics between Planck and Mach entailed emotional costs, as we shall see, but failed to disrupt the common constitution of the scientific community. No school of scientific thought is ever annihilated or liquidated as a class in the manner in which an aristocracy is eliminated in a middle-class revolution or a bourgeois in a proletarian one. The basic tenet of each school endures as a temporarily recessive theme in the evolution of science, a surviving minority opinion with adherents of a deviant temperament or bent of mind. Under unforeseeable conditions the recessive theme may emerge once more to dominance and contribute insight in the solving of some scientific problems.

In the historical retrospect, an impersonal, sociological determinism rather than an act of generational free will seems to govern the cumulative advancement of science. Thus Mendeleev could affirm in a retrospective spirit in his Faraday Lecture of 1889: "The law of periodicity was thus a direct outcome of the stock of generalizations and established facts which had accumulated by the end of the decade 1860–1870."[161] As we have seen, Mendeleev himself had been touched by the Decembrist revolutionary message that "youth must revolt against traditional authority."[162] This spirit animates the quest for novel schemes, facts, and systems which challenge the traditional conceptions. Yet a generation after his revolutionary discovery, his thinking took on the aspect of what the Fabians used to call the "inevitability of grandualism."

Invariably, the emotions of the generational revolutionist are found recorded in the personal documentary histories of the great discoverers. Hermann von Helmholtz, the discoverer of the principle of conservation of energy and a Prussian military physician, was very much a man of the establishment. Yet he was driven in his scientific work by the need to triumph over his disapproving father, Ferdinand von Helmholtz. The father, a teacher of classical languages, had served as a student volunteer in the war of liberation, and retained the enthusiasm of his generation for *Naturphilosophie*, apriorist speculative cosmology. Hermann recalled his father and his friends as "always ready to plunge into a metaphysical discussion." But their talk, as they discussed perpetual motion, seemed a sterile play with words compared to the concrete experiments of the science classroom. The father remained hostile to experimentation and inductive reasoning, and "missed no opportunity" to try to "shake his son from such directions."[163] As Hermann's scientific standpoint matured, however, "the more irreconcilable became the contrast with the wholly speculative philosophy of his father."[164] Their personal relations grew strained; the son finally felt it wiser not to discuss his work with his father. Thus, the principle of the conservation of energy was conceived, one infers, through energies liberated in Helmholtz's generational rebellion.

Sigmund Freud is a remarkable example of one who perhaps consciously diverted the energies of generational rebellion to find expression in scientific originality rather than in political revolution. The motto he placed at the beginning of *The Interpretation of Dreams* was *acheronto movebo*, which he translated as "stir up the underworld." "I had borrowed the quotation from Lassalle," wrote Freud, "in whose case it was probably meant personally and relating to social—not psychological classifications. In my case it was meant merely to emphasize the most important part in the dynamics of the dream."[165] Thus Freud, arousing his unconscious to consciousness, regarded himself as akin to Ferdinand Lassalle, the revolutionist, agitating the proletariat to unrest and revolt. As a schoolboy, his classmate, Heinrich Braun, later a socialist leader, was his "inseparable friend"; not only did he encourage Freud in what the latter described as "my aversion to school and what was taught there" but he also "aroused a number of revolutionary feelings within me. . . .[166] Freud later decided, however, that the outcome of his friend's political leftism was rather negative; Freud found a more constructive expression for generational revolt in his scientific work. Nonetheless, ideology of generational protest and activism persisted in Freud's unconscious, even in his later years, affecting the content of his dreams. In one such dream, Freud narrates, a fantasy took him back "to the year of revolution, 1848"; the dream was evidently evoked by an excursion Freud had made in the jubilee year of 1898, when he visited the place of refuge of Adolf Fischhof, the Jewish student leader of the Academic Legion in the Revolution of 1848. Several features of the manifest day-dream-content, said Freud, evidently referred to the student revolutionary chief. The dream also drew on the incidents of a "coup d'etat" which Freud had led as a student "against an unpopular and ignorant teacher," the tyrant of the school, as well as on Freud's defense of materialistic doctrines, in an "extremely one-sided" fashion, at a debate in a German students' club. In the dream, moreover, noted Freud, "I put myself in the place of a certain eminent gentleman of the revolutionary period, who had an adventure with an eagle (German, *Adler*)...."[167]

This revolutionary temperament is inherent in the scientific discoverer; without it the implications of his discovery for scientific thought remain unexplored. Of the joint discoverers of psychoanalysis, Breuer and Freud, for instance, it was Freud who was moved by the emotions of a scientific revolutionist and fought the battles which made for its impact on contemporary culture. In one generous moment Freud indeed declared that "it was not I who had brought psycho-analysis into existence: the credit for this was due to someone else, to Josef Breuer. . . ."[168] What Breuer entirely lacked, however, was the revolutionary impulse that stirred Freud. As a student Freud had joined with the later socialist politician Heinrich Braun in developing as we have seen, "a number of revolutionary feelings."[169] This was not the case for Breuer who wrote: "I never experienced, either during my schooldays or later, the heat and bitterness against school with which our youth seems filled even today and perhaps even more before the recent war." His father was, in Breuer's eyes, a man of tremendous force of character, one of "that generation of Jews who were the first to emerge from the intellectual ghetto into the free atmosphere of Western civilization." Leopold Breuer as a young man lived by what his son called the "bread of affliction," earning his livelihood by giving private lessons but in due time receiving an appointment as teacher of religion in the Jewish community of Vienna. Where Sigmund Freud felt a certain shame concerning his father, Josef Breuer called for "the most reverent honor from their sons and grandsons" to the "pathfinders." Breuer gave unstinting admiration to Freud's achievement: "I still regard Freud's work as magnificent," he wrote in 1907. Together with Freud he had observed the prominent place of sexuality in the neuroses. But Breuer was aware that he was not driven by the kind of compulsion that possessed Freud and moved him to the maximal inferences he could draw for the reconstruction of science rather than the minimal modifications of its existing system. Breuer wrote: "Freud is a man given to absolute and exclusive formulations: this is a physical need which in my opinion leads to excessive generalization. There may be in addition a desire *d'épater le bourgeois.*" Insofar as Freud's views moved beyond experience, "it is merely fulfilling the law of the swing of the pendulum, which governs all development." The bourgeois indeed bears the features of one's own father, and every romantic who has set out to confound the bourgeois was seeking a surrogate object for his generational revolt. And the pendulumlike oscillations in modes of thought arise from the underlying generational process. Freud charged that Breuer lacked the courage to probe sexuality and to endure the discomforts of a patient's transferences. Breuer conceded "that the plunging into sexuality in theory and practice is not to my taste." No doubt he had heard his father transmit Maimonides' teaching that the Hebrew language was so imbued with the values of sexual modesty that it lacked words for the sexual organs. What is impressive however is that Breuer experienced no discontent with his father's teachings; he felt no restless urge to disrupt them. In return he had the gift of personal happiness. "If anything is capable of averting the jealousy of the gods, it is my profound conviction that I have been fortunate far beyond my deserts."[170] The Promethean ingredient of the scientific revolutionist was not one which Breuer imbibed. Yet in his latter years Freud recognized that the revolutionary ingredient had only a subjective significance. He wrote in *The Future of an Illusion*: "The transformation of scientific ideas is a process of development and progress, not of revolution."

Freud in his old age barely survived the Nazi revolution. The French scientist Lavoisier, on the other hand, lost his life in 1794 at the hands of the Revolution. Yet Lavoisier too was a child of the revolutionary mode of thought and moved in its cultural milieu. That Lavoisier as a young man setting out in 1773 on his "long series of experiments" was filled with a revolutionary psychology we know from his famous memorandum dated February 20, 1772, in which he wrote: "The importance of the end in view prompted me to undertake all this work, which seemed to me destined to bring about a revolution in physics and chemistry. I have felt bound to look upon all that has been done before me merely as suggestive: I have proposed to repeat it all with new safeguards. . . ." And toward the end of his life, writing a letter to a friend in 1791, he could say: *Toute la jeunesse adopte la nouvelle théorie et j'en conclus que la révolution est faite en chimie.* (All the young are adopting the new theory from which I conclude that the revolution is accomplished in chemistry.)[171]

Lavoisier had in his youth imbibed deeply the revolutionary ideology of the Encyclopaedists Diderot and D'Alembert, and Abbé de Condillac, the godfather of "ideology." Born in 1743, Lavoisier, as a student, gave his time to the sciences rather than to his formal legal studies. He studied chemistry under Guillaume François Rouelle, a distinguished teacher and crystallographer, who was "enthusiastically admired by Diderot."[172] When in his maturity Lavoisier wrote his memoir, *Nomenclature Chimique* (1787), his philosophy of language was explicitly that of the ideologist Condillac.[173] With him, Lavoisier said that an advancing science aiming to banish the prejudices of its infancy demanded a reformed nomenclature.[174] The metaphors of the medieval alchemists and the preconceptions of the systematic chemists were to be eradicated.

The preface to Lavoisier's *Elements of Chemistry* began and ended with quotations from Abbé de Condillac. Lavoisier noted that his work on nomenclature, ostensibly designed only to improve the chemical language, had gradually transformed itself into a treatise on chemistry. This was in keeping with Condillac's statement in his *Logic*:[175] "Languages are true analytical methods. . . . The art of reasoning is nothing more than a language well arranged." And he concluded with observations of Condillac "made on a different subject" but nonetheless applicable to the recent state of chemistry.

Instead of applying observation to the things which we wished to know, we have chosen rather to imagine them. Advancing from one ill founded supposition to another. . . . These errors becoming prejudices, are, of course, adopted as principles, and we thus bewilder ourselves more and more. The method, too, by which we conduct our reasonings is as absurd; we abuse words. . . . There is but one remedy by which order can be restored to the faculty of thinking; this is to forget all that we have learned, to trace back our ideas to their source . . . to frame the understanding anew.

The cited passage came from Condillac's *Logic* which had been published posthumously in 1780.[176] Condillac was admired by Diderot and hailed by Voltaire in his *Philosophical Dictionary* for having rendered a very great service to the human spirit by showing the falsehood of all systems.[177] He was also the close friend of Jean Jacques Rousseau. Such was the philosophical spirit with which Lavoisier felt kinship; he too would destroy a system with its unverifiable, self-contradictory substance. Lavoisier was thus associated through his intellectual, revolutionary psychology with those who were moved by similar emotions in directions more political than scientific.

A child of the Encyclopaedists, Lavoisier's young manhood was lived during a time when D'Alembert, then secretary of the French Academy, was lobbying tirelessly to increase the representation of the *philosophes* in its ranks. There were only four in 1762, but between 1760 and 1770 nine were added, including Condillac in 1768.[178] D'Alembert and Condillac shared a hostility to the traditional metaphysics. They rejected, in Lockean fashion, the "innate ideas" of Descartes and the "being" of Aquinas.[179] With Condillac too, D'Alembert insisted that the sense of touch, which conveyed to us a body's essential property—impenetrability—was the one most important for affirming realities in the external world. Condillac in his *Traité des Systemes* (1749) had warned against the systems built on abstract principles: "As for suppositions, the imagination makes them with so much pleasure and so easily. . . . All this costs no more than a dream, and philosophers dream easily. One ignores details, the things under one's eyes; one takes flight into unknown hands, and one constructs systems."[180]

Lavoisier's chemical revolution was a fulfillment of Condillac's platform for ideological revolution. Condillac's theory of language became part of the official ideology of the French Revolution in 1795 when the Institut National des Sciences et Arts was founded.

"Linguistic empiricism, as expounded by Condillac, is the epistemology of egalitarianism," writes a historian of this episode.[181] The ideologists regarded themselves as the spokesmen of ordinary men, the vulgar, who could analyze better than the philosophers who abused language. Condillac had criticized the Newtonian philosophers who without any empirical evidence invented explanations in terms of attraction for such diverse phenomena as solidity, fluidity, elasticity, and fermentation. Lavoisier's critique of phlogiston was in the spirit of a disciple of ideology.[182] That his own political views, like those of Condillac, stopped far short of the revolutionary aims of the Convention, did not obviate their coinvolvement in the same underlying emotional-intellectual wave.[183]

The generational revolutionist in science does not necessarily share the specific political aims of the contemporary political revolutionist. Often his energies of generational rebellion are wholly sublimated, transformed, or transmuted into the effort of scientific reconstruction while his political and social views remain most conventional. Marxist sociologists always hope to find a close union of the revolutionary temperament in science with that in contemporary politics. They ignore the import of the sublimation of energies into alternative channels. Leon Trotsky was thus discomfitted as he wrote an essay celebrating Mendeleev's centenary. Mendeleev, he acknowledged ruefully, had been hostile in 1871 to the Paris Commune: "The attitude of our great chemist to this event can be gathered from his general hostility towards 'Latinism', its violence and revolutions." Trotsky also observed that Mendeleev had written of his fear for the future of science under socialism: "I especially fear," wrote Mendeleev, "for the quality of science and of all enlightenment, and general ethics under 'State Socialism.' "[184] His words had perhaps a prescience that Trotsky, the political revolutionist, imperiously ignored.

Julius Robert Mayer, the first discoverer of the principle of conservation of energy, was an unusual example of the rechanneling of revolutionary energies from the political to the scientific. A youthful enthusiast of the July Revolution in France, Mayer wrote to a friend in 1832: "Oh, if only I were among the students at Tübingen who, they say, have been expelled in part because they joined the Association for German Freedom of the Press; oh, if only I were among those who fell in the month of July. . . ."[185] Then Robert matriculated as a medical student at Tübingen. He helped found a corps fraternity, Guestphalia, which ran afoul of the political restrictions instituted by the Carlsbad Decrees. Suspended and jailed before he could take his examinations, he went on a hunger strike; after six days, his imprisonment was commuted to house arrest. Then a change in the direction of underlying emotion began to take place in Mayer. He attended university clinics, became a physician, and despite the disapproval of his liberal father, resolved in 1837 to go to the East Indies for several years as a military physician in the Dutch Colonial Service. He knew the risks. As he wrote a friend: "At least half of the newly arrived Europeans die within the first year. In this regard, East India is like the lion's well-known cave—of the great number of Europeans who go there annually, only a few come back." He wondered whether he would end dying in a garrison, surrounded by wild natives "at the lowest level of cultural development."[186] Thus it was that in Java this young doctor, puzzling over the heated blood of his patients, discovered the principle of the conservation of energy. Evidently, the energies of his generational rebellion were completely transformed from a political to a scientific form. During the Revolution of 1848, Robert Mayer became hostile to the revolutionary party, and spent much time endeavoring, as he said, "to uphold the order of the state through word and deed," thereby incurring "the hate of a fanaticized crowd."[187] He became estranged from his two older brothers who supported the Revolution. Then in the spring of 1849, Robert was arrested by revolutionists who thought he was a spy; he narrowly escaped execution by a firing squad. He was released on the grounds that nothing more could be proved about him except that he was known as "the chief reactionary in Heilbronn." Mayer had during one decade helped initiate the most profound scientific revolution of the nineteenth century, and then fought against its most far-reaching political and social revolution. Perhaps the law of conservation applies to the transformation of revolutionary political into revolutionary scientific energies, with the generational direction unchanging in the vectors.

Emotions of generational rebellion were also underlying in the direction of Darwin's achievement during these same years. Always reticent, Darwin indicated something of the nature of these emotions in his unusual avowal to Hooker in 1844:

> I have been now ever since my return engaged in a very presumptuous work, and I know no one individual who would not say a very foolish one. . . . At last gleams of light have come, and I am almost convinced

(quite contrary to the opinion I started with) that species are not (it is like confessing a murder) immutable.[188]

Darwin's spiritual parricide, his revolt against his father's views and authority, was long delayed; but in the measure to which he accomplished his long gestative rebellion, his energies were liberated for his supreme act of intellectual originality.[189] There were clear signs of an intense, repressed conflict with his father's authority. The father, "who was by far the best judge of character whom I ever knew" (in Darwin's estimate) had felt his son was cut out to be a physician. He erred grievously in this respect as well as others. Enrolled in the medical course at Edinburgh University, Charles was disgusted by the subject of human anatomy and the attendant dissections (a symptom which Freud found exhibited by medical students affected by a parental conflict). Nor could Darwin endure the sights, sounds, and smells of the operating theater: "I rushed away before they were completed." The father proposed that Charles become a clergyman. Charles thereupon dutifully overcame the scruples he had about the dogmas of the Church of England. He studied several books on divinity, and wrote: "As I did not then in the least doubt the strict and literal truth of every word in the Bible, I soon persuaded myself that our Creed must be fully accepted." His incipient revolt against his father's authority took two forms when he went to Cambridge University. In the first place Darwin became involved through his "passion for shooting and for hunting" with "a sporting set including some dissipated low-minded young men." They sang, played cards, and sometimes drank too much. Darwin knew that he "ought to feel ashamed of days and evenings thus spent," but he confessed, "I cannot help looking back to these times with much pleasure." A student crowd for gambling and sporting was a far cry from Einstein's Olympia Academy, Bohr's *Ekliptika* or Heisenberg's youth activists. Charles seemed to be confirming his father's words which had so mortified him: "You care for nothing but shooting, dogs, and rat-catching, and you will be a disgrace to yourself and all your family." Charles sought to supplant the parental authority through a second means. As with many such cases, he found an adult figure whom he both respected intellectually and whose interests he shared, and who in turn encouraged him in their pursuit. Such a man was Professor John Stevens Henslow whom Darwin had heard of before coming to Cambridge "as a man who knew every branch of science, and I was accordingly prepared to reverence him." A friendship grew between the professor and student which influenced Charles's career more than any other circumstance. Darwin became known as "the man who walks with Henslow."[190] With him, Darwin, the devoted collector of beetles, could feel a kinship of spirit. For a third time, however, Darwin's father came near to damaging his son's scientific development. In the fall of 1831 Professor Henslow informed Darwin that the ship *Beagle*, soon to sail for the South Seas, had a place for a young naturalist. Darwin's father "strongly objected," and Charles promptly wrote declining the offer. Luckily, the father had also said that if "any man of common-sense" were to give contrary advice, he would allow Charles to go. His mistrust of Charles's judgment was clear. By good fortune, Charles's uncle intervened with a positive recommendation; the father withdrew his objection while the son kindly consoled him by reminding him how hard it would be for him to spend more than his allowance aboard the *Beagle*. The father remained skeptical even of that.

Slowly, Charles Darwin began to doubt whether the logic of Paley's *Natural Theology* was as impeccable as he had thought. Aboard the *Beagle*, he was still regarded as religiously orthodox, but after his return to England, during the years from 1836 to 1839, he came to doubt that the Scripture were more sacred than, for instance, the Hindu books. He began to doubt the evidence for the Christian miracles, but "very unwillingly"; often he had daydreams that some archaeological discoveries would vindicate the truth of the Gospels. Gradually, however, unable "to invent evidence" that would convince himself, he wrote, "disbelief crept over me at a very slow rate, but was at last complete."[191] The old argument for God's existence from the evidence of His design in nature, which he had formerly regarded as conclusive, had collapsed.

Some strange guilt-feeling affected Charles Darwin with respect to his father. "The origin of this fear," surmises Gertrude Himmelfarb, was "personal: a sense of guilt at not earning his keep, and an acute sense of inadequacy at not being able to do so."[192] Darwin's wife, Emma, however, in her reminiscence of Darwin's father, indicated a deeper source of this guilt-feeling. "Doctor Darwin did not like him or understand or sympathize with him as a boy."[193] And Charles in his diffident way wrote: "I think my father was a little unjust to me when I was young, but afterwards, I am thankful to think I became a prime favorite with him." But as Charles Darwin experienced in 1837–1839 a burst of creative, original powers, he found that the unusual hypothesis of natural selection assuaged his own guilt feelings. He confided to his notebook:

> Let man visit ourang-outang in domestication, hear expressive whine, see its intelligence when spoken, as if it understood every word said—see its affection to those it knows,—see its passion & rage, sulkiness & very extreme of despair; let him look at savage, roasting his parent, naked, artless, not improving yet improvable, and then let him dare to boast of his proud preeminence.
>
> Man in his arrogance thinks himself a great work worthy the interposition of a deity. More humble & I believe true to consider him created from animals.[194]

Charles Darwin, who felt as he developed his hypothesis that he was committing murder, wrote at its inception his imagery of the savage roasting his parent. Though it was not a phenomenon he had observed in his years' voyaging on the *Beagle*, the image came to his consciousness as he thought of the arrogance of men. He was kin with the savage roasting his parent, no better than him, affected with the same animality that spoke in the despairing eyes of the orang-outang. His guilt was not personal; it was shared by all the creation of which he was part. He was at best "improvable." He renounced the sin of pride in his human status, but in doing so, felt he could no more be blamed for his inner feelings than the savage. Thus the hypothesis of natural selection, which universalized the lot of man and his struggles in all animate nature, may have assuaged his guilt about his deep feelings of generational revolt, thereby liberating the emotional basis of Darwin's burst of intellectual originality in 1837–1839.[195]

Generational emotions were involved not only in the origin of Darwin's theory; they also delineated its reception by his contemporaries. The young felt themselves joined to Darwin by a common isoemotional bond. As Gertrude Himmelfarb writes: "In the battle of the generations, the 'ancients' versus the 'moderns,' Darwin gave them the opportunity to assert themselves over their elders and superiors."

The intellectual war of the generations was fought most *à outrance* in geology. Archibald Geikie, the later noted geologist, twenty-four years old when the *Origin* appeared, was thus impatient with his teachers for not seeing it as a "new revelation." Nathaniel

Shaler, a student at Harvard, discussed Darwin's theory secretly with a friend, "for to be caught at it was as it is for the faithful to be detected in a careful study of heresy." He dared ask Professor Agassiz how species had originated, and was told: by a "thought of God." The biological statistician Karl Pearson recalled "the joy we young men then felt" as they saw the tight Biblical time-span dilated into broad evolutionary eons. The geneticist August Weismann, then twenty-five years old, recalled: "Darwin's book fell like a bolt from the blue; it was eagerly devoured, and while it excited in the minds of the younger students delight and enthusiasm, it aroused among the older naturalists anything from cool aversion to violent opposition."[196]

Withal, this was a struggle that expressed itself entirely within the constitutional workings of the scientific community. No one ventured to appeal to an epistemology of generational relativism, to argue that the perceptions of the generations are incommensurable, so that there is no common criterion of decision between them, or even to articulate a doctrine that the insights and perceptions of the young are superior to those of the old. In every social revolution doctrines of class relativism do flourish. With the Bolshevik Revolution, for instance, came the doctrine that the bourgeois mind was somehow incapacitated and could not discern objective truth, and that only the proletarian mind had privileged access to realities. Sometimes the epistemology of class relativism omitted the notion that an objective truth existed; then it simply affirmed the incommensurability of the different class standpoints. Every social revolution has tried to disrupt the epistemological community. But for all the generational feelings ignited by Darwin's theory of natural selection, the scientific community retained its unifying bond. From a sociological standpoint, it remained a constitutional system, resolving conflicts through its intrasystematic workings.

The significance of generational rebellion as the primary motivating force giving rise to new intellectual movements is now gradually being recognized by social scientists. At first it met with a certain resistance among young revolutionists, but even they have come to acknowledge its workings. At the convention of the American Sociological Association in September 1967, a leading "younger radical sociologist" informed *The New York Times* correspondent: "You have three kinds of sociologists these days. There's the old minister's son style—very decorous. These are the people at the top of the profession. Next there was the Jewish-radical generation. Following that you have the hipsters—that's us." The so-called Jewish-radical group were the founders of the Society for the Study of Social Problems in 1950 as a protest against the conservatism of the American Sociological Association. However, the correspondent was informed that the generation of 1950, now affluent and influential, was "viewed by younger, more radical men as part of the same establishment."[197] The young radicals were also in rebellion against the so-called value-free sociology of the generation of 1950.

During previous movements in sociology the nuclear generational energies have been enveloped by orbital doctrines of vicarious class spokesmanship; generational rebels professed to be proletarian sociologists seeking to challenge bourgeois sociology, the ideology of the ruling class. The most recent movement, however, is more overtly generational since the new revolutionary sociologists have not found allies among the working class; nor can they reasonably claim to be spokesmen for other races. Therefore the generational source of novel intellectual movements emerges in a more isolated and purer specimen. The newer revolutionary group has proportionately as many Jews as its forebears of 1950; hence, its rebellion can not be explained by the emergence of sociologists from a formerly unrepresented ethnic group. Themes of generational revolt have indeed defined the periodicities in the history of sociological theory, beginning, in America, with ministers' sons who sought to displace social religion with social science. Today, on the contrary, a new generation of revolutionary sociologists, finding that observed fact and statistical methods do not confirm its biases and longings, is constrained to revolt against the notions of an objective science and objective methods of verification, and to announce itself opposed to value-free sociology. The generational waves have thus taken sociologists from religion to science to ideology.

Curiously, as underlying generational motives have become more overt, a novel language for defining one's protest has come into being. From 1930 to 1950, young sociologists and their presumable allies among the working class called their enemy the "ruling class." Then in 1955 the word "establishment" began to appear, first among the Angry Young Men in Britain, then in America. The establishment first signified the elite that survived amidst all the redistributions of power and income among the classes, a Paretian elite that was invariant among all revolutions including Marxian ones.[198] A vector of generational resentment against the stodgy, domineering, crafty old men became overt. By the mid-sixties, establishment primarily signified the gerontocratic elite, and a new social science was sought to achieve a revolution against it.

Each group of revolutionists arriving within the establishment tends in turn to regard the next group of would-be revolutionists as essentially counter-revolutionists. Thus, a leading Keynesian economist observed that the "monetarist counter-revolution" in the late sixties "was also sparked by the same kind of social force that had touched off the Keynesian revolution—the burning desire of young economists to overthrow their profession's aging establishment." Keynes' doctrine, wrote Harry G. Johnson, having become the regnant orthodoxy, depended "on the authority and prestige of senior scholars which is oppressive to the young." The latter awaited the appropriate moment when they could do as the Keynesians had done in the thirties:

> A whole generation of students [was enabled] to escape from the slow and soul-destroying process of acquiring wisdom by osmosis from their elders and the literature into an intellectual realm in which youthful iconoclasm would quickly earn its just reward (in its own eyes at least) by the demolition of the intellectual pretensions of its academic seniors and predecessors.

The young monetarist Turks waited for a time when their counter-revolution would have strong political support. "That moment came with the escalation of the Vietnam war in 1965 and with the failure of the New Economists—the Keynesians—to control inflation."[199] Each generation desires to fashion its own theoretical ideas and language to declare the uniqueness of its own experience and to define a world different from that of its fathers; it awaits a strategic moment for its theoretical offensive.[200]

In the usual pattern of the development of science, the bold original thinker has the support of a generational circle; Einstein, Bohr, and Heisenberg were thus sustained by the emotional sympathies of comrades in their student circles. By contrast, such thinkers as Darwin and Freud worked for many years in relative isolation. To be sure, in his maturity Darwin sought the intellectual sympathy of close friends and Freud at the outset collaborated with Josef Breuer and later sought encouragement from Wilhelm Fliess. But Darwin at Cambridge or on H.M.S. *Beagle* had no experience of generational comradeship in revolutionary ideas comparable to Einstein's

Olympia, while Freud's student associations never seemed to have had that degree of closeness. Where a generational circle does not support the original thinker, his contribution tends to be postponed; he procrastinates, suffers extremes of self-doubt, allows his manuscripts to accumulate, and evidently develops illness and neurotic symptoms. Darwin labored for more than twenty years on a multivolumed manuscript whose end always receded as his pages multiplied; only a letter from Alfred Russel Wallace in 1858 from the Malay Archipelago relating his astonishingly similar hypothesis stirred Darwin to consult with friends about possible publication and led to the rapid appearance of *The Origin of Species.* In Freud's case, it was the impact of his father's death in 1896 that liberated his analytic powers for the discovery of his theory of the Oedipus complex and the writing of *The Interpretation of Dreams.* As Freud wrote in the preface to the second edition:

> For this book has a further subjective significance for me personally—a significance which I only grasped after I had completed it. It was, I found, a portion of my own self-analysis, my reaction to my father's death—that is to say, to the most important event, the most poignant loss, of a man's life.[201]

The lack of a supportive generational circle was like the absence of a base at which one could always refuel: "During the long years in which I have been working at the problem of the neuroses I have often been in doubt and sometimes been shaken in my convictions." *The Interpretation of Dreams*, born out of generational crisis, restored his confidence in himself. When Charles Lyell informed Darwin in late 1859 that he intended, after many years of skepticism, to declare himself in favor of the theory of evolution, Darwin confessed his lonely ordeal: "thinking of so many cases of men pursuing an illusion for years, often and often a cold shudder has run through me, and I have asked myself whether I may not have devoted my life to a phantasy. Now I look at it as morally impossible that investigators of truth like you and Hooker, can be wholly wrong, and therefore I rest in peace."[202]

Notes

1. Ignazio Silone, *Emergency Exit* (London, 1969), p. 118.
2. Leo Koenigsberger, *Hermann von Helmholtz*, trans. Frances A. Welby (Oxford, 1906), pp. 38, 43.

3. Francis Darwin, *The Life and Letters of Charles Darwin* (New York, 1898), vol. 2, p. 85.

4. Ibid., p. 147.

5. *More Letters of Charles Darwin*, ed. Francis Darwin and A. C. Seward (New York, 1930), vol. 1, p. 187; *Charles Darwin*, ed. Francis Darwin (London, 1908), pp. 221, 234; Robert H. Murray, *Science and Scientists in the Nineteenth Century* (London, 1925), p. 192.

6. Denis Diderot, *The Encyclopedia*, in *Rameau's Nephew and Other Works*, trans. Ralph H. Bowen (Indianapolis, 1964), pp. 286, 289.

7. Diderot wrote, for instance, "We are at the present time living in a great revolution of the sciences." But he feared that the revolution was one that would lead thinkers away from geometry "towards morals, fiction, natural history and experimental physics." "I feel almost certain," he wrote, "that before 100 years are up, one will not count three great geometers in Europe. The science will very soon come to a stop where the Bernoullies, the Eulers, the Maupertuis, Clairauts, Fontaines and D'Alemberts have left it. . . . We shall not get beyond that point. . . ." (*L'Interprétation de la nature*, in *Oeuvres de Denis Diderot* [Paris, 1818], vol. 1, pp. 420–422, 425).

8. Immanuel Kant, *Critique of Pure Reason*, trans. Norman Kemp Smith (1929; London, 1963), pp. 20–21. Though Kant never used the expression "Copernican revolution" to characterize his philosophy, he wrote of the "single and sudden revolution" which had been achieved in mathematics and natural science, and urged that metaphysics emulate the procedure of "the first thought of Copernicus"; rather than intuition conforming to the constitution of objects, it was to be the other way round. Ibid., pp. 21–22. Norwood Russell Hanson, "Copernicus' Role in Kant's Revolution," *Journal of the History of Ideas*, vol. 20 (1959), pp. 274–276.

9. D'Alembert, *Discours Preliminaire de l'Encyclopedie*, ed. F. Picavet (Paris, 1894), p. 99.

10. Ibid., p. 11.

11. *Oeuvres de D'Alembert: Sa Vie—Ses Oeuvres—Sa Philosophie*, par Condorcet (Paris, 1853), p. 24.

12. *Oeuvres de D'Alembert*, p. 22.

13. *Oeuvres de D'Alembert*, p. 217.

14. François Arago, *Biographies of Distinguished Scientific Men*, trans. W. H. Smyth, B. Powell, and R. Grant (Boston, 1859), 1st ser., p. 309.

15. A. A. Cournot, *Considérations sur la marche des idées et des Evènements dans les Temps Modernes* (Paris, 1872), vol. 1, p. 259; vol. 2, p. 12. The circumstances of Cournot's life are told in the translator's introduction to Antoine Augustin Cournot, *An Essay on the Foundations of our Knowledge*, trans. Merritt H. Moore (New York, 1956), pp. xxix–xxxvi.

16. Romain Rolland, *I Will Not Rest*, trans. K. S. Shelvankar (New York, 1937), p. 244. The centenary of 1789 was celebrated with an exhibition in Paris. Many, however, refused to join the festivities; there was even a countercelebration dedicated to the theme that the Revolution was a failure (D. W. Brogan, *The Development of Modern France, 1870–1939* [New York, 1940], p. 261).

17. Brogan, *The Development of Modern France*, pp. 102–103, 322.

18. M. Berthelot, *La Révolution Chimique: Lavoisier* (Paris, 1890), pp. 1–2. Marx's disciple, Edward Aveling, who was also a zoologist, popularized at this time the idea that Darwin and Marx had resembled each other, that each advanced "a generalization that not only *revolutionized* that branch, but is actually revolutionizing the whole of human thought, the whole of human life." *The Students' Marx* (London, 1892), pp. viii, ix.

19. Hans Christian Oersted, *The Soul in Nature*, trans. Leonora and Joanna B. Horner (London, 1852; reprint ed., London, 1966), pp. ix, 322.

20. John Frederick William Herschel, *Preliminary Discourse on the Study of Natural Philosophy* (London, 1830), p. 339.

21. Justus von Liebig, "An Autobiographical Sketch," trans. J. Campbell Brown, *Chemical News*, vol. 63, no. 164 (June 12, 1891), p. 277; W. A. Shenstone, *Justus von Liebig· His Life and Work 1803–1873* (London, 1895), p. 17.

22. William Whewell, *History of the Inductive Sciences, From the Earliest to the Present Time*, 3d ed. (New York, 1890), vol. 1, p. 339.

23. Ibid., p. 414.

24. John Theodore Merz, *A History of European Thought in the Nineteenth Century* (Edinburgh, 1896), vol. 1, p. 81.

25. Frederick Engels, *Herr Eugen Dühring's Revolution in Science*, 2d ed., trans. Emile Burns (New York, n.d.), pp. 16–18.

26. Frederick Engels, *Dialectics of Nature*, trans. Clemens Dutt (New York, 1940), p. 158.

27. Ibid., p. 159.

28. Ibid., p. xii. See also J. B. S. Haldane, *The Marxist Philosophy and the Sciences* (New York, 1939), pp. 53–54.

29. Henri Poincaré, *The Foundations of Science: Science and Hypothesis, The Value of Science, Science and Method*, trans. George Bruce Halsted (Lancaster, 1913), p. 303.

30. Poincaré, *The Foundations of Science*, p. 457. The metaphor of the "bankruptcy of science" was originated by the conservative literary critic Brunetière. (Romain Rolland, *Péguy* [Paris, 1944], p. 21). It was introduced by his article in January 1895, "After a Visit to the Vatican" which described his transition from determinist scientism to voluntarist idealism. Brunetière himself affirmed only the partial failure of science, not its bankruptcy. Cf. Harry W. Paul, "The Debate over the Bankruptcy of Science in 1895," *French Historical Studies*, vol. 5 (Spring 1968), pp. 303–306.

31. V. I. Lenin, *Materialism and Empiro-Criticism: Critical Notes Concerning a Reactionary Philosophy*, trans. David Kvitko (New York, 1927), pp. 213–214.

32. Abel Rey, *La Théories Physique chez les Physiciens Contemporains*, cited in Lenin, *Materialism and Empiro-Criticism*, p. 261.

33. Ibid.

34. Delegates of the USSR, *Science at the Crossroads* (London, 1931), pp. 84, 215.

35. Christopher Caudwell, *The Crisis in Physics* (London, 1939), reprinted in *The Concept of Freedom* (London, 1965), pp. 193–196.

36. Ibid., p. 198.

37. A. S. Eddington, *The Nature of the Physical World* (New York, 1928), pp. 1, 4, 352, 353.

38. In April 1971, a Conference on Socialist Studies was held at the University of Toronto on the theme "Marxism and the Philosophy of Science." Two of the papers were entitled "The Implication of T. Kuhn's Theory for the Marxist Sociology of Scientific Knowledge" and "Generalized Dialectics in Kuhn." Marxist ideologists were attracted to the model of "scientific revolution," its espousal of a "dialectical structure."

39. Rolland, *I Will Not Rest*, p. 315. Lovejoy noted in 1908 that the cry of "crisis," familiar for more than a century, was having a renewed vogue. Arthur O. Lovejoy, "Religious Transition and Ethical Thinking in America," *The Hibbert Journal*, vol. 6 (1907–08), p. 500.

40. V. I. Lenin, "*Left-Wing*" *Communism: An Infantile Disorder* (New York, 1934), p. 74.

41. *The Autobiography of Robert A. Millikan* (New York, 1950), pp. 23–24.

42. *The Autobiography of Robert A. Millikan*, pp. 269–270.

43. Robert Andrews Millikan, *Evolution in Science and Religion* (New Haven, 1928), pp. 7–11.

44. A. S. Eve, *Rutherford* (Cambridge, 1939), pp. 59–60.

45. Bertrand Russell, "H. Poincaré: Science and Hypothesis," *Mind*, vol. 14 (1905), p. 416.

46. Arthur Schuster, "H. Poincaré: Science and Hypothesis," *Nature*, vol. 73 (February 1, 1906), p. 315.

47. William Cecil Dampier Whetham, *The Recent Development of Physical Science* (London, 1904), pp. vi, 271.

48. Ibid., pp. 278, 279, 281.

49. Dampier Whetham, *The Recent Devolopment of Physical Science* (London, 1924), pp. 3, 253.

50. Emile Meyerson, *Identity and Reality*, trans. Kate Loewenberg (London, 1930), p. 60.

51. Emile Meyerson, *Du Cheminement de la Pensée* (Paris, 1931), vol. 2, p. 706.

52. Lawrence J. Henderson, *The Fitness of the Environment* (Boston, New York, 1913), pp. 20–21. Henderson's colleague, the mathematician E. V. Huntington, had already published in 1912 a paper on the logical structure of the theory of relativity. In 1913, the psychologist L. T. Troland reported to Josiah Royce's famous seminar in the logic of science on Einstein's "new" theory of relativity. Henderson was an active member of this seminar (*Josiah Royce's Seminar, 1913–1914: As Recorded in the Notebooks of Harry T. Costello*, ed. Grover Smith [New Brunswick, 1963], pp. 46–47, 51, 53–56, 150, 191).

53. Thomas S. Kuhn, *The Structure of Scientific Revolutions* (Chicago, 1962), p. 67. Professor Kuhn has recently amended his analysis in various respects in *The Structure of Scientific Revolutions*, 2d ed. (Chicago, 1970). The present author's concern, however, is with a sociological evaluation of the concepts of "revolution" and "scientific revolution" as they have become embedded in contemporary thought. Professor Kuhn's book is the most well-known example in recent years of the use of the image, model, or isomorpheme of "revolution." His new "Epilogue" gives a full account of the discussions of his thesis by philosophers.

54. Max Born, "Max Karl Ernst Ludwig Planck: 1858–1947," *Obituary Notices of Fellows of the Royal Society*, vol. 6, no. 17 (London, November 1948), p. 171.

55. Russell McCormmach, "Henri Poincaré and the Quantum Theory," *Isis*, vol. 58 (1967), p. 39.

56. Ibid. See also Martin J. Klein, *Paul Ehrenfest*, vol. 1, *The Making of a Theoretical Physicist* (Amsterdam, 1970), p. 252.

57. Silvanus P. Thompson, *The Life of William Thomson: Baron Kelvin of Largs* (London, 1910), vol. 1, p. 261. A few eminent men, according to Joule, did, however, express an interest in his work. (James Prescott Joule, *Joint Scientific Papers* [London, 1887], vol. 2, p. 215). See also vol. 1, pp. 156, 202.

58. Joule, *Joint Scientific Papers*, vol. 2, p. 215.

59. Thompson, *The Life of William Thomson*, vol. 1, p. 265.

60. *Autobiography of Charles Darwin* (London, 1931), pp. 60–61. Edmund Gosse, however, on the basis of his boyhood recollections, described that time as a "period of intellectual ferment, as when a great political revolution is being planned," and when "many possible adherents were confidentially tested with hints and encouraged to reveal their bias in a whisper" (*Father and Son* [1907; reprint ed. London, 1958], p. 98). Leonard Huxley, *Life and Letters of Thomas Henry Huxley*, (London, 1903), vol. 1, p. 245.

61. See T. E. Thorpe, *Essays in Historical Chemistry* (London, 1894), p. 136; Sir William Ramsay, *The Life and Letters of Joseph Black, M.D.* (London, 1918, p. 93.

62. Anatole G. Mazour, *The First Russian Revolution: 1825: The Decembrist Movement* (Berkeley, 1937), pp. 212, 214, 249, 252. Michael Zetlin, *The Decembrists*, trans. George Panin, New York, 1958), pp. 52, 264, 283, 321, 324.

63. Daniel Q. Posin, *Mendeleev: The Story of a Great Scientist* (New York, 1948), pp. 9, 23, 118; Benjamin Harrow, *Eminent Chemists of Our Time* (New York, 1920), p. 37. O. N. Pisarzhevsky, *Dmitry Ivanovich Mendeleyev: His Life and Work* (Moscow, 1954), p. 9. Mendeleev told Sir William Ramsay, the discoverer of the inert gases, that "he was raised in East Siberia and knew no Russian till he was 17 years old" (Sir William A. Tilden, *Sir William Ramsay*, London, 1918, p. 89).

64. O. N. Pisarzhevsky, *Dmitry Ivanovich Mendeleyev: His Life and Work* p. 76.

65. D. Mendeleeff, "The Periodic Law of the Chemical Elements" (1889), in *Readings in the Literature of Science*, ed. William C. Dampier and Margaret Dampier (New York, 1959), p. 117.

66. Cited in Pisarzhevsky, *Dmitry Ivanovich Mendeleyev*, p. 74.

67. F. P. Venable, *The Development of the Periodic Law* (Easton, Pa., 1896), p. 107. Mendeleev's "Natural System," as he called it, as well as Meyer's, aroused in 1869 "no great stir" and "little comment." It

was threatened with the same fate of dust and oblivion which had befallen the systems of earlier writers. After several years of neglect, even on the part of its authors, attention was drawn to the system once more by the fortunate fulfillment of certain bold predictions made by Mendeleev in his table. The discovery of scandium and gallium and their fitting into the predicted places with atomic weights and other properties coinciding with those predicted for them, gave a new impetus to the study of these tables, and their use in the classroom" (ibid., pp. 8–9).

68. Ibid., p. 109.

69. Harrow, *Eminent Chemists of Our Time*, p. 29.

70. George Sarton, "*The Discovery of the Law of Conservation of Energy*," *Isis*, vol. 13, no. 40 (September 1929), p. 21.

71. John Stuart Mill, *Principles of Political Economy with Some of Their Applications to Social Philosophy*, bk. 3, chap. 1, sec. 1, in *Collected Works of John Stuart Mill* (Toronto, 1965), vol. 3, p. 456.

72. J. A. Hobson, *Confessions of an Economic Heretic* (London, 1938), p. 23.

73. W. Stanley Jevons, *The Theory of Political Economy*, 3d ed. (London, 1888), pp. 162–163.

74. See Vilfredo Pareto, *A Treatise on General Sociology*, trans. Andrew Bongiorno and Arthur Livingston (New York, 1935), vol. 3, pp. 1,410–1,412; *Les Systèmes Socialistes: Oeuvres Complètes* (Geneva, 1965), vol. 2, pp. 348, 368; vol. 5.

75. Jevons, *The Theory of Political Economy* (London, 1871), pp. 266–267.

76. Joseph A. Schumpeter, *History of Economic Analysis* (New York, 1954), p. 826. Schumpeter writes that the new theory was linked to "the call for social reform." This was untrue in Jevons' case, though in Lausanne, the strange genius, Leon Walras, also a founder of marginal analysis, advocated various schemes of collectivization.

77. Shaw exaggerated Jevons' misgivings concerning state enterprises. Cf. Jevons, *Methods of Social Reform* (London, 1883), pp. 277 ff, 324 ff.

78. *Letters and Journal of W. Stanley Jevons* (London, 1886), pp. 343, 385, 409. According to Alfred Marshall, Jevons "was impressed by the mischief which the almost pontifical authority of Mill exercised on young students; and he seemed perversely to twist his own doctrines so as to make them appear more inconsistent with Mill's and Ricardo's than they really were" (*Memorials of Alfred Marshall*, ed. A. C. Pigou, [London, 1925], p. 99).

79. R. F. Harrod, *The Life of John Maynard Keynes* (London, 1951), pp. 462–463. Also cf. Lord Robbins, *Autobiography of an Economist* (London, 1967), p. 153. Sir William Beveridge, the director of the London School of Economics, spoke for many of the elder economists when he found fault with Keynes's comparing his intent with Einstein's. The theory of relativity was testable by crucial experiment whereas Keynes's views, he argued, were "without verification of any kind." *Power and Influence* (New York, 1955), p. 253. In economics and sociology, the relative lack of crucial experiments leads to the use of another criterion for accepting a given theory over its rivals—namely its heuristic power for suggesting programs of action for averting undesired events and realizing desired ones. A theory which tells one to act, and how to act, will be preferred over a theory which has no advice in the way of interventionist measures.

80. M. S. Kedrov, *Book Publishing under Tzarism* (New York, 1932).

81. Gertrude Himmelfarb, *Darwin and the Darwinian Revolution*, (New York, 1959), p. 291.

82. *Les Prix Nobel en 1921–1922*, (Stockholm, 1923), p. 6.

83. Sigmund Freud, *The History of the Psychoanalytic Movement*, in *The Basic Writings of Sigmund Freud*, ed. A. A. Brill, (New York, 1938), p. 960.

84. Ibid., pp. 943–944.

85. Ibid., p. 956.

86. Freud, *The History of the Psychoanalytic Movement*, in *The Basic Writings*, pp. 962, 963. During the nineteenth century, research on mesmerism (later called hypnotism) had incurred an official ban. John Elliotson, a physician and member of the faculty of University College, London, began mesmerizing patients at the Uni-

versity College Hospital with beneficial therapeutic results. He was urged to discontinue such work on the ground that he was hurting the hospital's reputation. Shortly afterward, the Council of University College enacted a resolution forbidding the practice of mesmerism within the college. Thereupon Elliotson resigned from both the college and hospital. Since no medical journal would publish his papers, he founded the magazine *Zoist* in which he could print his work. *Zoist* described itself as "a journal of cerebral psychology and mesmerism, and their application to human welfare." It lasted from 1843 to 1856. See Edwin G. Boring, *History, Psychology and Science*, ed. R. I. Watson and D. T. Campbell (New York, 1963), p. 72, ff; *A History of Experimental Psychology* (New York, 1929), pp. 120–121.

87. Zhores A. Medvedev, *The Rise and Fall of T. D. Lysenko*, trans. I. Michael Lerner (New York, 1969); Julian Huxley, *Memories* (London, 1970), p. 284.

88. Freud, *The Question of Lay-Analysis*, in *The Standard Edition of the Complete Psycholanalytical Works*, trans. James Strachey et al. (London, 1959), vol. 20, pp. 207–208.

89. Marthe Robert, *The Psychoanalytic Revolution*, trans. Kenneth Morgan (New York, 1968), p. 241.

90. As Ernst Mach wrote: "The fate of all momentous discoveries is similar. On their first appearance they are regarded by the majority of men as errors. J. R. Mayer's work on the principle of energy (1842) was rejected by the first physical journal of Germany; Helmholtz's treatise (1847) met with no better success; and even Joule . . . seems to have encountered difficulties with his first publication (1843)" (*Popular Scientific Lectures*, trans. Thomas J. McCormack [Chicago, 1910], p. 138).

91. George Sarton, "The Discovery of the Law of Conservation of Energy," *Isis*, vol. 13, no. 40 (1929), p. 20.

92. Koenigsberger, *Hermann von Helmholtz*, p. 43.

93. Alex Wood, *Joule and the Study of Energy* (London, 1925), pp. 52–55; Murray, *Science and Scientists in the Nineteenth Century*, p. 89.

94. Ernst Mach, *History and Root of the Principle of Conservation of Energy*, trans. P. E. B. Jourdain (Chicago, 1911), pp. 10, 13, 80, 87, 88.

95. "Jubelband," *Annalen der Physik und Chemie* (Leipzig, 1874), pp. xii–xiii.

96. "J. C. Poggendorff," *Annalen der Physik und Chemie*, 6th ser., vol. 10 (Leipzig, 1877), pp. vii–xiv.

97. A. Vibert Douglas, *The Life of Arthur Stanley Eddington* (London, 1956), p. 44.

98. Ibid., p. 42.

99. McCormmach, "Henri Poincaré and the Quantum Theory," p. 52.

100. Louis de Broglie, "Recherches sur la Théorie des Quanta," *Annales de Physique*, vol. 3 (1925), pp. 28–29.

101. Klein, *Paul Ehrenfest* vol. 1, p. 249.

102. J. H. Jeans, *Report on Radiation and the Quantum Theory*, 2d ed. (London, 1924), p. 19.

103. Henri Poincaré; "Sur la Theorie des Quanta," *Journal de Physique*, vol. 2 (January 1912), p. 37, cited in ibid., p. 24.

104. Jeans, *Report on Radiation and the Quantum Theory*, pp. 28, 29.

105. Klein, *Paul Ehrenfest*, p. 253.

106. Jeans, "Discussion on Radiation," *Report of the Eighty-Third Meeting of the British Association for the Advancement of Science, Birmingham, 1913* (London, 1914), p. 378.

107. McCormmach, "Poincaré and the Quantum Theory," pp. 53–54.

108. Jeans, "Discussion on Radiation," p. 379. E. N. Da C. Andrade, *Rutherford and the Nature of the Atom* (New York, 1964), p. 136.

109. E. A. Milne, *Sir James Jeans: A Biography* (Cambridge, 1952), p. 17. See also, Klein, *Paul Ehrenfest*, p. 253.

110. Jeans, *Report on Radiation and the Quantum-Theory*, p. 25 (translated by J. H. Jeans).

111. L. C. Dunn, *A Short History of Genetics: The Development of Some of the Main Lines of Thought: 1864–1939* (New York, 1965), p. ix.

112. R. C. Punnett, "Early Days of Genetics," *Heredity*, vol. 4, part 1 (April 1950), p. 8.

113. Dunn, *A Short History of Genetics*, p. 5.

114. Punnett, "Early Days of Genetics," p. 4.

115. Ibid., p. 5.

116. Ibid., p. 7.

117. Theodor Gomperz, *Essays und Erinnerungen* (Stuttgart, 1905), pp. 14–15. Adelaide Weinberg, *Theodor Gomperz and John Stuart Mill* (Geneva, 1963), pp. 9–10. Hugo Iltis, *Gregor Johann Mendel: Leben, Werk und Wirkung* (Berlin, 1924), pp. 24, 64.

118. See Mendel's original paper, "Experiments in Plant-Hybridization," reprinted in James A. Peters, ed. *Classic Papers in Genetics* (Englewood Cliffs, 1959), pp. 1–20.

119. Hugo Iltis, *Life of Mendel*, trans. Eden and Cedar Paul (New York, 1937), pp. 35, 39, 58, 103, 105, 282. Dr. Louis Büchner, *Force and Matter*, ed. J. Frederick Collingwood (London, 1870), p. 240. The first edition had eleven admiring references to Ludwig Feuerbach.

120. Bentley Glass, "The Long Neglect of a Scientific Discovery: Mendel's Laws of Inheritance," *Studies in Intellectual History* (Baltimore, 1953), pp. 148–160. Mendel, observes Elizabeth B. Gasking, was not interested either in the origin or the producing of a new species. He was solely concerned with the laws governing the inheritance of particular traits. "Why Was Mendel's Work Ignored?" *Journal of the History of Ideas*, vol. 20 (1959), pp. 68, 71.

121. Erik Nordenskiöld, *The History of Biology*, trans. L. B. Eyre (New York, 1928), p. 507.

122. Ibid., p. 506.

123. Nordenskiöld. *The History of Biology*, pp. 554–555, 557. Daniel Gasman has challenged Nordenskiöld's linkage of Haeckel's view to liberalism, and has argued that he was rather an intellectual father of national socialism. At the same time, Gasman indicates the considerable extent to which Engels, Marx's coworker, incorporated Haeckel's notions into what came to be called dialectical materialism. Haeckel's works, even in our time, are still sold in socialist bookshops. All of this shows that a scientific idea is multipotential in its ideologcial uses; in its career, it traverses contrary phases. See Daniel Gasman, *The Scientific Origins of National Socialism: Social Darwinism in Ernst Haeckel and the German Monist League* (London, 1971), pp. xxviii, 108.

124. Conway Zirkle, "Mendel and his Era," in *Mendel Centenary: Genetics, Development and Evolution*, ed. Roland M. Nardone (Washington, D.C., 1968), pp. 123–124. When Hugo Iltis, Mendel's later biographer, was a student, he discovered in 1899 Mendel's paper. He took it "in great excitement" to his professor. The latter said: "Ah! I know all about that paper, it is of no importance. It is nothing but numbers and ratios, ratios and numbers. It is pure Pythagorean stuff; don't waste any time on it, forget it."

125. Albert Einstein, "The Mechanics of Newton and Their Influence on the Development of Theoretical Physics," *The World As I See It*, trans. Alan Harris (New York, 1934), p. 57.

126. Philipp Frank, *Philosophy of Science* (Englewood Cliffs, N.J. 1957), pp. 145–146. Cf. Bertrand Russell, *The ABC of Relativity* (New York, 1925), p. 160.

127. *The Autobiography of Robert A. Millikan*, p. 271.

128. Engels, *Herr Eugen Dühring's Revolution in Science*, pp. 167, 346.

129. T. L. Kuhn, *The Structure of Scientific Revolutions*, pp. 100–101.

130. Bertrand Russell, *Introduction to Mathematical Philosophy*, 2d ed. (London, 1920), p. 12.

131. R. F. Harrod, *The Life of John Maynard Keynes*, p. 462.

132. John Maynard Keynes, *The General Theory of Employment, Interest and Money* (London, 1936), p. 3.

133. Kuhn, The *Structure of Scientific Revolutions*, pp. 65, 53.

134. Ibid., p. 24.

135. Kurt Lewin wrote: "At the beginning of this century . . . the experimental psychology of 'will and emotion' had to fight for recognition. . . . Although every

psychologist had to deal with these facts realistically in his private life, they were banned from the realm of 'facts' in the scientific sense. . . . Like social taboos, a scientific taboo is kept up not so much by a rational argument as by a common attitude among scientists . . ." (*The Philosophy of Ernst Cassirer*, ed. P. A. Schilpp [Evanston, 1949], pp. 278–279).

136. Edwin G. Boring, *A History of Experimental Psychology* (New York, 1929).

137. Joseph Priestley, *The History and Present State of Electricity, with Original Experiments*, 5th ed., corrected, (London, 1794), pp. vi, x. The experiments on electricity were also described as "romantic." "Romantic as these things may seem, they should not be absolutely condemned without a fair trial, since we all, I believe remember the time, when these phenomena in electricity, which are now the most common and familiar, would have been deserving as little credit as the cases under consideration may seem to do" (ibid., p. 156).

138. Max Planck, *A Survey of Physics*, trans. R. Jones and D. H. Williams (London, 1925), p. 132.

139. Boring, *A History of Experimental Psychology*, pp. 133 ff.

140. Otto Glasser, *Dr. W. C. Röntgen* (Springfield, Illinois, 1945), p. 116.

141. W. Robert Nitske, *The Life of Wilhelm Conrad Röntgen: Discoverer of the X Ray* (Tucson, 1971), pp. 4, 50.

142. Ibid., p. 95.

143. Glasser, *Dr. W. C. Röntgen*, p. 39.

144. Nitske, *The Life of Röntgen*, p. 5.

145. Glasser, *Dr. W. C. Röntgen*, pp. 84–85.

146. Ibid., p. 117.

147. Nitske, *The Life of Röntgen*, p. 19.

148. Ibid., p. 200.

149. Ibid., p. 215.

150. Ibid., p. 153.

151. Ibid., p. 129.

152. In a brilliant but neglected essay, George Boas wrote that "problems are concerned with deviations from the rule. . . ." Several amendments must, however, be made. Given deviations or anomalies will pose a problem to some scientists, but not to others. The degree of "problematicity" that attaches to a deviation is dependent primarily on the extralogical motivation of the thinker; it depends on the emotions that lead him to wish to see a certain kind of theory realized. Also, the existence of rules, not deviations, can itself constitute a "problem" if the scientist's personality has the wayward emotions of Charles Peirce. His particular temperament led him to desire a universe in which chance was primordial. As Boas writes: "The determination of the problematic is not itself logical" ("What Is a Problem?" *The Journal of Philosophy*, vol. 34 [1937], p. 200).

153. *The Autobiography of Robert A. Millikan*, p. 106. The other five discoveries that Millikan enumerates were at the outset revolutions in theory: Planck's quantum (1900), Einstein's special relativity (1905), relativistic transformation of mass into energy (1905), completion of the kinetic theory and equipartition principle (1905), and photoelectric law. (1905).

154. The psychology of the revolutionary experimenter remains to be explored. The documentary evidence is usually of the most fragmentary kind. Niels Bohr, for instance, tells of C. T. R. Wilson, the inventor of the cloud chamber method for tracing the paths of alpha rays, that "his interest in these beautiful phenomena had been awakened when as a youth he was watching the appearance and disappearance of fogs as air currents ascended the Scottish mountain ridges and again descended in the valleys." One would have wished to know more of Wilson's youthful feelings and the source of their deep impression (Niels Bohr, *Essays 1958–1962 on Atomic Physics and Human Knowledge*, p. 31). Rutherford's propensity for a "joke" may have affected his experimental direction as it did Einstein's in his theoretical work. Cf. E. N. da C. Andrade, *Rutherford and the Nature of the Atom* (New York, 1964), pp. 72–73.

155. Einstein, *The World As I See It*, pp. 57, 69.

156. Werner Heisenberg, *Physics and Beyond: Encounters and Conversations*, trans. Arnold J. Pomerans (New York, 1971), pp. 147–148.

157. Daniel J. Boorstin, *The Genius of American Politics* (Chicago, 1953), p. 68 ff; J. Franklin Jameson, *The American Revolution Considered As A Social Movement* (1926; reprint ed., Boston, 1963), pp. 8–9; Peter Viereck, *Conservatism from John Adams to Churchill* (New York, 1956), pp. 26–27.

158. Born, "Max Karl Ludwig Planck: 1858–1947," p. 170.

159. Ibid., p. 167.

160. Max Planck, *The Philosophy of Physics*, trans. W. H. Johnston (New York, 1936), p. 96.

161. D. Mendeleeff, "The Periodic Law of the Chemical Elements," p. 117.

162. Daniel Q. Posin, *Mendeleyev*, p. 118.

163. Koenigsberger, *Hermann von Helmholtz*, p. 3.

164. Ibid., p. 30.

165. *Letters of Sigmund Freud*, ed. Ernst L. Freud, trans. Tania and James Stern (New York, 1964), p. 375.

166. Ibid., p. 379.

167. Freud, *The Interpretation of Dreams*, in *The Basic Writings of Sigmund Freud*, pp. 270–273. The life of Adolf Fischhof is told in my *The Conflict of Generations* (New York, 1969), pp. 72–73. Werner J. Cahnman, "Adolf Fischhof and his Jewish Followers," *Publications of the Leo Baeck Institute of Jews from Germany, Year Book IV* (London, 1959), pp. 111–139.

168. Sigmund Freud, *On the History of the Psycho-Analytic Movement*, trans. Joan Riviere (New York, 1966), pp. 7–8.

169. *The Letters of Sigmund Freud*, p. 379.

170. "Autobiography of Josef Breuer (1842–1925)" trans. C. P. Oberndorf, *The International Journal of Psycho-Analysis*, vol. 34 (1953) pp. 65–57; Paul F. Cranefield, "Josef Breuer's Evaluation of his Contribution to Psycho-Analysis," ibid., vol. 39 (1958), pp. 319–322.

171. Douglas McKie, *Antoine Lavoisier: The Father of Modern Chemistry* (Philadelphia, n.d.), pp. 121–122.

172. Ibid., p. 21.

173. Ibid., p. 248.

174. Ibid., pp. 249–50. Language, wrote Lavoisier, should be regarded as simply

> a collection of representative signs. . . . The word should give birth to the idea and the idea should portray the fact. . . . And, since it is the words that preserve the ideas and transmit them, it follows that it is impossible to improve the science without perfecting its language, and that, however true the ideas that they give rise to, they will still transmit only false impressions if we have no exact expressions to convey them. The perfection of chemical nomenclature, considered in this respect, consists in rendering the ideas and the facts in their exact truth, . . . it must be nothing but a faithful mirror . . .

175. Lavoisier, *Elements of Chemistry*, trans. Robert Kerr (Edinburgh, 1790; reprint ed., New York, 1965), pp. xiii, xiv, xxxvi.

176. Baguenault de Puchesse, *Condillac: Sa Vie, Sa Philosophie, Son Influence* (Paris, 1910), p. 205. The passage cited is from Condillac, *La Logique* (1780), chap. XXII, pp. 100, 107.

177. Ibid., p. 46.

178. Ronald Grimsley, *Jean d'Alembert (1717–83)* (Oxford, 1963), pp. 94–95.

179. Ibid., p. 225.

180. Condillac, *Traïté des Systèmes*, cited in R. J. White, *The Anti-Philosophers: A Study of the Philosophers in Eighteenth Century France* (London, 1970), pp. 44.

181. H. B. Acton, "The Philosophy of Language in Revolutionary France," *Proceedings of the British Academy*, vol. 45 (1959), p. 217.

182. Robert McRae, *The Problem of the Unity of the Sciences: Bacon to Kant* (Toronto, 1961), p. 102.

183. Isabel F. Knight, *The Geometric Spirit: The Abbé de Condillac and the French Enlightenment* (New Haven, 1968), pp. 287–289.

184. Leon Trotsky, "Dialectical Materialism and Science," *The New International*, vol. 6, no. 1 (February 1940), pp. 26, 28, 29. Though nihilist and Marxist modes of thought flourished in Russia during Mendeleev's lifetime, he repudiated the notion that revolution was a source of either natural or social change. See Alexander Vucinich, "Mendeleev's Views on Science and Society," *Isis*, vol. 58 (1967), p. 348.

185. Wilhelm Schütz, *Robert Mayer* (Leipzig, 1969), p. 28 f.

186. Ibid., pp. 30–31.

187. Schütz, *Robert Mayer*, p. 75 f.

188. *More Letters of Charles Darwin: A Record of his Work in a Series of Hitherto Unpublished Letters*, ed. Francis Darwin and A. C. Seward (New York, 1903), vol. 1, p. 45.

189. "An inner revolutionary process of a special kind," observed Ernest Jones, is crucial to "the perceiving of a truth to which the world had previously been blind." Darwin had to triumph in an unconscious conflict before he could " 'see' an answer previously invisible." For his unconscious parricide, "he paid the penalty in a crippling and lifelong neurosis" (*Free Associations: Memoirs of a Psycho-Analyst* [New York, 1959], p. 203).

190. *Autobiography of Charles Darwin*, pp. 7, 11, 12, 20, 23, 26, 33, 34. "No man could be better formed to win the entire confidence of the young and to encourage them in their pursuits," wrote Darwin of Henslow. In 1848 Darwin told him: "the only thing at Cambridge which did my mind any good, were your lectures, & still more your conversation" (*Darwin and Henslow: The Growth of an Idea, Letters 1831–1860*, ed. Nora Barlow [London, 1967], pp. 161, 222).

191. Ibid., pp. 143–145.

192. Himmelfarb, *Darwin and the Darwinian Revolution*, p. 136.

193. Ibid., p. 19.

194. *Darwin's Notebooks on Transmutation of Species: Part II. Second Notebook (February to July 1838)*, ed. Sir Gavin de Beer, *Bulletin of the British Museum (Natural History), Historical Series*, vol. 2, no. 3 (London, 1960), pp. 91, 106.

195. Himmelfarb, *Darwin the Darwinian Revolution*, p. 151.

196. Ibid., pp. 281–282.

197. *The New York Times*, September 3, 1967.

198. Charles Hussey, "And Now There's the Anti-Establishment," *The New York Times Magazine*, May 3, 1964, p. 42 f.

199. *The New York Times*, December 30, 1970; Harry G. Johnson, "Revolution and Counter-Revolution in Economics," *Encounter*, vol. 36 (April 1971), pp. 26, 28.

200. Even the expression "industrial revolution" was first used in the context of historical thinking to interpret a series of industrial changes as a vehicle for emotions of generational rebellion. The idea of associating industrial changes with revolutionary political changes came into being at the time of the French Exposition of 1798. To young General Bonaparte just beginning to rule France, industrial renovation seemed to offer a field for youthful energies as political revolution had before. French writers in 1804, 1806, and 1810–1837 used the word "revolution" to describe the technical innovations in industry. As Anna Bezanson observed, only a minority of the French working population was affected by these changes: "One must not take too seriously the actual changes which led to the large phrases of enthusiasts" ("The Early Use of the Term Industrial Revolution," *The Quarterly Journal of Economics*, vol. 36 [1922], p. 348). But clearly, industrial innovation was becoming a domain for expressing sublimated energies of political revolution, especially so among the Saint-Simonians; the two shared a common source in emotions of generational rebellion. Curiously, in Britain, where industrial changes proceeded on a truly massive scale, the term "industrial revolution" did not come into use until a half-century later, for Britain did not experience those outbreaks of generational tension characteristic of the succession of French revolutions. Then in 1884, a year after Arnold Toynbee's death, his *Lectures on the Industrial Revolution* were published. Toynbee, interestingly enough, had had the spirit of a student activist; he had been "anxious to utilize for political

reform the ferment of thought at the Universities." As his biographer, F. C. Montague, writes: "Had Arnold Toynbee lived in the thirteenth century, he would probably have entered or founded a religious order, unless he had been first burnt for a heretic." (*Arnold Toynbee: Johns Hopkins University Studies in Historical and Political Science*, 7th ser., [Baltimore, 1889], vol. 1, pp. 31, 35). As it was, he was led, in a back-to-the-people spirit, to give himself to the impoverished London workingmen in Whitechapel. The use of "industrial revolution" was thus expressive of a perspective shaped by the emotions of generational revolt. Rapid changes, in and of themselves, had not been a sufficient condition for the origin and diffusion of the term.

201. *The Standard Edition of the Complete Psychological Works of Sigmund Freud*, vol. 4, p. xxvi.

202. *Charles Darwin*, ed. Francis Darwin (London, 1908), p. 213.

4

THE Conflict OF Scientific Schools

The Notion of Isomorpheme

The word *isomorpheme* has been used throughout this book to refer to those basic ideas or forms of laws whose appearances, disappearances, alterations, and alternations constitute the basic wave movement in the history of thought. Various schools have used different terms to denote this central philosophical idea. Thomas L. Kuhn has borrowed the word *paradigm* from the linguistic philosophy so fashionable today, whereas a half-century ago *model* was still preferred. Almost a hundred years ago Lord Kelvin could say with authority that "the test of *Do we or do we not understand a particular point in physics?* is *Can we make a mechanical model of it?*"[1] The Austrian physicist Ludwig Boltzmann wrote an influential article on *model* tracing its origin from such usages as the sculptor's wax model to the more generic one of "theory whether actually existing or only mentally conceived of, whose properties are to be copied."[2] Josiah Royce, the idealist philosopher, liked to speak of the *canonical forms* of the diverse theories, emphasizing, for instance, how the same statistical forms recurred in such separate fields as the distribution of molecular velocities and the distribution of observations according to the magnitude of their errors.[3] Max Planck, bred on Kantian philosophy, simply preferred to speak of controlling *world-pictures*, the most basic of which was "a fixed world-picture independent of the variation of time and people"; such principles as those of conservation and least action were again aspects of world-pictures.[4] The historian of science, John T. Merz wrote of the different *views of nature*: the astronomical, mechanical, physical, morphological, genetic, statistical, and psychological. Heinrich Hertz asserted that certain basic *images* were central to the scientist's thinking: "We form for ourselves images or symbols of external objects. . . . The images which we here speak of are our conceptions of things. . . . Various images of the same objects are possible, and these images may differ in various respects." Whereas the classical mechanical image allowed for action at a distance between masses, Hertz was guided by an image of continuous connections between hidden masses.[5] Ernst Mach used *theoretical idea* to denote the common property of a "great system of resemblances." Such a system of resemblances was operative when, for instance, the scientist Joseph Black employed the notion of a substance, caloric, to explain the phenomenon of heat. The caloric was conceived as flowing from hot to cold bodies. Thus, too, a stream of water suggested to Joseph Fourier the image of flowing currents of heat.[6] Mach called for a "universal physical phenomenology" which would investigate the central theoretical ideas or images that recurred in all domains, in the manner, for example, in which a common idea was involved in physical falls of all kinds, whether of pressure, temperature, or electrical potential. Mach knew, however, that any image could also evolve into an agent of misperception. The notion of substance, for example, "blinded the eyes of the successors of Black, and prevented them from seeing the manifest fact, which every savage knows, that heat is produced by friction."

Every sociologist of ideas and logician of comparative science has required a conception similar to that which Arthur O. Lovejoy called *unit-ideas* whose combinations and recombinations result in the variations in philosophies much as the combination of elements in varying proportions yield different compounds.

A linguistic usage, however, can bring with it the constraint of some narrowing doctrine. The use of *paradigm* tends to suggest that what is involved in scientific originality is primarily a new way of speaking not unlike a new way of conjugating verbs or declining nouns, which used to be enumerated as paradigm cases in old-fashioned Latin textbooks. A linguistic paradigm is a convention; whatever the inflections of verbs, whether they conform to the first, second, or third conjugation has nothing to do with the objective realities they are used to describe. Similarly, linguistic variation such as most cases of genders in nouns has no correspondence to differences in physical facts. Einstein's aim was to get closer to reality; he was trying to discover hitherto unknown relations. The theory of relativity introduced a new way of speaking only because its underlying conception of nature was different.

The *model*, a favorite of many nineteenth-century scientists, had mechanical overtones that were uncongenial to the entire line of physical investigators who developed the profound phenomenalistic method; the latter aimed to describe the observable relations of phenomena without venturing hypotheses concerning unobserved, hypothetical entities. This was the achievement of such French scientists as Ampère and Fourier early in the nineteenth century. They were scientists who had what we might call a postideological frame of mind; they cheerfully discarded electrical fluids and calorics. Ampère had seen his father guillotined in the Revolution and then suffered a nervous breakdown. As a scientist, he preferred to avoid the pitfalls studied in "ideogeny." He wanted purely mathematical formulations of observed physical phenomena; these, synthesized with mathematical skill, emerged as beautiful systems of the formal relationships of occurrences. Whoever renounced speculation about hypothetical entities could at least be sure of one thing, Ampère told Fourier: their theories would not be overthrown.[7] Here was a method that made scientific revolutions less likely. Even if there were a restoration of caloric, the validity of Fourier's mathematical equations for heat would remain unimpaired.

Central to all these usages is the notion that a "theoretical idea" is fertile to the extent that the forms or relational structures which it sets forth are replicated in various independent domains. The purely mathematical pattern, for instance, of Newton's law of gravitation, the structure of the inverse variation of a force between entities in accordance with the square of the distance between them, and its variation directly with their product, could be exemplified by different interpretations of the variables in the fields of electrostatics and magnetism (in Coulomb's law, for instance, for the strength of electrostatic attraction). Within the field of mechanics itself, the generality of the law of gravitation embraces a class of partially isomorphic corollaries; the special cases that are solved for different bodies in the heavens and earth alike, the solution for the moon's motion and that of a falling object on the earth, apart from their initial conditions, share an isomorphism with regard to the law of gravitation. That is why one could say that the moon is "falling" toward the earth.

The number of interpreted physical laws or subsystems that are isomorphic with each other is the measure of the fertility of a "theoretical idea." For this reason, *isomorpheme* would appear to be the most descriptive word for such key, root ideas. We will define *isomorpheme* as the logical pattern that is invariant through the repeated uses of an analogy. Every analogy is based on some degree of isomorphism. Two systems or sets of axioms are said to be isomorphic when their terms and relations can be put into a mutual, one-to-one correspondence; each statement affirmed in one system will have its counterpart in the other. In other words, two systems are isomorphic when their logical structures are alike.

The "structure" of a relation, as Russell wrote, "does not depend on the particular terms that make-up the field of the relation. The field may be changed without changing the structure. . . ."[9] A philosophy, however, is something more than a structural form which finds many exemplifications. A philosophy usually also aims to state what, in its view, is the intrinsic nature of reality, or the character of the entities among which the structural relations hold. From a purely structural standpoint, the intrinsic nature of the terms in an equation is of no moment. They may be taken as subjective sensations, or external events, or thoughts in God's mind. As Russell says, "What matters in mathematics, and to a very great extent in physical science is not the intrinsic nature of our terms, but the logical nature of their interrelations."[9] In this sense, however, no philosophy rests content with a purely structuralist standpoint. Every philosophy proposes how the terms of the scientific system shall finally be interpreted. Ernst Mach identified his world-elements with perceptual data; Bertrand Russell constructed a world of perspectives, perceived and unperceived; Whitehead proposes that "the key notion from which such construction should start is that the energetic activity considered in physics is the emotional intensity entertained in life."[10] Every philosopher is engaged, in Eddington's phrase, in "world-building," and in building a world, a philosopher describes not only what he takes its structure to be, but also the intrinsic nature of its terms. A philosophical theory, from this standpoint, proposes not only a controlling isomorpheme for the sciences; it also enunciates what might be called an *intrinsotheme*, its notion of the intrinsic nature of the terms. The different intrinsothemes and isomorphemes are the unit-ideas, as Lovejoy uses that term.

The philosophical associations surrounding the word "structure" are apt to mislead us, however, especially today when an ideology of "structuralism" exerts an attraction on many thinkers. It was probably Arthur Eddington who, following Russell's definition of structure, first developed in *Space, Time and Gravitation* a whole philosophy of structuralism. Eddington wrote of the work of unification achieved by the theory of relativity.

And yet, in regard to the nature of things, this knowledge is only an empty shell—a form of symbols. It is knowledge of structural form, and not knowledge of content. All through the physical world runs that unknown content, which must surely be the stuff of our consciousness. Here is a hint of aspects deep within the world of physics, and yet unattainable by the methods of physics.[11]

Thus, according to both Eddington and Russell, the scientist was inevitably a structuralist; scientific knowledge, whether of the atom or the universe, was always knowledge of abstract structure because all it gave us was equations correlating the pointer-readings of scales and clocks. "Results of measurement are the subject-matter of physics,"[12] its terms; the relational structure is provided by the mathematical construction. Thus the scientist is engaged in world-building, but the inner character of the building material, the terms, the relata, remains unknown.

But no social scientist remains content with knowledge that is limited to the structure of social events or movements. For much of social science rests precisely on a knowledge of felt emotions and conflicts, on the character of the intrinsic experience. When we speak of the revolutionary emotions, the feelings of generational revolt that were experienced by scientists, the unconscious pressures that predisposed them to a given mode of thought, we are speaking not merely of structural relations but of intrinsic, qualitative contents. We are not dealing solely with the logical structure of the null-result of the Michelson-Morley experiment, the structure of classical theories, the auxiliary hypotheses of the Lorentz-Fitzgerald contraction, and the contrasting logical elegance of Einstein's special theory. These may constitute the structure of the scientific revolution, but what concerns social scientists are the non-structural, impelling, emotional motivations—the rebellions, the yearnings for a new type of world—for a new definition of reality. We inquire into conditions and choices, into the flesh and moving blood, not merely the structures of events, toward a realistic theory.

When Eddington explicated the structure of the physical world, he thus prepared the way for his mystical intrinsotheme; for a purely formal structuralism, like the frame of a building without the bricks, was empty without a qualitative content. Eddington's emotions as a Quaker mystic then provided his solution for the unknown terms of reality. A distinguished religious thinker wrote aptly:

> Even if I had known it otherwise I could have guessed from his writings that he belonged to the Society of Friends, who might well be called the Friends of Humanity, essentially a mystical movement. . . . As man of science he stands in the tradition of Newton; as mystic, in the tradition of George Fox—two men who were almost contemporary, and both much occupied with the Light; Fox with the Inner Light that never was on land or sea, and Newton with the outer light that illuminates them both. . . .[13]

A strong, recurrent emotion among scientists generates the belief that the dualism between isomorpheme and intrinsotheme, between structure and content, form and matter, can be overcome. Eddington in his last twenty years labored to show how from the "stuff of consciousness," pure reason itself, the mind of God, there were mathematically deducible the values for the empirical constants of nature. As social scientists, our theory, however, confines itself to the qualitative workings of the finite consciousness of human beings. That the gropings of these finite consciousnesses in different directions, the trials and errors with rival isomorphemes, the intellectual experiments with different modes of thought sometimes so tragic in their consequences for their adherents are all themselves episodic alternations within the workings of a Transcendent Consciousness, might perhaps constitute a final isomorpheme and intrinsotheme, a final model for the human history of science.

The Emotional Character of Scientific Schools

A school of thought arises as a sociological phenomenon when there is an emotional commitment to a controlling intrinsotheme and isomorpheme. The emotions that underlie the formation of a school of thought are powerful and largely unconscious; they generally involve more than a conception of what is admissible as science. Often the founder of the school is internalized by its members as an inner moral authority. Therefore, a scientific school usually advocates a distinctive way of life and ethical conception. Ernst Mach, for instance, thought that eliminating the ego, and constructing a world exclusively of sensory elements would result in a higher ethic:

The ego must be given up. . . . We shall then no longer place so high a value upon the ego. . . . We shall then be willing to renounce individual immortality, and not place more value upon the subsidiary elements than upon the principal ones. In this way we shall arrive at a freer and more enlightened view of life, which will preclude the disregard of other egos and the overestimation of our own. The ethical ideal founded on this view of life will be equally far removed from the ideal of the ascetic, which is not biologically tenable for whoever practices it . . . and from the ideal of an overweening Nietzschean "superman," who cannot, and I hope will not be tolerated by his fellow-men.[14]

Darwin's theory likewise involved a distinctive way of life for his ardent young followers. Frederick Pollock, in his biography of the mathematician William Kingdon Clifford, recalled:

For two or three years the knot of Cambridge friends of whom Clifford was the leading spirit were carried away by a wave of Darwinian enthusiasm: we seemed to ride triumphant on an ocean of new life and boundless possibilities. Natural selection was to be the master-key of the universe. . . . Among other things it was to give us a new system of ethics, combining the exactness of the utilitarian with the poetical ideals of the transcendentalist. We were not only to believe joyfully in the survival of the fittest, but to take an active and conscious part in making ourselves fitter. At one time Clifford held that it was worth our while to practise variation of set purpose; not only to avoid being the slaves of custom, but to eschew fixed habits of every kind, and to try the greatest possible number of experiments in living. . . .[15]

As the Italian mathematician, Federigo Enriques, once said, the rise of a "heterodox science" is founded on "psychological motives." "All the rebellious spirits panting for a new light" band together. They proclaim that

the problems of orthodox science are out of place, meaningless, and worthless . . . that a new attitude of thought is tending to turn men's minds in a radically opposite direction. . . . To one body of doctrines problems, and methods, there is opposed another quite different, inspired by a strong spirit of negation, and announcing itself as a renovating school.

Without the challenge of heterodox science, orthodox science, based on the normal "tendency to consolidate methods," would "lose itself in Byzantinism," and with a fixated attachment to the sole legitimacy of its own methods, become indifferent to the facts, and forget the ends of science.[16]

Wilhelm Ostwald, who regarded the law of the conservation of energy as the all-embracing model for science, founded the school of energetics. He went to the extreme of proclaiming the energetic imperative: "Not only the human but the cosmic is governed in the same way by this law of nature. . . . Since the phrase should especially characterize its ethical side, I shall call it the energetic imperative. And this energetic imperative which governs our whole life from technics to ethics means: 'Waste no energy'."[17] Ostwald also formulated the ethical maxim: "So act that the crude energy is transformed into the higher with the least possible loss."

The Copenhagen spirit of the school of Niels Bohr and the relativistic spirit of the Zurich-Berne circle of Einstein are later examples of ways of life linked to respective conceptions of science. This phenomenon of schools of thought has hitherto been much less characteristic of scientific life in England and America than in Europe.[18] In France, divergences between schools reached to such an extent that people spoke of the political struggle between the Normalians and the Polytechnicians as reflecting the metaphysical loyalties of the one and the positivism of the other.[19]

Emotions, often of a kind unexpected in scientists, condition not only the apperception of certain traits of reality but also the nonperception of others. Neither of the two discoverers of the theory of natural selection, for instance, could endure working at dissecting organisms. Alfred Russel Wallace, like Darwin, wrote of his "positive distaste for all forms of anatomical and physiological experiment."[20] It was precisely this type of naturalist-traveler, however, with rich and abundant experience of animals and plants in distant places, immersed in their lives and virtually empathizing with their struggles, who was the most likely to envisage the theory of natural selection.

Given the emotional bases of schools of scientific thought, we can understand why their methodological and philosophical conflicts engender animosities. Each school is characterized by an affirmation that its mode of thought is the only valid one; each school denies that what the others are doing is truly science. The proliferation of schools, however, does not necessarily signify that emotional involvements will predominate over a common loyalty to scientific method. For the competition of schools engenders a contribution of diverse insights; the different emotional perspectives can thus provide competing energies and multiple directions for scientific fruitfulness. Schools, however, become deleterious insofar as their animosities provoke fixations on their respective isomorphemes and intrinsothemes; they then become one-ideaed and blind to the others' achievements. A psychologist studying this phenomenon in his own discipline wrote: "With regard to the cause for the existence of schools—each of which regards itself as exclusively valid when they all actually complement each other—can it fairly be said that the exponents of the schools have been committing a 'fallacy of arrogation'."[21] Eclecticism, however, can be a sterile scholastic formula, for the real question concerns the degree, the extent, and the way in which diverse factors complement each other.

A unit-idea, an isomorpheme, often has its independent career, its own intellectual world-line, diffusing from one philosophical school to another, so that the map of its diffusional route defies any notion of purely logical connection. Herbert Spencer's evolutionary mode of analysis, for instance, had a lineage that derived from Schelling's *Naturphilosophie*; the embryologist von Baer was the intervening link. Spencer told of how he conceived his evolutionary formula in 1851: "I became acquainted with von Baer's statement that the development of every organism is a change from homogeneity to heterogeneity. The substance of the thought was not new to me, though its form was. . . ."

The notions of function and structure, as Spencer had previously conceived them, had been limited to organic phenomena.

> It was otherwise with the more generalized expression of von Baer. Besides being brief it was not necessarily limited to the organic world; though it was by him recognized only as the law of evolution of each individual organism. Added to my stock of general ideas, this idea did not long lie dormant. It was soon extended to certain phenomena of the super-organic class. At the close of the essay on "The Philosophy of Style," published in October, 1852, it made an unobtrusive first appearance in supplying a measure of superiority in style. Change from homogeneity to heterogeneity, began to be recognized as that change in which progress

other than organic, consists. . . . The extension of von Baer's formula expressing the development of each organism, first to one and then to another group of phenomena, until all were taken in as parts of a whole, exemplified the process of integration.[22]

Von Baer, famous for his discovery of the mammalian egg, was nurtured on Schellingian ideas during his student days at Würzburg. Following Schelling, he rejected the notion of epigenesis, according to which an embryo, for instance, was conceived as arising *de novo* from a curdling of fluid substance; rather he held that what took place was a series of transformations in the egg which made its structure more differentiated and specialized. Through the process of impregnation "a part becomes whole." "Every development strives from the center toward the periphery." "Every animal, which is begotten by a sexual union, develops out of an egg; none out of a formative liquor."[23] Von Baer, to be sure, was saved from Schellingian vagaries because he had the temperament of an empirical investigator. But the example illustrates the important truth that an isomorpheme pursues an independent life, and that it can find lodgings in the most diverse schemes of thought. The notion that evolution was a differentiation prefigured in some original potency was attached first to Schelling's philosophy of identity, then provided a model for the discovery and study of the mammalian egg, and lastly stimulated Spencer's formulation of a law of evolution (embedded in the unknowable), which profoundly affected the social sciences at the end of the nineteenth century.

The attachment to a particular mode of thought, a particular isomorpheme or intrinsotheme, is no dispassionate decision; for the founder, disciples, and exponents of a school, the choice of its mode of thought is expressive of deepest personal strivings and emotional needs. In the later stages, however, of an idea's diffusion, when it has become something of a cliché used by all wings, left, right, and center, its use will be invested with little emotion, and may indeed be quite peripheral to the emotional needs of its users. This is not so, however, during the stages of formation and controversial strife of the given idea. As Mendeleev wrote:

Since the scientific world view changes drastically not only from one period to the other but also from one person to another, it is an expression of creativity. . . . Each scientist endeavors to translate the world view of the school he belongs to into an indisputable principle of science.[24]

The Tragedy of Ludwig Boltzmann: A Case Study in the Consequences of Generational, Methodological Fanaticism

A scientific school, as a socioemotional group, can at times engender a species of fanaticism which we scarcely associate with the pursuit of scientific method. For to the young disciple, the choice of a school, the emotional identification with a master, and the adherence to a particular mode of explanation are frequently vehicles for the resolution of his personal conflicts. His aggressive feelings are then directed against those who question the authority of his master or the supremacy of his method. The fanatical disciple is emotionally a totalitarian; instead of adhering to the constitutional, republican criterion for due process of verification, binding on all schools, he insists on an allegiance to his school's model for explanation. The school then acquires the emotional characteristics of a totalitarian sect or party. As such, it does violence to the scientific spirit.

Methodological fanatics can impose an intolerable strain on even the most sensitive creative mind. Jean Racine, France's great dramatist, renounced his art for twelve years because of vindictive critics. Under similar circumstances, Ludwig Boltzmann, the powerful intellect of molecular physics, took his own life in 1906. He was too vulnerable to the criticism of methodological fanatics. Such persons, always represented in scientific schools, delight in using their favorite isomorpheme as an instrument for assailing those who pursue another method or vision; they forget that the alternation and competition of theoretical ideas has been the circulatory life-blood of science.

Ludwig Boltzmann, professor of physics at Vienna, had among other achievements derived the law of entropy, the second law of thermodynamics, from calculations concerning the value of the probability for the distribution of the velocities of gas molecules. His bold and far-reaching work met with some initial doubts even from Max Planck who later followed in his path. Planck was troubled by the fact that in Boltzmann's analysis the law of the increase of entropy became a probable one; the entropy might then conceivably decrease, and the law would not be "as general and free from exception as the law of conservation of energy."[25] The most unbridled criticism, however, came from young Machians—not the genial Mach himself—but fanatics among his followers who ridiculed all atomists as if they were obsolete survivors of a prehistoric era. This experience led Boltzmann to some incisive reflections on the nature of methodological controversies in science. There "is a process," he wrote, "which is by no means restricted to theoretical physics, but to all appearances recurs in the history of every field of intellectual activity." The motives that led to a new style in physical theory were analogous to those involved in the formation of a new artistic school:

In like manner in art the Impressionists and Secessionists stand arrayed against the old schools of painting, and the Wagnerian school of music against the schools of the ancient classical masters. There is accordingly no occasion for surprise that theoretical physics does not form an exception to this general law.[26]

In effect, the "general law" underlying the strife of scientific schools was in Boltzmann's view, the conflict between generations of scientists, the old and the new; hence Boltzmann felt himself a lonely survivor of the old, fighting off the tireless, endless onslaughts of a host of young. This was the primary, underlying wave in the polemics of methodology and the law of the succession of methods. Every method, said Boltzmann, goes through a series of stages in its development. At first it yields "the most beautiful results"; many persons are then

tempted to believe that the development of science to the end of all time would consist in the systematic and unremitting application of them. But suddenly they begin to show indications of impotency, and all efforts are then bent upon discovering new and antagonistic methods. Then there usually arises a conflict between the adherents of the old method and those of the new.

The latter ridicule the adherents to the old "as antiquated and obsolete"; the old "in their turn look down with scorn upon the innovators as perverters of true classical science." Boltzmann described his own isolation in poignant words:

> I seem to myself like a veteran on the field of science; nay, I might even say that I alone am left of those who embraced the old doctrines heart and soul; at least I am the only one who is still sturdily battling for them. . . . I appear before you, therefore, as a reactionary and belated thinker, as a zealous champion of the old classical doctrines as opposed to the new.[27]

Behind the alternation of intellectual standpoints, Boltzmann discerned an underlying alternation of psychological mood. The classical physics, with its attraction to Newtonian mechanical models, had led to a fervor that multiplied unverifiable, mechanical, intervening entities. "Every Tom, Dick, and Harry felt himself called upon to devise his own special combinations of atoms and vortices, and fancied in having done so that he had pried out the ultimate secrets of the Creator."[28] Frequent "extravagances" had taken place against which "some sort of reaction was imperative." Thus arose the new schools of "energeticians" and "phenomenologists" whom Boltzmann called "the moderate secessionists." Their "extreme wing" was the "mathematical wing," especially militant psychologically against mechanical models. The standpoint of the insurgents as a whole was "a reaction against the predominating tendency of the old view to regard the hypotheses concerning the composition of atoms as the real aim of scientific enquiry. . . ." The extreme wing of "mathematical phenomenology" argued that what was important were the equations for the observed phenomena, not the models that might have been instrumental in their discovery. "The hypotheses by which the equations were originally reached were found to be precarious and subject to change; but the equations themselves, after they had been put to experimental test . . . remained intact. . . ."[29] Though physicists acknowledged that as a matter of historical fact nearly all the equations were first derived from molecular assumptions, the insurgent phenomenologists held that "the molecular theory was superfluous. . . . All vowed its annihilation."[30]

Did this desire of the phenomenological school for the "annihilation" of its elders' standpoint arise solely from motives of scientific prudence and concern for verifiability? Or rather was its manifest zeal for an extremist degree of verification the emotional expression of a new generation looking for a scientific method that would declare its independence from the old? Boltzmann acknowledged that he was "only too well aware that like others, I also see the things of this world as colored by the glasses of my own subjectivity."[31] A revolutionary emotion, however, he observed, was strong among the phenomenologists. They became so enthusiastic for instance, over developments in the application of the principle of conservation of energy, that they argued that it was the canonical form for the solution of all problems: "A revolution which affected the entire domain of physics . . . was inaugurated with the rapid growth in import and scope of the principle of energy." Energeticians "declared for absolute rupture with the old ideas."[32] They even decreed for themselves, as a new chosen elect, a new scientific commandment: "Thou shalt make unto thee no mental image or mental likeness whatsoever."[33] To Boltzmann it seemed strange that images that had been so helpful in the derivation of laws should arouse such intense animosity. One might have thought that the scientific community would wish to preserve in its heritage of instrumentalities those ideas that had been proven to contribute to the advancement of science. But the revolutionary spirit has an ingredient of negation. The classical models of the kinetic theory of gases, the atomistic hypotheses, were all compulsively adjudged in the tribunal of the young as the invalid fantasies of the elders. The phenomenologists declared that the pattern of revolution governed the progress of science; they "appealed to the historical principle that frequently opinions which are held in the highest esteem have been supplanted within a very short space of time by totally different theories. . . ." Boltzmann pondered the notion of revolution as the law of scientific advance and concluded that it did violence to an important aspect of the scientific spirit. For "historical principles are sometimes double-edged weapons," he wrote. "History does no doubt often show revolutions which have been unforeseen . . . but it should also be borne in mind that some achievements may possibly remain the possessions of science for all time, though in a modified and perfected form." The revolutionary isomorpheme did indeed fail to do justice to the cumulative incorporation of the generalizations of one generation into the limiting laws of those of its successors. And if history testified to successive annihilations by schools of their predecessors, did this not condemn all schools, including the current revolutionary one, which proclaimed its standpoint the ultimate one? "Indeed, by the very historical principle in question a definitive victory for the energeticians and phenomenologists would seem to be impossible, since their defeat would be immediately required by the fact of their success."[34]

Personal tragedy intervened in this drama of the conflict of schools and generations in science. Boltzmann's dedication to the life of the spirit was total. He affixed quotations from Faust to the prefaces of his first books, and said: "As Faust before the world spirit, so mortal man stands trembling before abstract science."[35] He felt the scientist was confronting existence in a spirit of storm and stress; without Schiller's poetic works, he said, he himself could never continue. The atoms that he described in his scientific work were part of the pattern of life and death. When a dear friend and a distinguished physicist died, Boltzmann said at the close of his memorial address: "Now Loschmidt's body has disintegrated into its atoms; into how many we can calculate on the basis of the principles he himself worked out. . : ."[36] Then to honor the dead experimental physicist, he had this number written on the blackboard. Boltzmann became sensitive to assaults on the symbols of manhood. He wore a long beard and gave a sardonic reply to Schopenhauer's argument that beards and hairiness are associated with animality. But he sank into melancholy when the Machians dubbed him "the last pillar" of atomistics. He felt that "an era of barbarism" was triumphing in physics which was overwhelming his lifework and would finally suppress all atomistic thinking. He could find no university setting that was not marred by these polemics. The labor of science in which he had once found a refuge from human pains and from his inner tendency to melancholy was now itself penetrated by strife and vindictiveness. Boltzmann killed himself on September 6, 1906. "That the opposition to atomic theory contributed to Boltzmann's depression," notes his biographer, "impresses us as especially tragic, since it was precisely in our century that the atomic theory won its greatest victories."[37] The work of J. J. Thomson, Ernest Rutherford, and Jean Perrin soon culminated in Niels Bohr's model of the atom; Ostwald, the chief of the energetical school, tried to explain Boltzmann's suicide by saying *Er war ein Fremdling dieser Welt* (he was "a stranger in this world").[38] This was, however, another way of saying that the strife of schools and generations could become inimical to the community of scientists.

Before he died, Boltzmann in a sense bequeathed his mantle to Max Planck. The younger Planck, prior to his momentous work on black-body radiation, was engaged in research in thermodynamics using classical conceptions. He sent his research to Boltzmann whom he greatly admired. Thereupon Boltzmann replied that he would never succeed in constructing a truly adequate theory of the statistical thermodynamics of radiation unless he introduced into the process of radiation an element of discontinuity still unknown. "It seems doubtless," writes Louis de Broglie, "that Boltzmann's remark contributed towards orienting Planck toward his immense discovery."[39]

In the conflict between phenomenologists and atomists, metaphorical utterances have recurred which suggest the character of the ultimate unconscious emotions underlying methodological controversies. The periodic phenomenological revolt against hypotheses, hypothetical entities, unperceived entities, absolutes, and hooks and eyes of atoms may indeed be founded on underlying emotional waves. The phenomenological standpoint has a kinship with certain sublimated romantic desires. Heinrich Hertz once said: "The rigor of science requires that we distinguish well the undraped figure of nature itself from the gay-colored vesture with which we clothe it at our pleasure." To this Ludwig Boltzmann responded: "But I think the predilection for nudity would be carried too far if we were to forgo every hypothesis."[40] The phenomenological method thus carried with it the emotional valences of a "predilection for nudity" and "the undraped figure of nature." The phenomenological mood seems to satisfy the longing to approach and grasp this figure of nature in "beauty bare," in the words of Edna St. Vincent Millay. Does the imagery of maternal nudity take us to the innermost source of the recurrent phenomenological motif in generational rebellions of young scientists?

Einstein, in his most productive years, oscillated remarkably between his attachment to the Machian standpoint and an attraction to Boltzmann's. He rose above schools and found pleasure in a duality of philosophical moods. Thus, his friend Friedrich Adler thought in 1908 that Einstein was then an adherent to Boltzmann's atomistic realism. In the letter to his father Victor Adler which we have already cited, Friedrich concluded his description of Einstein by saying: "Today he is one of the most authoritative and recognized of Boltzmann's tendency. And this trend, not that of Mach, is the fashion today."[41] Einstein had indeed followed along Boltzmann's lines when in 1905 he had developed the photoelectric law and the theory of Brownian movement; indeed even earlier in 1902 he had restated Boltzmann's theory of random movements, though Professor Kleiner at the University of Zurich in 1901 rejected Einstein's essay on the kinetic theory of gases for the doctorate degree because he felt it was too critical of Boltzmann. Nevertheless, when the French physicist Jean Perrin subsequently confirmed Einstein's theory of the Brownian movement, it was at the same time a confirmation of Boltzmann's theory of molecular motion. Each of the great rival schools of thought seemed to Einstein in his most creative years to be privileged for grasping some central truth. In his most spirited Machian days he was still a "deviationist" in the Boltzmann direction, as Friedrich Adler, the orthodox Machian, no doubt sadly observed. As a child Albert had read in the *Pirke Aboth* (The Sayings of the Fathers) of the strife between the schools of Hillel and Shammai over the meaning of the Law.[42] He had evidently imbibed the Talmudic wisdom which acknowledged to all schools in honest search, their perceptions of truths whose total reality transcended them all.

The Planck-Mach Controversy

The suicide of Ludwig Boltzmann did not dampen the methodological polemics waged between the phenomenological and atomic schools. For the strife of schools has been founded on four recurrent, nonlogical factors: the revolt of the younger generation against the old, a periodicity in underlying unconscious emotions, the distribution of personality traits within the same generation, and diversities of national tradition and rivalries. Only a few years after Boltzmann's death, a great controversy took place between Max Planck and Ernst Mach. It illustrated that ultimate methodological differences cannot in the last analysis be resolved; as Planck himself acknowledged, he could not "refute" Mach's standpoint. Each school, founded on its respective emotion, with its respective isomorpheme, wishes to regard physical reality in a certain way; each school is a gamble with the unknown, an act of faith at its maximum in the character of the universe, and at its minimum, in the next stage of science; each school hopes to objectify in the physical reality that it would discover an embodied counterpart of the emotional longings of its subjective life.

Max Planck had strong emotions about the Machian theory of

knowledge. "I have occupied myself thoroughly with this theory for years," he wrote in 1910. Then he told how he had come to his own intellectual maturity by ridding himself of its incubus:

> Did I not count myself among the convinced followers of Mach's philosophy during my years in Kiel (1885–1889), which philosophy—as I voluntarily acknowledge—had a strong influence on my thinking about physics? But later I turned from it, principally because I came to see that Mach's philosophy of nature is in no way capable of fulfilling its most dazzling promise: the elimination of all metaphysical elements from the theory of physical knowledge.[43]

Planck stood with Boltzmann in his controversy with the phenomenalists. With the memory of his death still keen, Planck raised the question: Why was Mach's standpoint "in great favor in scientific circles?" He conceded that there was "no contradiction in Mach's system, if correctly carried through." But he found it utterly sterile; it lacked some guiding picture concerning the nature of the world; it was "only formalistic," an affair of arranging words after the discoveries had been made, but not expressive of the inner intuition that always guided the creative researcher, "the finding of a *fixed* world-picture independent of the variation of time and people." Therefore, wrote Planck: "It still remains to ask how it comes about that Mach's theory has won so many disciples among students of Nature." Planck's answer invoked an unformulated law of human psychology:

> If I am not mistaken it represents, fundamentally, a reaction against the proud expectations of previous generations associated with special mechanical phenomena following the discovery of the energy principle. . . . Physics has renounced the development of the mechanics of the atom. . . . Mach's positivism was a philosophic example of the inevitable disillusionment. He fully deserved the credit for having found again, in the sense-perceptions, the only legitimate starting-point of all physical research, and this in the face of threatened scepticism, but he overshot his mark, for he lowered the standard of the physical world-picture to that of the mechanical world-picture.[44]

We may use the conceptions of contemporary psychology to elucidate further Planck's explanation of the vogue of Machian ideas. The younger generation of scientists had grown up under teachers who had experienced the traumatic impact of repeated defeats in the attempt to arrive at a working mechanical hypothesis of nature; consequently, in a temper of disillusionment, the younger scientists chose to fall back narcissistically on the certainty of their own sense-perceptions. Their sensations were at any rate relatively unquestionable, and in a narcissistic, vaguely solipsistic spirit, the younger physicists elevated those sensations to the status of world-elements. All this, however, was tantamount to choosing cultural regression; it seemed to Planck that the way of reason was not to abandon world-pictures but to try to construct one rich enough to encompass physical reality.

Ernst Mach replied to Planck in combative, personal terms. He felt that Planck was trying to excommunicate him from the scientific community.

After having with an even Christian charity (*mit christlicher Milde*) admonished me to show consideration for my adversary, Planck brands (*brandmarkt*) me, finally with the well-known Biblical term, with the mark of a false prophet (*als falschen Propheten*). One sees that the physicists are well along toward becoming a church, and they avail themselves of current practices. To such I shall simply reply: If faith in the reality of atoms is essential for you, I declare myself liberated from the physical mode of thought, I don't wish to be a true physicist, I renounce all scientific reputation, finally I render immeasurable thanks through you to the community of believers (*die Gemeinschaft der Glaubigen*). Freedom of thought seems to me preferable.

It was wrong, said Mach, to try to explain the appeal of his ideas in their time by reference to a psychology of disillusionment. There were "Lorentz, Einstein, Minkowski" dealing with the problems of space, time, and matter in a spirit akin to his; the kinetic theory did not exhaust reality. Indeed, suggested Mach, it was likely that non-logical, emotional factors were prompting Planck to a hostility toward phenomenalist physics: "What reasons have motivated the attacks of Planck against my theory of knowledge and what goal he has pursued, I shall not enquire into here." The great dividing issue was faith in the reality of atoms: it became the touchstone for amity or animus: "This is also why Planck cannot find for such a perversion (*Verkehrtheit*) terms degrading enough. If one wished to indulge in psychological conjectures, one must read the text itself. . . ."[45]

As a psychologist of ideas, Mach readily affirmed that each scientist worked with that world-picture (*Weltbild*) congenial to his personality:

It is also necessarily dependent on each one's individuality, and consequently in a certain sense equally arbitrary. . . . Who could prevent investigators from directing their special attention to different aspects of facts? . . . Naturally the image which man makes of the universe, an image which society preserves throughout the change of inquirers in a manner which seems independent of the individuality of each, that image provides an expression of the facts which purifies itself progressively. But in general, in every observation, in every opinion, the surrounding environment, as well as the observer, finds its expression.[46]

What we may call methodological pluralism is the natural expression of the diversity of human temperaments. This Mach acknowledged. Nonetheless, he still contested on behalf of his own distinctive standpoint; while recognizing the historically demonstrated "utility" of atomic theories, he steadfastly remained opposed to them. He could not accept Boltzmann's argument that the second law of thermodynamics was purely a probabilistic one founded on the likeliest distribution of molecules. He felt that science would absorb the observable consequences of atomic theories, while discarding their hypothetical premises.

Thus the controversy raged between Planck the realist and Mach, the positivist.[47] Planck, in his sardonic rejoinder, cast doubts as to the cogency and heuristicity of Mach's standpoint. The principle of economy of thought was used vaguely, he said, to cover both a biological utility as well as any scientific satisfaction. Mach's phenomenalism, he reiterated, was devoid, of any suggestive world-image. He felt that Mach's underlying psychology of fear inhibited the scientific imagination. For example, Mach regarded the notion of an absolute zero in temperature, a lowest limit in cooling at which a body became devoid of heat energy, as "a much too daring extrapolation." Would Mach have felt the same about Einstein's notion that the velocity of light was an upper limit for velocities generally? Planck, moreover, challenged Mach's relativism: If Mach thought it equally valid to say that the fixed stars rotate with respect to a stationary earth as the other way round, would he grant an infinite angular velocity to infinitely distant bodies (an unpleasant consequence for him indeed)? Planck, with the feelings of a Kantian absolutist, found a certain unreality in Mach's relativistic logic: "Mach's theory is incapable of doing justice to the awesome progress which is connected with the introduction of the Copernican world-view,—a circumstance which alone would suffice to put Mach's theory of knowledge in a somewhat dubious light."[48]

The controversy simmered down. Both Planck and Mach realized that neither could refute the other. Both standpoints were "confirmed" during the next half-century. The Bohr model of the atom was developed in a spirit congenial with Planck's ideas; but Heisenberg's observational standpoint was a reversion, as he himself affirmed, to the Machian emotions. The cumulative growth of science proceeds through alternating waves of generational emotion; each in its cycle tends to become as intolerant as the other, and each in turn is provisionally superseded during the next primary wave. Perhaps all that sociological understanding can hope to contribute is to reduce the amplitude, the animosity, and the severity in the wave transitions of generational emotion.

Generational Movements and the Scientific Community

Each generation, whether in the sciences, arts, or politics, aspires to actualize its own standpoint toward reality. In the arts the rhythm of generational styles has been most evident, so that the history of literature takes the form of generational waves. Emotional motivations express themselves most directly in works of art because the artistic imagination can reconstruct realities most freely.[49] In scientific work, however, a hypothesis has to be tested against experimental fact; and in politics, existing institutions cannot easily be ignored lest one tilt at windmills. Nonetheless, generational waves are the underlying source of the alternation of modes of thought in the sciences.

Relativists, absolutists, classicists, and romantics are thus emotional-intellectual types found in both the sciences and the arts. No generation is cut to one pattern; every generation has its distribution of diverse personalities. Nevertheless, among its intellectual leaders, at a given time a certain type will tend to be dominant. Bernard Shaw, who himself led the revolt of the Fabian classicists against romantic idealism, wondered therefore how one could transcend the unending alternations: "One generation sets up duty, renunciation, self-sacrifice as a panacea. The next generation . . .

that defrauded generation foams at the mouth at the very mention of duty, and sets up the alternative panacea of love, their deprivation of which seems to have been the most cruel and mischievous feature of their slavery to duty."[50] But these wave motions are inherent in human existence; one can hope only to mitigate the intolerance and destructiveness that have characterized some of them.

Wilhelm Ostwald first enunciated the truth that common emotions were underlying in both scientific and literary work. The psychological types of the great scientists, he wrote, could be classified, like the great artists, as either classicists or romantics. The makers of revolution in science, he believed, were romantics.[51] They responded to new ideas rapidly and went from one subject to another, filled with enthusiasm. They made the best teachers because they conveyed this enthusiasm to others; they became the founders of distinctive schools because besides inspiring their students, they always had a surplus of ideas, plans, and problems which they cheerfully shared with their students. By contrast, the classicists were withdrawn, single-minded, and rather egotistical men. They had no emotional need to surround themselves with a school of disciples. The classicist was in no hurry to publish. Like the American Josiah Willard Gibbs, the classicist is content to devote his entire life to the intensive investigation of one or two problems. Newton, for example, waited patiently for nearly twenty years to announce the law of gravitation until he had satisfactorily resolved all mathematical and experimental difficulties. The classicist tends to regard his work as a private, personal calling. Reserved, retiring men, working slowly and laboriously, they write treatises which endure longer than the contributions of the romantics. The latter, however, activate many minds with more immediate influence.[52]

But Ostwald misperceived the psychology of the would-be scientific revolutionists. For scientists of classical temperament can be moved by revolutionary impulses as much as their romantic colleagues. The emotions of generational revolt can take, under different initial conditions, either a classical or romantic form. In the nineteenth century, during which Ostwald lived, political revolutionists were indeed usually romantics. This has not always been the case in science. Charles Darwin was, according to Ostwald's classification, a classicist; a slow, painstaking researcher given to the study of a few life-problems, he shunned social life, and had no wish for disciples. He procrastinated for a long time before publishing his results. Had it not been for the letter he received from young Alfred Russel Wallace in the Malay Archipelago formulating the theory of natural selection, Darwin would have been content to add to the many years he had already devoted to accumulating evidence for projected volumes on his hypothesis. Nonetheless, the classicist Darwin was the prime maker of a scientific revolution. Isaac Newton also had the psychology of a classicist, "reserved and retiring" (in Ostwald's words), shrinking from social relations, secretive, indifferent to publication, and, as noted above, content to wait many years for some needed observations before publishing his own analysis. Yet Newton put his stamp on physical conceptions for more than two centuries. According to Ostwald, the romantic gathers students around himself and thus creates the nuclear revolutionary circle. Einstein, however, had no impulse to surround himself with disciples. He preferred to work alone. His situation grew so solitary that it became difficult in later years even to find an assistant for him, with the result that his assistants were a motley group unlikely to effect any insurrection.[53] Einstein, however, had fashioned his revolution without the support of a scientific school. People having classical temperaments can be revolution-makers as well as romantics. Even the French Revolution had its enthusiasts for a cult of classical antiquity.[54]

In the nineteenth century, however, the generations avowedly engaging in political and artistic revolt were predominantly romantics. The century of Byron, Shelley, Lamartine, and Hugo was that of the romantic revolutionaries of 1830 and 1848.[55] And this mood made itself felt in science. The first Nobel Laureate in chemistry, Van't Hoff, thus regarded Lord Byron as the guiding star of his life. He wrote his mother: "If I had not been captured by the highly emotional Byron then I would soon have shriveled to a dried out scientific conglomerate. . . ."[56] Even the Sandemanian Michael Faraday spoke of himself as a "very lively imaginative person" who "could believe in the 'Arabian Nights' as easily as in the Encyclopaedia."[57] But it was Byron who was the major figure of Western European romanticism. Brooding on nameless sins, dwelling in self-isolation, he was the rebel against God; his fate was that of Manfred and Cain. And the nineteenth century, opening with Malthus' gloomy generalization, was in the mood to impart a touch of the diabolical to God. The romantic physicists, like the poets, reveled in the diversities of things; Victor Hugo in his preface to *Cromwell* proclaimed the *mélange des genres*, and von Helmholtz enunciated the multiplicity of the forms of energy. But suffusing the cosmological drama was the melancholy outcome, symbolized in the title of Lord Kelvin's paper, "On a Universal Tendency in Nature to the Dissipation of Mechanical Energy." God, whom the eighteenth century conceived of as a Perfect Designer, vanished from mid-nineteenth-century science as from nineteenth-century literature. Kelvin's final vision was as somber as Manfred:

> Within a finite period of time past the earth must have been, and within a finite period of time to come the earth must again be, unfit for the habitation of man as at present constituted, unless operations have been or are to be performed which are impossible under the laws to which the known operations going on at present in the material world are subject.[58]

Generational Relativism and Intergenerational Invariance

Although each generation seeks its distinctive idiom, images, vision, and forms of laws, a common allegiance to truth and method still binds the generations into a scientific community. No relativism sets up its incommensurable perspectives and impermeable walls between the scientific generations. The source of its cohesive tenacity is the emphasis on the criterion of verification, the evidence of perception, and the confirmation of the predicted consequences of a theory.[59] The subcommunities among scientists of generation, class, religion, nation, race, and school have their respective cultural, aesthetic, and psychological preferences as to mode of theory and isomorpheme; each may have its favored aesthetic criterion as to what constitutes elegance, beauty, or simplicity in a theory. But again, what transcends all such cultural variation is the common allegiance to verification. We may affirm the following generalization in the sociology of the scientific community: Resistances to a novel theory which are founded on generational or cultural aesthetic motives will persist and prevent its common acceptance until some novel prediction derived from the proposed theory is confirmed. For this reason Einstein attached crucial significance to the British expedition of 1919 and its observations during the eclipse of the sun. Einstein also emphasized the importance of a further test, the displacement of solar absorption lines in the spectrum toward the red, saying: "If it were proved that this effect does not exist in nature, then the whole theory would have to be abandoned."[60]

Older scientists such as Sir Oliver Lodge frequently expressed their aversion to the special theory of relativity with its "extraordinary postulate" so contrary to common sense, of the absolute constancy of the velocity of light as perceived by all observers moving with uniform velocities relative to each other. To Lodge, the explanation of this phenomenon proposed earlier, the so-called Lorentz-Fitzgerald contraction, seemed to be convincing and physically "true." Whatever the superior mathematical elegance of Einstein's theory, many saw this as outweighed by its rift with common sense. As long as an explanation consistent with common sense might still be forthcoming, they were reluctant to quit its territory: "I still clung, and cling, to the idea that the Fitzgerald contraction is a reality," wrote Lodge. Nevertheless, despite his emotional preferences, he acknowledged in later years that the theory of relativity with its prediction of novel consequences had emerged supreme: "The later generalised theory of relativity of 1915, however, was applied to many other phenomena, and proved itself so consistent with every observation—both those that had been made and some that were going to be made in the future—that it has now secured all but universal acceptance."[61] He retained the hope that posterity might look at the matter "a little differently." The whole episode showed that what is decisive in the emergent ascendancy of one theory over another is not any cultural-aesthetic criterion but its heuristic achievement, that is, its prediction of novel, hitherto unforeseen consequences.

The revered Lorentz did indeed aver frankly his own generational resistance to the theory of relativity; nonetheless, he abided by the criterion of prediction and verification. He said in 1917:

> It will certainly depend to a considerable degree on the type of thought to which one is accustomed, whether one feels himself drawn to one or the other view. I personally still find a certain satisfaction in older views that the ether still preserves a certain substantiality, that space and time are sharply distinguished, and that one can speak of simultaneity without a closer specialization.[62]

Withal, he awaited the "judgment" rendered with all the "thoroughness" of the practice of science. The intergenerational community founded on scientific method transcended the relativism of generational perspective. The philosophy and style of a generation were not transmuted into a generational a priori.

The distinguished experimentalist, A. A. Michelson, had also retained a preference for the ether hypothesis, and therefore did not welcome the theory of relativity. He too would not, however, allow his emotional preferences to veto the experimental and observational confirmations, culminating in the measurements of the photographs of the eclipse in 1919. The deflection of a light-ray in the sun's neighborhood, inversely proportional to its distance from the sun's center, was an observational fact which he would not allow his feelings to repress. Michelson wrote:

Thus the theory of relativity has not only furnished an explanation of known phenomena, but has made it possible to predict and to discover new phenomena, which is one of the most convincing proofs of the value of a theory. It must therefore be accorded a generous acceptance notwithstanding the many consequences which may appear paradoxical. . . .[63]

The constitutional law of the scientific community was thus upheld. The heuristic prediction of novel consequences has thus been the decisive criterion in the acceptance of theories. Maxwell's theory of the electromagnetic field had elegance, mathematical beauty; on reading Maxwell's equations Boltzmann said: *War es ein Gott, der diese Zeichen schrieb?*[64] (Was it a God who wrote these symbols?) Nevertheless, as Einstein writes, "it took physicists some decades to grasp the full significance of Maxwell's discovery, so bold was the leap that his genius forced upon the conceptions of his fellow-workers. Only after Hertz had demonstrated experimentally the existence of Maxwell's electromagnetic waves did resistance to the new theory break down."[65] Indeed when Maxwell proposed his theory of the electromagnetic field, it produced the usual generational division. The older men would not adopt it. As J. J. Thomson described it:

They had been working for years with the old theory, they had been led to it by general discoveries, they knew it was not inconsistent with any known electrical phenomena. For them to forsake this theory for one which introduced a new principle for which there was no direct evidence was quite a different thing from the adoption of a new theory by young men who were just beginning to study electricity and who had no theories to give up.[66]

Maxwell had virtually no support "outside a small group of young Cambridge men," and the status of his theory was in doubt when he died. In Germany, his concept of a field of force "found no foothold . . . and was scarcely even noticed," according to Max Planck. But all that changed when Heinrich Hertz in his experiments generated electromagnetic waves. Where mathematical elegance and aesthetic symmetry had failed to win the scientific community, intergenerational and international, a radically new experimental consequence did.

Similarly, the resistance on the continent to the Newtonian system vanished when in 1758 the French mathematician Clairaut predicted accurately the reappearance of Halley's comet of 1682. Clairaut calculated that its period of revolution would be increased through the perturbational action of Jupiter and Saturn by a total of 618 days. Then, as François Arago vividly narrates:

> Never did a question of astronomy excite a more intense, a more legitimate curiosity. All classes of society awaited with equal interest the announced apparition. A Saxon peasant, Palitzch, first perceived the comet. Henceforward, from one extremity of Europe to the other, a thousand telescopes traced each night the path of the body through the constellations. The route was always, within the limits of precision of the calculations, that which Clairaut had indicated beforehand . . . astronomy had just achieved a great and important triumph. . . .[68]

Generational, ethnic, religious, and cultural traits do exert an influence on the way we perceive our everyday reality resulting in emphases and obliterations. But in scientific work, the impact of external reality in perception is the umpire among competing extralogical circumpressures, and contravenes those which persist in being nonlogical. W. W. Campbell, the astronomer, hoped (as we have mentioned) that his observations of the sun's eclipse would infirm Einstein's theory of relativity. But his wishes did not shape his perception. The distinction remains between what have been justly called the reality principle and the pleasure principle.

Our categories and our theories do affect our perceptions of external reality, but only within limits of tolerance set by reality itself. Where those limits are exceeded, where the theory-holder tries to impose his wishes on reality, trying to bend reality in a strained quasi-perceptual act, then the hold on reality breaks under the strain. The resulting misperception is what Irving Langmuir has called "pathological science." The misperceivers in such cases shared certain common properties: appeal to effects almost undetectable, claims of extraordinary accuracy in observation, and a readiness to invoke ad hoc auxiliary hypotheses to outflank objections.[69] In every instance of pathological science some subjective, generational, or cultural a priori is impelled to do violence to the intersubjective, intergenerational, intercultural reality. Therefore, Einstein insisted upon the great truth of the absolute character of reality: "The belief in an external world independent of the percipient subject is the foundation of all science."[70]

It is of course inevitable that what a younger scientific generation experiences as emotionally liberating is felt by its elders to be emotionally traumatic. The generations regard the turning points in the history of science from different emotional perspectives.[71] Niels Bohr, as we have seen, characterized the emotions involved in the later quantum physics as those of renunciation. Such emotional descriptions, however, have their own generational complementarity. For what in older years seems an act of renunciation, a melancholy leave-taking from the world that one has known, is for the young man a joyous act of adventure, of asserting his individuality, and achieving manhood. The history of science has been marked by a series of such heroic renunciations—most severe of all was the ascetic abandonment of the desire for teleological explanations, the "why" of things, in favor of a modest satisfaction with descriptive uniformities, the "how" of things. Other examples of such renunciation have been: the discarding of Aristotelian qualitative categories in favor of Galilean mechanical properties; the Copernican heliocentric hypothesis as against the more cozy, geocentric Ptolemaic; the Newtonian law of gravitation with its departure from action through contact; the relativist theory with its renunciation of the absolute space and time and simultaneity of common sense; the Bohr theory of the atom with its renunciation of continuous motion; and the Heisenberg principle of indeterminacy with its renunciation of complete causal explanation.

But what is renunciation? It is a surrender of some component of the world of childhood, the world that we first experienced when our mothers and fathers cradled and fed us. In the deepest psychological sense, the controlling primitive model for explanation, the primordial isomorpheme, is the structure of our childhood world and its emotions. Our fathers' purposive ordering of things, our mothers' feeding us through contact, are primary traits of the world; no phenomenon seems explained unless it has been shown to conform to these characters. Every act of scientific renunciation is thus a rebellion against ties that bind us to the older generation. Darwin, destroying the argument from His design for God's existence, could then really feel as if he were committing murder. The paramount moments of renunciation were thus the maxima of generational revolt and liberation.

Yet the word "renunciation" imposes an emotional interpretation that actually belies another aspect of the experience of generational revolt. The young Einstein, playing his "joke" on the elders and overturning the familiar world of absolute simultaneity, enjoyed his revolutionary experience. The young relativistic scientists felt a sense of their liberated manhood. They could depart from their fathers' houses, venturing forth into a strange world, living with heroic risks, not quite knowing what they would encounter. That which to the conservative mood seems an act of renunciation was to their complementary radical mood an act of liberation. What is described in the Kierkegaardian vocabulary of Christian ascetic renunciation was not thus experienced by the younger scientist; rather he was exhilarated by the discomfiture of the old, and enthralled by the sense of a universe opened to unforeseeable, uncharted dangers which he would dare to navigate. Each act of the younger generation's self-liberation was however, a renunciation enforced on the old. The same events in the history of science can thus be viewed from two generational standpoints: that which is renunciation from the standpoint of the old is adventure for the young; and that which is ascetic deprivation for the old is hedonistic liberation for the young. What is a lost familiar world to the old is a cheerfully scrapped, rejected world to the young. The saturnine Pareto, as he appraised the theories of the revolutionary Marxists, argued that throughout the history of societies, one could observe the workings of a "law of the circulation of elites." There is a similar law of the circulation of ideas throughout the history of science; interwoven are what are called the scientific revolutions.

The generational mentality is apt, of course, to exaggerate the rout of its predecessors and to regard the preceding scheme of ideas as utterly vanquished and consigned to the dust-bin of history. The revolutionary-minded Newtonians, for example, scorned the Cartesian school and regarded adherence to any Cartesian ideas as evidence of backwardness or a corrupt spirit. The Encyclopaedist D'Alembert characterized the Cartesian as "almost ridiculous." Something of the Cartesian standpoint persisted, however, as an undercurrent in scientific theory. In 1690, Christian Huygens, Newton's great Dutch contemporary, wrote to Leibniz in a Cartesian spirit that Newton's principle of attraction seemed to him "absurd." Together with Leibniz, he was troubled by the notion of gravity acting at a distance; there were misgivings too about the conception of absolute space. The nineteenth-century theories of the ether represented, from one aspect, a return to the Cartesian way of conceiving physical reality. And although Einstein dismantled absolute space and time, his theory of gravitation, with its geometrizing of physics and rejection of action at a distance, evoked still another Cartesian theme. Einstein persisted in his rejection of the standpoint of quantum theory and toward the very end of his life, he wrote to Max Born: "For the time being, I am alone in my views—as Leibniz was with respect to the absolute space of Newton's theory."[72] The alternation of ideas makes contemporaries of fathers, sons, their ancestry, and their posterity.

Generational Evolution from Empirical Contingency to Rational Necessity

Within the lifetime of a single generation it seems to be a general law that what was perceived in one's youth as an empirical, experimental discovery tends in one's later years to be seen as having been the rationalistic intuition of a logically necessary truth. We might call this the law of generational evolution from the empirically contingent to the rationally necessary. Erwin Schrödinger, the discoverer of the "wave equation," remarked upon this uniformity:

The following process is recurrent in physical science. A certain amount of special knowledge, empirically accumulated and asserted, is tentatively cast into a comprehensive theoretical aspect. The theory, after having been gradually corrected by further experiments . . . tends to acquire an unforeseen general validity. But, strangely enough . . . the knowledge which its propositions are supposed to convey turns out to be more and more tautological.

Paul Dirac thus had the feeling that while lacking in rigor, Eddington's a priori derivations were "probably substantially correct" for at least some of the universal cosmological constants, while P. W. Bridgman, the polemicist of the operational standpoint, in his latter years likewise joined in the hope that a rationalistic derivation of the universal constants would some day be achieved.[72] A scientific theory reflecting the perspective of the scientist's advancing age acquires the character of a rational, timeless necessity, in a sense distinct from the analytic tautologies of logic, but partaking indeed of the intrinsic necessity of Spinoza's Substance. When Max Born vigorously attacked Eddington's apriorism in his lecture "Experiment and Theory in Physics," Einstein replied ironically in October 1944 that he was "confident that 'Jewish Physics' is not to be killed." What was "Jewish Physics"? It had been defined by Born as the endeavor "to get hold of the laws of nature by thinking alone," something in his view that resulted in "pure rubbish" when "average people" tried to do it.[74] But only the Nazis would have reduced the recurrent aspiration of pure reason to the privileged standpoint of a single people. The scientific community also excludes privileged frames of reference.

Thus, too, the experimental foundation for the theory of relativity hard-earned in Einstein's youth tended in his old age to be superseded by the logical a priori; what had been empirically contingent evolved over the years into the rationally necessary. The noted cosmologist Hermann Bondi, for instance, wrote: "With our modern outlook and our modern technology, the Michelson-Morley experiment is a mere tautology." At first Einstein, an old Machian, cavilled at any attempt to expound the theory of relativity as akin to logically necessary truth. When Eddington set forth the theory in an almost aprioristic way, Einstein wrote in 1926 that his book was

really extraordinarily witty. But still I cannot accept the aim to understand the laws of nature only as ordering-schemata for dividing the cases that are subsumable from those which are not. Also to be challenged is that he presents the theory of relativity altogether too much as logically necessary. God could also have decided to create instead of a relativistic ether an absolute one at rest.[75]

By his later sixties, however, Einstein's faith in a rational universe became akin to Eddington's. He affirmed

a theorem which at present can not be based upon anything more than upon faith in the simplicity, i.e., intelligibility of nature: there are no *arbitrary* constants of this kind; that is to say, nature is so constituted that it is possible logically to lay down such strongly determined laws that within these laws only rationally determined constants occur (not constants, therefore, whose numerical value could be changed without destroying the theory).[76]

Thus in different moods and at different times of his life Einstein varied in his evaluation of the relative importance of experimental as compared to purely theoretical considerations. He described these two competing standpoints in later years as those of "external confirmation" as against "inner perfection."[77] In recent years, it has thus been said that experiment had little to do with the formation of the theory of relativity, that in 1905 Einstein had not even heard of the Michelson-Morley experiment.[78] Yet we have seen that Einstein's primary effort as a student in the Federal Polytechnic physical laboratory was precisely to design an apparatus that would measure the earth's movement in the ether.[79] The first page of his 1905 paper alluded briefly to "the unsuccessful attempts to discover any motion of the earth relatively to the 'light medium,'" and said this lent support to the notion that "the phenomena of electrodynamics as well as of mechanics possess no properties corresponding to the idea of absolute rest."[80] The reference to the Michelson-Morley experiment was indeed elliptical, for by this time, especially through the writings of Lord Kelvin, its null result had been posed as one of the residual "clouds" in physical theory.[81] This was now part of the intellectual climate to which a revolutionary-minded young student of physics in 1905 would have been especially sensitive. Einstein once remarked good-naturedly to his assistant Leopold Infeld that he had often been remiss in making acknowledgments in his papers to predecessors.[82] The elliptical reference in his relativity paper of 1905 was probably one such case. For the importance of the Michelson-Morley experiment in the formation of Einstein's theory was indicated by him in 1916 when in his memorial to Ernst Mach he wrote that "Mach would have arrived at the theory of relativity" if when he was still young "the question of the velocity of light had already engaged the attention of the physicists."[83]

To be sure, the motive for "inner perfection" was always present in Einstein's theorizing; he felt the imperfection in the existence of "asymmetries," in the asymmetry which obtained in the fact that whereas the laws of mechanics were invariant for the ordinary (Galilean) frames of reference, Maxwell's laws of electrodynamics were not, and presumably required some privileged inertial frame. The emotionally revolutionary opposition to absolutes would incline one to seek an "inner perfection" in the elimination of asymmetries.

Nevertheless, during the years immediately following World War I, when the theory of relativity evoked sharp, often bitter controversies, Einstein placed the emphasis on the experimental foundation of his theory. At the Collège de France, for instance, on March 31 and April 3, 1922, in a historic confrontation with the French scientific elite, Einstein emphasized that the theory of relativity was born of problems posed by experience, that the Michelson experiment and others had led to the need for reformulating the principles of relativity and the velocity of light, and abandoning the conceptions of absolute space and time. To the astronomer Nordmann, Einstein recalled that Michelson had said ruefully: "If I had been able to foresee the consequences that have been derived from my experiment, I really think I never would have done it."[84] Mach's influence, said Einstein, had been foremost in evolving his own standpoint.

Einstein continued to recognize the significance of crucial experiment for the theory of relativity. As Karl Popper tells, when D. C. Miller announced that he had convincing evidence that contravened Michelson's null result, "Einstein at once declared that if these results should be substantiated he would give up his theory. . . . Einstein did not try to hedge . . ."[85] When his friend Ehrenfest complained that he was having difficulty teaching relativity to his students, Einstein told him that he got into trouble because he based his exposition on epistemological rather than experimental grounds. "The 1905 innovation," Einstein wrote, rested on the "empirical" ground of the "equivalence of all inertial systems compared with light. The epistemological requisite begins first in 1907.[86] Yet it is certainly true that as Einstein grew older the "inner perfection" of a theory, its "naturalness," in a sense which he confessed he could not further explain, acquired the controlling role for Einstein. For, granted that a theory must not contradict empirical fact, it was always possible for its rivals to accommodate themselves to the experimental situation "by means of artificial additional assumptions."[87] God, however, from Einstein's standpoint, could not be as arbitrary or clumsy as that.

In truth, however, it was hard in practice to separate the criterion of "inner perfection" from "external confirmation." The famed historian and philosopher of science and scientist, Pierre Duhem, invoked his immense erudition and philosophical authority to contravene Einstein's theory of relativity. For as Duhem insisted, no experiment could be "crucial" for a decision in favor of the relativity theory; in scientific work one can never confirm or invalidate an isolated proposition; one is testing a complex theory, all of whose components conjointly lead to the testable consequence; but then one has a margin of choice in evaluating which of the component propositions shall be regarded as falsified by the experiment. And Duhem's feeling of "inner perfection" was such that he felt a theory lost that quality when it exorcised the classical conceptions of space and time; he preferred to make minor compensatory alterations in associated components and thereby preserve the divine-like character of absolute space and time.

Duhem spoke indeed for the conservative older generation in France which was antidemocratic and antirepublican. Duhem was an anti-Dreyfusard during that crisis in France, an admirer of the anti-Semitic editor Edouard Drumont (author in 1886 of the obsessed *La France juive*), and a reader of his anti-Semitic daily newspaper *La Libre Parole*.[88] All the violence of the Paris Commune had not kept Pierre Duhem as a young student from faithfully attending his catechism classes; his loyalties remained unswerving in later life.[89]

Duhem once said that a cosmological theory could be conceived that would scientifically explain the miracle of Joshua. Naturally he found Einstein's theory of relativity repugnant. Michelson's experiment, he felt, showed that "certain retouches" of optical theory were required, but nothing like what the German physicists were doing "to overturn the notions which common sense provides us concerning space and time." He felt that these physicists were taking a positive delight in overthrowing "the intuitions of common sense." The notion that the velocity of light was a limiting one seemed to him to violate the imagination of common sense. He contended against "the principle of relativity as it has been conceived by an Einstein, a Max Abraham, a Minkowski, a Laue. . . ."[90] Three of the four names that Duhem mentioned were Jews, and curiously the nationalist Germans who attacked Einstein used arguments identical with Duhem's.

Thus Pierre Duhem's God and Einstein's God spoke with different voices as to the "inner perfection" of a theory. And the allegiance to an "inner perfection" could determine one's view as to what propositions were being falsified by "external confirmation." An experimental decision would then depend on an arbitrary notion of "inner perfection". All sorts of arbitrary sociological premises would enter physical theory to read into it alternative, incommensurable physicosocial reflections. Clearly no such sociotheological relativism entrapped the younger generation. For them their relativism, felt, to be sure, as an emotional generational postulate, was nonetheless decisively reinforced by the heuristic power of the theory of relativity. It led to new work, new laws, new predictions, in ways which the old "common-sense" physics did not. Youthful generational energies are attracted to theories that open up new possibilities; they are not drawn to projects for the repair or alteration of old edifices. With many modifications, one might render the old doctrine consistent with the novel experimental findings; but it was not the old doctrine that had led the way to the latter. The younger generation of scientists is invariably attracted to the heuristic, and this alone is their common platform of agreement with their elders. Sometimes the novel consequence will emerge in a crucial experiment designed to test a theory, as was the case with Eddington's observations in 1919 to test Einstein's predictions of the deflection of light during the solar eclipse. On other occasions a novel phenomenon educed without regard to a given theory is seen as unexpectedly confirming the latter; in his experiments that produced an electromagnetic wave Heinrich Hertz had actually not set out to test Maxwell's theory.[91]

The principle of heuristicity, we might say, is the form that the criterion of verifiability takes when it is impressed with a generational pragmatism. The heuristic theory acts unaided as the pioneer in the derivation of hitherto unforessen consequences; it thus appears to correspond with the grain of nature, to unlock new doors to fresh facts. At critical junctures and fortunate conditions, it coincides with an emotional-intellectual generational postulate for the younger group of scientists. But for the scientific community, inter-generational, inter-religious, and international, it is the purely scientific heuristicity which is its unwritten constitutional law for amendments.

Conclusion

Each generation renews the wave of scientific intuition and discovery with its search for a new "generational postulate"; thereby it defines its own emotional character in philosophical terms. The generational revolutionists, Einstein, Bohr, de Broglie, and Heisenberg, drew inspiration from philosophers who proposed bold new conceptions that projected underlying unconscious emotional responses to reality. Einstein read Mach, Spinoza, and Schopenhauer; Bohr drew on Kierkegaard, Heisenberg turned to Plato, and de Broglie listened to Bergson, whose writing similarly aroused the enthusiasm of Paul Ehrenfest.[92] Without exception, they tended to ridicule academic philosophers and positivists. Max Born expressed what was indeed the common feeling of the generational revolutionists in science when he wrote to Einstein in 1948: "I am annoyed that you reproach me for my positivistic ideas; that really is the very last thing I am after. I really cannot stand those fellows."[93] The scientific revolutionists were repelled by the adroit verbalism of academic philosophers. For them a philosophy had to propose, if it was worth a hearing, some guiding image of reality which was at once emotionally satisfying and scientifically heuristic. As a student Max Born thus found the lectures of Professor Edmund Husserl uninteresting; on the other hand, he was stirred by the classic of anarchistic rebellion, Max Stirner's *The Ego and his Own*, a book that was "widely read" in his youth and conveyed to him a " 'solipsistic' " message.[94] Analytic linguistic philosophy, even when expounded by such a master as Bertrand Russell, failed to evoke interest from the workers in scientific theory. When Russell was at Princeton during World War II, he went to Einstein's house every week for discussions with him, Kurt Gödel, and Wolfgang Pauli. But Russell found these meetings "disappointing," for "although all three of them were Jews and exiles and, in intention, cosmopolitans, I found that they all had a German bias towards metaphysics, and in spite of our utmost endeavors we never arrived at common premises from which to argue."[95] What the scientists still sought in their self-made philosophizing was fructifying world-images and world-ideas—precisely the ingredient expelled in the universities' analysis.[96]

Any essay, finally, in the understanding of the activities of scientists as human beings also involves its general conception of reality. For the science of science, or more accurately speaking, the sociology of truth-seeking men, is itself a branch of theoretical sociology and its central standpoint is be shaped by the inquirer's philosophy.[97] If psychologically this is my preconception, it is also a logical conclusion.

Reality consists of many strata of existence which come into view when different methods of investigation are employed. Each generation has its favored insight and method which in time, as it reaches a region of low diminishing returns, becomes exhausted. The history of science with its manifold skirmishes, revolts, rebellions, and revolutions is Nature's way of reinstating periodically a law of increasing returns. The new generation comes forward with a new method which, under optimal conditions, will be in harmony with a hitherto inaccessible stratum of reality. But there is no pre-established harmony, and a generation, like a god, may fail. Each revolutionary generation moreover is destined to find that its revolution was an illusion, that its standpoint was only partial and remained circumscribed by a reality that is perhaps transgenerational. However, in legend science has always been a revolution against the gods, against the fathers. Daring to eat of the Tree of Knowledge Adam defied the paternalistic Jehovah to learn the law of sexuality; Prometheus rebelled against Zeus to bring man the secret of fire, the law of the transmutation of energy. The creative scientist remains energized by these same Promethean drives against the established viewpoint. No scientific standpoint proves to be valid for more than a chapter of existence, which never yields its plot's entirety. The way of wisdom is to understand how "scientific revolution," in its illusory character, is Nature's method for liberating human energies for scientific progress. Thereby we may secure from the gods their gift of serenity.

Notes

1. W. Thomson, *Notes of Lectures on Molecular Dynamics* (Baltimore, 1884), p. 132. Cited in Silvanus P. Thompson, *The Life of William Thomson*, vol. 2, p. 830.

2. Ludwig Boltzmann, "Model," *Encyclopaedia Britannica*, 11th ed. (New York, 1911), vol. 18, p. 638.

3. Josiah Royce, "The Mechanical, the Historical and the Statistical," *Science*, n.s. vol. 39 (April 17, 1914), p. 11.

4. Max Planck, *A Survey of Physics: A Collection of Lectures and Essays*, trans. R. Jones and D. H. Williams (London, 1925), pp. 38, 40, 126.

5. Heinrich Hertz, *The Principles of Mechanics*, trans. D. E. Jones and J. T. Walley (London, 1900), pp. 1–2.

6. Ernst Mach, *Popular Scientific Lectures*, trans. Thomas J. McCormack (Chicago, 1910), pp. 241, 244, 249.

7. E. Littré, "M. Ampère: Physique," *Revue des Deux Mondes*, vol. 9 (1837), p. 432.

8. Bertrand Russell, *Introduction to Mathematical Philosophy*, 2d ed. (London, 1920), pp. 60–61.

9. Ibid., p. 59.

10. Alfred North Whitehead, *Modes of Thought* (New York, 1938), pp. 231–232.

11. A. S. Eddington, *Space, Time and Gravitation*: (Cambridge, 1920), p. 200; cited in A. Vibert Douglas, *Arthur Stanley Eddington*, p. 49. There was an earlier use of "structuralism" which was associated with the standpoint of the introspective psychologist, Edward Bradford Titchener. His school emphasized the structural universals of the human mind as against the pluralistic functionalists who stressed not the "mind" but "minds" in different situations. See Titchener, *Systematic Psychology: Prolegomena* (New York, 1929), p. 178.

12. Douglas, *Arthur Stanley Eddington*, p. 53.

13. L. P. Jacks, *Sir Arthur Eddington: Man of Science and Mystic* (Cambridge, 1949), pp. 3–4.

14. Ernst Mach, *The Analysis of Sensations and the Relation of the Physical to the Psychical*, trans. C. M. Williams, ed. Sydney Waterlow 5th German ed. rev. (Chicago, 1914), pp. 24–25.

15. William Kingdon Clifford, *Lectures and Essays*, ed. Leslie Stephen and Frederick Pollock, 2d ed. (London, 1886), pp. 24–25.

16. Federigo Enriques, "Heterodox Science and Its Social Function," *Revista di Scienza*, 1 (1907), pp. 326–331.

17. Wilhelm Ostwald, "Efficiency," *The Independent* (October 19, 1911), p. 871; Edwin E. Slosson, *Major Prophets of To-day* (Boston, 1914), p. 203.

18. John Theodore Merz, *A History of European Thought in the Nineteenth Century* (Edinburgh, 1896), vol. 1, p. 250: "No great master in scientific research in this country can point to a compact following of pupils—to a school which undertakes to finish what the master has begun, to carry his ideas into far regions and outlying fields of research, or to draw their remoter consequences."

19. Harald Høffding, *A History of Modern Philosophy*, trans. B. E. Meyer (London, 1900), vol. 2, p. 319.

20. Alfred Russel Wallace, *My Life: A Record of Events and Opinions*, vols. 1, 2 (London, 1905), p. 39.

21. Saul Rosenzweig, "Schools of Psychology: A Complementary Pattern," *Philosophy of Science*, vol. 4 (1937) p. 104.

22. Herbert Spencer, *An Autobiography*, vol. 2 (New York, 1904), pp. 9–10, 13–14.

23. George Sarton, "The Discovery of the Mammalian Egg and the Foundation of Modern Embryology," *Isis*, vol. 16 (1931) p. 322; Erik Nordenskiöld, *A History of Biology*, trans. L. B. Eyre (New York, 1928), pp. 363–366.

24. Alexander Vucinich, "Mendeleev's Views on Science and Society," *Isis*, vol. 58 (1967), p. 349.

25. Born, "Max Karl Ernst Ludwig Planck: 1858–1947," p. 166.

26. Ludwig Boltzmann, "The Recent Development of Method in Theoretical Physics," (1899) trans. T. J. McCormack, *The Monist*, vol. 11 (1901), pp. 229–230.

27. Ibid., p. 229.

28. Ibid., p. 247.

29. Ibid., pp. 248, 250.

30. Ibid., p. 252.

31. Ibid., p. 233.
32. Ibid., pp. 245–246.
33. Boltzmann, "On the Necessity of Atomic Theories in Physics," trans. T. J. McCormack, *The Monist*, vol. 12 (1902), p. 66.
34. Boltzmann, "The Recent Development of Method in Theoretical Physics," pp. 245–246, 252–253.
35. Engelbert Broda, *Ludwig Boltzmann: Mensch, Physiker, Philosoph* (Vienna, 1955), pp. 15–18.
36. Ibid., p. 25.
37. Ibid., p. 27.
38. Wilhelm Ostwald, *Grosse Männer: Studien Zur Biologie des Genies*, 5th ed. (Leipzig, 1919), pp. 404–405. Professor Martin J. Klein observes that Boltzmann felt himself to be, with some justice, "scientifically isolated" and fighting a "rearguard action" against "the victorious barbarian hordes of the energeticists." Although he rejects what he calls the "oversimplified view" that Boltzmann's suicide was "caused" by the attacks of his critics, he notes that it was "inseparable" from his scientific sacrifices. Boltzmann, he argues, was not broken by his critics any more than John Keats was. But the vulnerability to generational onslaught in a man of sixty-two is scarcely to be equated with that of one in his twenties (Klein, *Paul Ehrenfest* [Amsterdam, 1970], pp. 76–77).
39. Louis de Broglie, *Sur les Sentiers de la Science* (Paris, 1960), p. 164.
40. Boltzmann, "On Certain Questions of the Theory of Gases," *Nature*, vol. 51 (February 28, 1895), p. 413..
41. Julius Braunthal, *Victor und Friedrich Adler: Zwei Generationen Arbeiter-bewegung* (Vienna, 1965), p. 196.
42. Anton Reiser, *Albert Einstein*, p. 69. Philipp Frank, *Einstein: His Life and Times*, p. 68.
43. Max Planck, "Zur Machschen Theorie der physikalischen Erkenntnis: Eine Erwiderung" (On the Machian Theory of Physical Knowledge: A Reply), *Physikalische Zeitschrift*, vol. 11 (1910), p. 1187.
44. Planck, *A Survey of Physics: A Collection of Lectures and Essays*, trans R. Jones and D. H. Williams (London, 1925), pp. 35–38.
45. Ernst Mach, "Die Leitgedanken meiner Naturwissenschaftlichen Erkenntnislehre und ihre Aufnahme durch die Zeitgenossen." (The Leading Ideas of my Theory of Knowledge in the Natural Sciences and their Reception from my Contemporaries), *Scientia*, vol. 7, (1910) pp. 233, 237, 240.
46. Ibid., p. 230.
47. Cf. Karl Gerhards, "Zur Kontroverse Planck-Mach," *Vierteljahrsschrift für wissenschaftliche Philosophie und Soziologie* (Leipzig, 1912), pp. 19–67.
48. Max Planck, "Zur Machschen Theorie der physikalischen Erkenntris," pp. 1189–1190.

Often one reads that in the light of the theory of relativity the Inquisition's preference for the Ptolemaic frame of reference was as valid as Galileo's choice of the Copernican. It is said, for instance, that the Inquisition could logically reject the Copernican hypothesis because the movements of Venus could be "accommodated equally" to the view that the sun moves around the earth (Cf. I. Bernard Cohen, review of *The Scientific Intellectual*, *The New York Times Book Review*, June 16, 1963). Insofar as such arguments have appealed to the theory of relativity, it should be noted that the earth does not constitute a frame of reference moving at a uniform velocity with reference to the sun; both its speed and direction change in its orbital motion. Moreover, the relativist standpoint is not a matter of choosing with complete arbitrariness a frame of reference. If it were that, it would have been applicable without further ado to rotational as well as translational motions, and all the problems that such men as Planck and Russell raised would have been regarded as otiose. A stationary, nonrotating earth would have required a cosmology of an incredible complexity with stars whirling at speeds exceeding the velocity of light. Planck was indeed making such a point in his warning that a Machian dogmatism could not dismiss the Copernican standpoint.

49. Henri Peyre, *Les Générations Littéraires* (Paris, 1948), pp. 174–176, 201; Detley

W. Schumann, "Cultural Age-Groups in German Thought," *Publications of the Modern Language Association of America*, vol. 51 (1936), pp. 1180–1207.

50. Bernard Shaw, *The Perfect Wagnerite: A Commentary on the Niblung's Ring* (1898; 2d ed. London, 1912), p. 69.

51. Wilhelm Ostwald, *Grosse Männer*, pp. 371 ff; "The Biology of the Savant: A Study in the Psychology of Personality," *Scientific American Supplement*, vol. 62, no. 1862 (September 9, 1911), p. 170.

52. Florian Cajori, "Newton's Twenty Years Delay in Announcing the Law of Gravitation," in *Sir Isaac Newton, 1727–1927: A Bicentenary Evaluation of his Work* (Baltimore, 1928), pp. 127–188.

53. "It was not easy for Einstein to find a suitable assistant. . . . The students at the University (Berlin) who were working for the doctorate or to pass examinations for physics teachers were busy enough trying to satisfy all the demands made on them. . . ." One of his first assistants was a Russian Jew who was so pathologically deformed that he made "a repulsive impression on people that no one wanted to engage him as an assistant, much less as a teacher" (Philipp Frank, *Einstein: His Life and Times*, p. 207. "Einstein was not the man to have many students working under him. Whatever he undertook was always so difficult that he alone was able to carry it through" (ibid., p. 117).

54. Harold Talbot Parker, *The Cult of Antiquity and the French Revolutionaries: A Study in the Development of the Revolutionary Spirit* (Chicago, 1937), pp. 37–46.

55. Literary and practical politics, however, sometimes made some curious blends in the nineteenth century. Blanqui, the famed professional revolutionist, was a romantic in politics but a classicist in literature. He opposed Hugo's romanticism. At the height of the July Revolution of 1830, bespattered with powder and blood, Blanqui cried: "The Romantics are done for!" (Alan B. Spitzer, *The Revolutionary Theories of Louis Auguste Blanqui* [New York, 1957], pp. 48–49).

56. J. H. Van't Hoff, *Imagination is Science*, trans. G. F. Springer (New York, 1957), pp. 4, 15. Norbert Wiener has written probingly on the relationship between the Romantic era and its mathematicians, speculating indeed that Evariste Galois was the original for Victor Hugo's Enjolras. See "Pure and Applied Mathematics," in *Structure, Method and Meaning: Essays in Honor of Henry M. Sheffer*, ed. P. Henle, H. M. Kallen, and S. K. Langer (New York, 1951), pp. 95–96.

57. John Tyndall, *Faraday As a Discoverer*, 2d ed. (London, 1870), p. 8.

58. Silvanus P. Thompson, *The Life of William Thomson: Baron Kelvin of Largs*, (London, 1910), vol. 1, p. 291.

59. As Roentgen said: "Only very slowly came the conviction that the experiment is the mightiest and trustworthiest lever through which we can overhear nature, and that the same must reflect the last resort for the decision of the question if a hypothesis should be retained or cast off" (W. A. Nitske, *The Life of Wilhelm Conrad Röntgen*, p. 154).

60. Douglas, *The Life of Arthur Stanley Eddington*, p. 41. On the other hand, Einstein showed little interest in D. C. Miller's report of a positive result in replication of the Michelson-Morley experiment. "I think that the Miller experiments rest on an error in temperature. I have not taken them seriously for a minute," wrote Einstein to Besso in 1925. An analysis thirty years later of Miller's experiments showed that Einstein's surmise was well-founded. Cf. Gerald Holton, "Influences on Einstein's Early Work in Relativity Theory," *Actes du XIe Congrès International d'Histoire des Sciences* (1965; Wroclaw, 1968), vol. 1, p. 105.

61. Sir Oliver Lodge, *Past Years: An Autobiography* (New York, 1932), pp. 204–207. See also Joseph Larmor, in James Clark Maxwell, *Matter and Motion*, (reprint, New York, n.d.), pp. 140–142.

62. Lorentz, cited in A. V. Vasiliev, *Space, Time, Motion*, p. 147.

63. A. A. Michelson, *Studies in Optics* (Chicago, 1927), p. 161.

64. Engelbert Broda, *Ludwig Boltzmann, Mersch, Physiker, Philosoph* (Berlin, 1957), p. 31. Also cited in Vasiliev, *Space, Time, Motion*, p. 132.

65. Einstein, *Out of My Later Years*, p. 99; cf. Philipp Lenard, *Great Men of Science: A History of Scientific Progress*, trans. Stafford Hatfield (London, 1933), p. 359.

66. J. J. Thomson, "James Clerk Maxwell," in *James Clerk Maxwell: A Commemorative Volume, 1831–1931* (Cambridge, 1931), pp. 25–26.

67. Max Planck, "Maxwell's Influence on Theoretical Physics in Germany," ibid., p. 58.

68. François Arago, *Biographies of Distinguished Scientific Men*, trans. W. H. Smyth, B. Powell, R. Grant (Boston, 1859), 1st ser., p. 320.

69. As examples of pathological science, Langmuir mentioned Blondlot's N-Rays, the Davis-Barnes Effect, Mitogenetic Rays, and the Allison Effect. Cf. J. H. Ziman, "Some Pathologies of the Scientific Life," *The Advancement of Science*, vol. 27 (September 1970), p. 10.

70. Albert Einstein, "Maxwell's Influence on the Development of the Conception of Physical Reality," *James Clerk Maxwell: A Commemorative Volume*, p. 66.

> 71. Each generation of men of science has to consider anew the relation of the newly current theories to the perennial problems of philosophy. Though much survives, in each generation, of the scientific achievements of its predecessors, the conclusions of physics, even about itself, have lately changed radically every twenty years or so (E. A. Milne, *Sir James Jeans* [Cambridge, 1952], p. 153).

72. Paul Mouy, *Le Développement de la Physique Cartésienne, 1646–1712* (Paris, 1934), p. 322; A. E. Bell, *Christian Huygens and the Development of Science in the Seventeenth Century* (London, 1947), pp. 84–85; *The Born-Einstein Letters*, ed. Irene Born (London, 1971), p. 189.

73. Ernst Schrödinger, "Sir Arthur Eddington: The Philosophy of Physical Science," *Nature*, vol. 145 (March 16, 1940), p. 402; P. A. M. Dirac, "The Cosmological Constants," *Nature*, vol. 139 (February 20, 1937), p. 323; P. W. Bridgman, *Reflections of a Physicist* (New York, 1950), p. 112.

74. *The Born-Einstein Correspondence*, pp. 149, 164.

75. Albert Einstein, Arnold Sommerfeld, *Briefwechsel* (Basel, 1968), p. 109.

76. P. A. Schilpp, ed., *Albert Einstein*, vol. 1, p. 63.

77. Ibid., p. 23.

78. Hermann Bondi, *Assumption and Myth in Physical Theory* (Cambridge, 1967), p. 24.

79. Anton Reiser, *Albert Einstein*, p. 52.

80. H. A. Lorentz et al., *The Principle of Relativity*, p. 37.

81. "Alas," exclaimed Lord Kelvin, in his famous lecture, "Nineteenth Century Clouds over the Dynamical Theory of Heat and Light," April, 1900, as he described the outcome of Michelson's "admirable experiment." Kelvin took heart from the "brilliant suggestion" of both Lorentz and Fitzgerald for compensatory contraction. (*Baltimore Lectures in Molecular Dynamics and the Wave Theory of Light* [London, 1904], Appendix B, p. 492; also p. 485). Kelvin hoped for further experiments that would establish, as against Michelson and Morley, the existence of an "ether at rest absolutely throughout the universe," but he acknowledged "their great experimental work" (ibid., p. vi).

82. Leopold Infeld, *Quest: The Evolution of a Scientist* (New York, 1941), p. 277.

83. A. Einstein, "Ernst Mach," *Physikalische Zeitschrift*, vol. 17, (April 1, 1916), p. 103; cited in A. V. Vasiliev, *Space, Time, Motion*, pp. 90–91.

84. Charles Nordmann, "Einstein Expose et Discute la Théorie," *Revue des Deux Mondes*, vol. 9 (1922), pp. 134–135, 138–139, 142–143.

85. G. J. Whitrow, ed., *Einstein: The Man and His Achievement* (London, 1967), pp. 26–27. When Emile Meyerson placed the gravamen for the origin and acceptance of the theory of relativity on the Michelson-Morley experiment, and when he added further that if the eclipses of 1919 and 1922 had contravened Einstein's predictions, Einstein would have lost "most of his partisans," Einstein in his review of *La Déduction Relativiste*, the only philosophical work which he so honored, did not dispute Meyerson's statements (*La Déduction Relativiste* [Paris, 1925], p. xii).

86. Martin J. Klein, *Paul Ehrenfest*, pp. 315–316.

87. P. A. Schilpp, *Albert Einstein*, p. 23.

88. Cf. Robert F. Byrnes, *Antisemitism in Modern France, Volume 1: The Prologue to the Dreyfus Affair* (New Brunswick, 1950), p. 148 f., 324 f.

89. Hélène Pierre-Duhem, *Un Savant Francais, Pierre Duhem* (Paris, 1936), pp. 17–18, 126, 131, 72–73.

90. Pierre Duhem, *La Science Allemande* (Paris, 1915), pp. 134–136. "Quelques Réflexions sur la Science Allemande," *Revue des Deux Mondes*, vol. 25 (1915), pp. 680–681.

91. Planck, "Maxwell's Influence on Theoretical Physics in Germany," *James Clerk Maxwell: A Commemorative Volume*, p. 63.

92. *The Born-Einstein Letters*, p. 166. "While Harald Høffding's objectivity and free search for harmony in existence seems to have been significant for his [Bohr's] development, he had otherwise little in common with the prevailing philosophical trends of this time." Oskar Klein, in *Niels Bohr*, ed. S. Rozental, p. 74.

93. Martin J. Klein, *Paul Ehrenfest*, vol. 1, p. 309.

94. Max Born, *My Life and My Views*, p. 165. Stirner's book, forgotten soon after its publication in 1843, enjoyed a revival in the 1890s in the wake of the Nietzschean cult of the free individual. Rebellious spirits, young intellectuals, and workers read it. The historian of anarchism calls it "the most tedious of all the libertarian classics," but in Born's youth, its call to reject all abstractions—Man, Humanity, Party—and to seek freedom in one's uniqueness, one's "ownness," could act as a generational manifesto. "*My own* I am at all times and under all circumstances, if I know how to possess myself . . ." Cf. George Woodcock, *Anarchism: A History of Libertarian Ideas and Movements* (Cleveland, 1962), pp. 98–105; R. W. K. Paterson, *The Nihilistic Egoist: Max Stirner* (London, 1971), p. 145. Wolfgang Pauli was especially drawn toward a Jungian theory of archetypes as underlying all the sciences. See "Naturwissenschaftliche und erkenntnistheoretische Aspekte der Ideen vom Unbewussten," *Dialectica*, Vol. 8 (1954), p. 301. Schrödinger's reaction to positivism was that it filled him with a "cold clutch of dreary emptiness," that eliminating metaphysics meant "taking the soul out of both art and science, turning them into skeletons incapable of any further development." William T. Scott, *Erwin Schrödinger: An Introduction to his Writings* (Amherst, 1967), p. 114.

95. *The Autobiography of Bertrand Russell: The Middle Years: 1914–1944* (New York, 1968), pp. 326–327.

96. Even Darwin, confronted with the criticisms of a professor of philosophy, said: "He is a metaphysician, and such gentlemen are so acute that I think they often misunderstand common folk" (*The Life and Letters of Charles Darwin*, ed. Francis Darwin [New York, 1896], vol. 2, p. 230). Darwin's grandson, an eminent physicist, wrote likewise: "Is it not the salient fact about the philosophy of science that no professional philosopher can write a book that a man of science wants to read?" Prof. C. G. Darwin, *Nature*, vol. 139 (June 12, 1937), p. 1008.

97. "Generalizations about natural science are generalizations in social science," said Lawrence J. Henderson (*Eleven Twenty-Six: A Decade of Social Science Research*, ed Louis Wirth [Chicago, 1940], p. 272). The expression "science of science" was already used by Friedrich Engels (cf. *Herr Eugen Dühring's Revolution in Science*, trans. Emile Burns [New York, n.d.], p. 70).

Index

Index